Pre-Calculus FOR DUMMIES®

by Krystle Rose Forseth, Christopher Burger, and Michelle Rose Gilman, with Deborah Rumsey

Wiley Publishing, Inc.

Pre-Calculus For Dummies®

Published by
Wiley Publishing, Inc.
111 River St.
Hoboken, NJ 07030-5774
www.wiley.com

Library of Congress Control Number: 2008922118

ISBN: 978-0-470-16984-1

Manufactured in the United States of America

10 9 8 7 6 5 4 3

About the Authors

Krystle Rose Forseth graduated from the University of California, Santa Cruz, where she majored in mathematics with an emphasis in education. She has been tutoring math for eight years and teaching for three. Currently, she is the head of the math department at Fusion Learning Center and Fusion Academy, where she teaches mathematics and oversees the math instructors. Teaching students math has made Krystle a more compassionate individual, and her enthusiasm for the subject makes learning fun.

Christopher Burger graduated with a Bachelor of Arts degree in mathematics from Coker College in Hartsville, South Carolina, with minors in art and theatre. He has taught math for more than 10 years and tutored math subjects ranging from basic math to calculus for 20 years. Currently, he is the director of independent studies for Fusion Learning Center and Fusion Academy in Solana Beach, California, where he not only teaches students one-on-one, but also writes curriculum, oversees a staff of 35 teachers, and maintains high levels of academic rigor within the school. Christopher takes teaching and connecting with his students very seriously, and he believes he makes a difference not only in their math education, but in their lives.

Michelle Rose Gilman is proud to be known as Noah's mom (Hi, Noah!). A graduate from the University of South Florida, Michelle found her niche early, and at 19, she was working with emotionally disturbed and learning-disabled students in hospital settings. At 21, she made the trek to California, where she found her passion for helping teenagers become more successful in school and life. What started as a small tutoring business in the garage of her California home quickly expanded and grew to the point where traffic control was necessary on her residential street.

Today, Michelle is the founder and CEO of Fusion Learning Center and Fusion Academy, a private school and tutoring/test prep facility in Solana Beach, California, serving more than 2,000 students each year. She is the author of *The ACT For Dummies* and other books on self-esteem, writing, and motivational topics. Michelle has overseen dozens of programs over the last 20 years, focusing on helping kids become healthy adults. She currently specializes in motivating the unmotivated adolescent, comforting their shell-shocked parents, and assisting her staff of 35 teachers.

Michelle lives by the following motto: There are people content with longing — I am not one of them.

Dedication

We would like to dedicate this book to every student we've ever taught — each one of you taught us something in return. To everyone who we've missed over the last few months while writing this book — see you soon!

Authors' Acknowledgments

Thanks to everyone who helped pick up the pieces; to Bill Gladstone for the opportunity to write this book; Virginia Highstone for her much needed advice; Kate Brutlag for lending a hand when no one else could; Kristin DeMint, Tracy Boggier, Joyce Pepple, and all the editors at Wiley; and finally, Nicholas Angelo for putting up with it all and supporting this endeavor (thanks for the food, bah!).

Publisher's Acknowledgments

We're proud of this book; please send us your comments through our Dummies online registration form located at www.dummies.com/register/.

Some of the people who helped bring this book to market include the following:

Acquisitions, Editorial, and Media Development

Project Editor: Kristin DeMint

Acquisitions Editor: Tracy Boggier

Copy Editor: Josh Dials

Editorial Program Coordinator: Erin Calligan Mooney

Technical Editor: Carol Spilker

Editorial Manager: Michelle Hacker

Editorial Assistants: Joe Niesen, Leeann Harney

Cover photo: © Wiley Publishing, Inc.

Cartoons: Rich Tennant (www.the5thwave.com)

Composition Services

Project Coordinator: Erin Smith

Layout and Graphics: Carrie A. Cesavice, Stephanie S. Jumper, Erin Zeltner

Proofreaders: Laura Bowman, Cynthia Fields, Caitie Kelly, Betty Kish, Jessica Kramer

Indexer: Potomac Indexing, LLC

Special Help

Victoria M. Adang; Chad R. Sievers

Publishing and Editorial for Consumer Dummies

Diane Graves Steele, Vice President and Publisher, Consumer Dummies

Joyce Pepple, Acquisitions Director, Consumer Dummies

Kristin A. Cocks, Product Development Director, Consumer Dummies

Michael Spring, Vice President and Publisher, Travel

Kelly Regan, Editorial Director, Travel

Publishing for Technology Dummies

Andy Cummings, Vice President and Publisher, Dummies Technology/General User

Composition Services

Gerry Fahey, Vice President of Production Services

Debbie Stailey, Director of Composition Services

Contents at a Glance

Table of Contents

Introduction

Welcome to *Pre-Calculus For Dummies*. This is a nondiscriminatory, equal-opportunity book. You're welcome to participate whether you are a genius or (like us) you need a recipe to make ice. Don't let the title throw you. If you've gotten this far in math, in no way are you a dummy! You may be reading this book for a few perfectly great reasons. Maybe you need a reference book that you can actually *understand* (we've never met a pre-calc text that we liked). Perhaps your guidance counselor told you that taking pre-calc would look good on your college application, but you couldn't care less about the subject and just want to get a good grade. Or, maybe you're contemplating buying this book and you want to check us out to see if we're a good match (not unlike looking at your blind date through the window before you walk into the restaurant). Regardless of why you opened up this book, it will help you navigate the tricky path that is pre-calc.

You may also be wondering, "When will I ever really use pre-calculus?" You're not alone. Some of our students have referred to it as pre-calc-uselessness. Well, they quickly found out how wrong they were. The concepts throughout this book are used in many real-world applications.

This book has one goal and one goal only — to teach you pre-calculus in as painless a way as possible. If you thought that you could never tackle this subject and you end up with a decent grade in this class, would you mind sending us a letter? E-mail's good too. We love to hear our students' success stories!

About This Book

This book is not necessarily meant to be read from beginning to end. It's structured in a way that you can flip to a particular chapter and get your needs met (those pesky needs we all have). Sometimes we may tell you to look in another chapter to get a more in-depth explanation, but we have tried to allow each chapter to stand on its own.

All vocabulary is mathematically correct and clear. We have taken liberties at some points throughout this book to make the language more approachable and likable. It's just more fun that way.

Pre-calc is its own special math topic. You see, some states, like California, don't have any set standards that students need to learn to officially master pre-calculus. As a result, the subject of pre-calc varies between districts, schools, and individual teachers. Because we don't know what your teacher is going to want you to take away from this course, we've covered pretty much every concept in pre-calc. We may have covered areas that you'll never be required to tackle. That's okay. Just use this book according to your individual needs.

If you use this book only to prop open a door or as a bug smoosher, you won't get what you need from it. We suggest two alternatives:

- Look up only what you need to know when you need to know it. This book is handy for this. Use the Index, the Table of Contents, or better yet, the quickie Contents at a Glance found in the very front of this book to find what you need.
- Start at the beginning and read through the book, chapter by chapter. This is a good way to tackle this subject because the topics sometimes build on previous ones. Even if you're a math god and you want to skim through a section that you feel you know, you may be reminded of something that you forgot. We recommend starting at the beginning and slowly working your way through the material. The more practice you have the better.

Conventions Used in This Book

For consistency and ability to navigate easily, this book uses the following conventions:

- Math terms are *italicized* to indicate their introduction and to help you find their definition.
- Variables are also *italicized* to distinguish them from common letters.
- The step-by-step problems are always **bold** to help you identify them more easily.
- The symbol for imaginary numbers is a lowercase *i*.

Foolish Assumptions

We can't assume that just because we absolutely love math that you share the same enthusiasm for the subject. We can assume, however, that you opened up this book for a reason: You need a refresher on the subject, need to learn it for the first time, are trying to relearn it for college, or have to help your kid understand it at home. We can also assume that you have been exposed, at least in part, to many of the concepts found in this subject because pre-calc really takes geometry and Algebra II concepts to the next level.

We also assume that you're willing to do some work. Although pre-calculus isn't the end-all to math courses out there, it's still a higher level math course. You're going to have to work a bit, but you knew that, didn't you?

We also are pretty sure that you're an adventurous soul and have chosen to take this class because pre-calculus is not necessarily a required high school course (in most U.S. high schools, anyway). Maybe it's because you love math like we do, or you have nothing better to do with your life, again like us, or because the course will enhance your college application. Obviously, you managed to get through some pretty complex concepts in geometry and Algebra II. We can assume that if you made it this far, you'll make it even farther. We're going to help!

How This Book Is Organized

This book is broken down into four sections dealing with the most frequently taught and studied concepts in pre-calc.

Part I: Set It Up, Solve It, Graph It

The chapters in Part I begin with a review of material you should already know from Algebra II. Then we review real numbers and how to operate with them. From there we cover functions, including polynomial, rational, exponential, and logarithmic, and graphing them, solving them, and performing operations on them.

Part II: The Essentials of Trigonometry

The chapters in Part II begin with a review of angles, right triangles, and trig ratios. Then we build the glorious unit circle. Graphing trig functions may or may not be a review, depending on the Algebra II course you've taken, so we show you how to graph the parent graph of the six basic trig functions and then explain how to transform those graphs to get to the more complicated ones.

This part also covers the harder formulas and identities for trig functions, breaking them down methodically so you can internalize each identity and truly understand them. We then move right along into simplifying trig expressions and solving for an unknown variable using those formulas and identities. Finally, this part covers how to solve triangles that are not right triangles using Law of Sines and Law of Cosines.

Part III: Analytic Geometry and System Solving

Part III covers a multitude of pre-calc topics. It begins with understanding complex numbers and how to perform operations with them. Next comes polar coordinate graphing and finally conics. Systems of equations live in this part, as do sequences and series, and binomial expansion. Finally, this part concludes with calculus and the study of limits and continuity of functions.

Part IV: The Part of Tens

After you've covered everything up to this point in the book, you may be eyeing the next big math challenge: calculus. (And if you decide to stop with pre-calc, that's okay, too.) But before you head off for even more complex concepts, you need to do two things: pick up some good math habits to take into calculus, and break any bad habits you've developed along the way. This part helps you with those tasks. Both ends of this spectrum are critical for success because the problems get longer and teachers' patience for algebra errors gets shorter.

Icons Used in This Book

Throughout the book you'll find little drawings (we call 'em *icons*) that are meant to draw your attention to something important or interesting to know.

Pre-calc rules are the basic rules of pre-calc. They must be observed every time to make problems come out correctly.

Math Mumbo Jumbo alerts you to information that is helpful but not required to gain full knowledge of the concept in that section.

We love Tips! When you see this icon, you know that it points out a way to make your life a lot easier. Easier is good.

You'll see this icon when we mention an old idea that you should never forget. We use it when we want you to recall a previously learned concept or a concept from a lower math course.

Think of Warnings as big stop signs. Its presence alerts you to common errors, or points out something that can be a bit tricky.

Where to Go from Here

If you have a really firm background in basic algebra skills, feel free to skip Chapter 1 and head right over to Chapter 2. If you want a brush-up, we suggest reading Chapter 1. In fact, everything in Chapter 2 is also a review except interval notation. So if you're really impatient or are a math genius, ignore everything until interval notation in Chapter 2. As you work through this book, keep in mind that many concepts in pre-calc are take-offs from Algebra II, so don't make the mistake of completely skipping chapters because they sound familiar. They may sound familiar but are likely to include some brand-spanking new material. We also didn't sit next to you when you took Algebra II, so we can't be sure what your teacher covered. So here's a brief list of sections that may sound familiar but include new concepts that you should pay attention to:

- Translating common functions
- Solving polynomials
- All the trig information
- Complex numbers
- Matrices

So where do you go from here? You get going straight into pre-calc! Good luck.

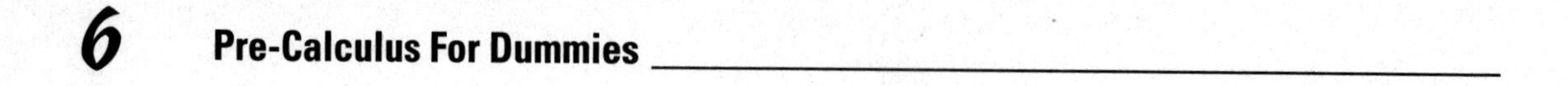

Part I

Set It Up, Solve It, Graph It

"David's using algebra to calculate the tip. Barbara– would you mind being a fractional exponent?"

In this part . . .

A major goal of pre-calculus is to bring the big ideas of algebra to the surface and sharpen the skills most needed for calculus. This part pulls together and expands on those algebra concepts. Perhaps most importantly, it identifies the most common mistakes students make in algebra so you can conquer those before moving on to higher-level concepts.

The chapters in Part I move through a review of working with real numbers, including the ever-elusive radicals. From there, we review functions — from how to graph them by transforming their parent graphs to how to perform operations on them. Then we move on to polynomials and review how to solve polynomials using common techniques, including factoring, completing the square, and the quadratic formula. We also explain how to graph complex polynomial and rational functions. Lastly, we show you how to deal with exponential and logarithmic functions.

Chapter 1

Pre-Pre-Calculus

In This Chapter

- Refreshing your memory on numbers and variables
- Accepting the importance of graphing
- Preparing for pre-calculus by grabbing a graphing calculator

Pre-calculus is the bridge (or purgatory?) between Algebra II and calculus. In its scope, you'll review concepts you've seen before in math but then quickly build on them. You'll see some brand-new ideas, but even those build on the material you've seen before; the main difference is that the problems get much harder (take going from systems to nonlinear systems, for instance). You keep on building until the end of the course, which doubles as the beginning of calculus. Have no fear! We're here to help you cross the bridge (toll-free).

Because you've probably already taken algebra, Algebra II, and geometry, we assume throughout this book that there are certain things you already know how to do. (We go over them briefly in the Introduction to this book.) Just to make sure, though, we go over each of them in this chapter in a little more detail before we move on to the pre-calculus that is pre-calculus.

If we cover any topic in this chapter that you're not familiar with, don't remember how to do, or don't feel comfortable doing, we suggest that you pick up another *For Dummies* math book and start there. Don't feel like a failure in math if you need to do that. Even the pros have to look up things from time to time. These books can be like encyclopedias or the Internet — if you don't know the material, you look it up and get going from there.

Pre-Calculus: An Overview

Don't you just love movie previews and trailers? Some people show up early to movies just to see what's coming out in the future. Well, consider this section a trailer that you see a couple months before the *Pre-Calculus For Dummies* movie comes out! (We wonder who will play us in the movie.) In the following list, we present some material you've learned before in math, and then we give you some examples of where pre-calculus will take you next:

- **Algebra I and II:** Dealing with real numbers and solving equations and inequalities.

  Pre-calculus: Expressing inequalities in a new way called *interval notation.*

  Before, your solutions to inequalities were given as set notation. For example, one solution may look like $x > 4$. In pre-calc, you express this solution as an interval: $(4, \infty)$. (For more, see Chapter 2.)

- **Geometry:** Solving right triangles, where all sides are positive.

  Pre-calculus: Solving non-right triangles, where the sides aren't necessarily always positive.

  You've learned that a length can never be negative. Well, in pre-calc you use negative numbers for sides of triangles to show where these triangles lie in the coordinate plane (they can be in any of the four quadrants).

- **Geometry/trigonometry:** Using the Pythagorean Theorem to find the length of a triangle's sides.

  Pre-calculus: Organizing information into one nice, neat package known as the unit circle (see Part II).

  In this book, we give you a handy shortcut to finding the sides of triangles, which is an even handier shortcut to finding the trig values for the angles in those triangles.

- **Algebra I and II:** Graphing equations on a coordinate plane.

  Pre-calculus: Graphing in a brand-new way, with the polar coordinate system (see Chapter 11).

  Say goodbye to the good old days of graphing on the Cartesian coordinate plane. You have a new way to graph, and it involves goin' round in circles. We're not trying to make you dizzy; actually, polar coordinates can make you some pretty pictures.

- **Algebra II:** Dealing with imaginary numbers.

  Pre-calculus: Adding, subtracting, multiplying, and dividing complex numbers gets boring when the complex numbers are in rectangular form ($A + Bi$). In pre-calc, you'll become familiar with something new called *polar form* and use that to find solutions to equations that you didn't even know existed.

All the Number Basics (No, Not How to Count Them!)

When entering pre-calculus, you should be comfy with sets of numbers (natural, integer, rational, and so on). By this point in your math career, you should also know how to perform operations with numbers. We quickly review these concepts in this section. Also, certain properties hold true for all sets of numbers; some math teachers may want you to know them by name, so we review these in this section, too.

The multitude of number types: Terms to know

Dorky mathematicians love to name things simply because they can; it makes them feel special. In this spirit, mathematicians attached names to many sets of numbers to set them apart and cement their places in math students' heads for all of time:

- **The set of *natural* or *counting numbers:* {1, 2, 3 . . .}.** Notice that the set of natural numbers doesn't include 0.
- **The set of *whole numbers:* {0, 1, 2, 3 . . .}.** The set of whole numbers does include the number 0, however.
- **The set of *integers:* {. . . –3, –2, –1, 0, 1, 2, 3 . . .}.** The set of integers includes positives, negatives, and 0.

  Dealing with integers is like dealing with money: Think of positives as having it and negatives as owing it. This becomes important when operating on numbers (see the next section).

- **The set of *rational numbers,* which are the numbers that can be expressed as a fraction where the numerator and the denominator are both integers.** The word *rational* comes from the idea of a ratio (fraction or division) of two integers.

  Examples of rational numbers include (but in no way are limited to) $\frac{1}{5}$, $-\frac{7}{2}$, and 0.23. If you look at any rational number in decimal form, you notice that the decimal either stops or repeats.

  Adding and subtracting fractions is all about finding a common denominator. And roots must be like terms in order to add and subtract them.

- **The set of *irrational numbers,* which are all numbers that can't be expressed as fractions.** Examples of irrational numbers include $\sqrt{2}$, $\sqrt{21}$, and π.
- **The set of *all real numbers,* which comprise all the sets of numbers previously discussed.** For examples of a real number, think of one number . . . any number. Whatever it is, it's real. Any number from the previous lists would work as an example. The numbers that aren't real numbers are imaginary.

  Like telemarketers and pop-up ads on the Net, real numbers are everywhere; you can't get away from them — not even in pre-calculus. Why? Because they include *all* numbers except the following:

  - **A fraction with a zero as the denominator:** Such numbers don't exist.
  - **The square root of a negative number:** These numbers are called *complex numbers* (see Chapter 11).
  - **Infinity:** Infinity is a concept, not an actual number.
- **The set of *imaginary numbers,* which are square roots of negative numbers.** Imaginary numbers have an imaginary unit, like i, $4i$, and $-2i$. Imaginary numbers were once made-up numbers, but mathematicians soon realized that these numbers pop up in the real world. We still call them imaginary because they're square roots of negative numbers, but they do exist. The imaginary unit is defined as $i = \sqrt{-1}$. (For more on these numbers, head to Chapter 11.)
- **The set of *complex numbers,* which are the sum and difference of a real number and an imaginary number.** Complex numbers appear like these examples: $3 + 2i$, $2 - \sqrt{2i}$, and $4 - \frac{2}{3}i$. However, they also cover all the previous lists, including the real numbers (3 is the same thing as $3 + 0i$) and imaginary numbers ($2i$ is the same thing as $0 + 2i$).

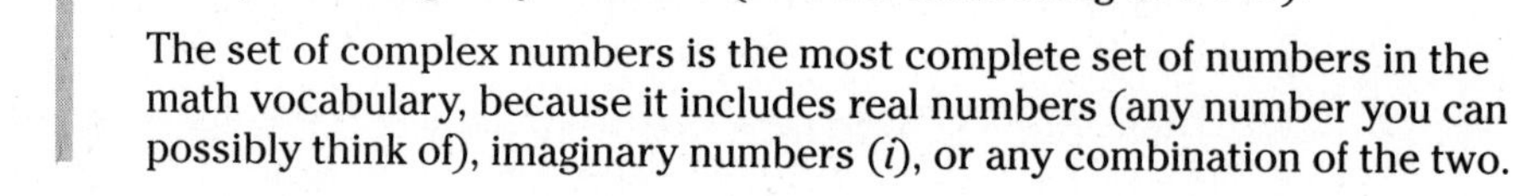
The set of complex numbers is the most complete set of numbers in the math vocabulary, because it includes real numbers (any number you can possibly think of), imaginary numbers (i), or any combination of the two.

The fundamental operations you can perform on numbers

From positives and negatives to fractions, decimals, and square roots, you should know how to perform all the basic operations on all real numbers. This means adding, subtracting, multiplying, dividing, taking powers of, and taking roots of numbers. The *order of operations* is the way in which you perform these operations.

The mnemonic device used most frequently to remember the order is PEMDAS, which stands for

1. **P**arentheses (and other grouping devices)
2. **E**xponents
3. **M**ultiplication and **D**ivision, whichever is first from left to right
4. **A**ddition and **S**ubtraction, whichever is first from left to right

One type of operation most of our students overlook or forget to include on the previous list: the absolute value. *Absolute value* is the distance from 0 on the number line. Absolute value should be included with the parentheses step, because you have to consider what's inside the absolute value bars first (because the bars are a grouping device). Don't forget that absolute value is always positive. Hey, even if you're walking backward, you're still walking!

The properties of numbers: Truths to remember

It's important to remember the properties of numbers because you'll use them consistently in pre-calc. However, you often won't see them used by name in pre-calc, but it's assumed that you know when you need to utilize them. The following list presents the properties of numbers:

- **Reflexive property: $a = a$.** For example, $10 = 10$.
- **Symmetric property: If $a = b$, then $b = a$.** For example, if $5 + 3 = 8$, then $8 = 5 + 3$.
- **Transitive property: If $a = b$ and $b = c$, then $a = c$.** For example, if $5 + 3 = 8$ and $8 = 4 \cdot 2$, then $5 + 3 = 4 \cdot 2$.
- **Commutative property of addition: $a + b = b + a$.** For example $2 + 3 = 3 + 2$.
- **Commutative property of multiplication: $a \cdot b = b \cdot a$.** For example, $2 \cdot 3 = 3 \cdot 2$.
- **Associative property of addition: $(a + b) + c = a + (b + c)$.** For example, $(2 + 3) + 4 = 2 + (3 + 4)$.
- **Associative property of multiplication: $(a \cdot b) \cdot c = a \cdot (b \cdot c)$.** For example, $(2 \cdot 3) \cdot 4 = 2 \cdot (3 \cdot 4)$.

- **Additive identity: $a + 0 = a$.** For example, $0 + -3 = -3$.
- **Multiplicative identity: $a \cdot 1 = a$.** For example, $4 \cdot 1 = 4$.
- **Additive inverse property: $a + (-a) = 0$.** For example, $2 + -2 = 0$.
- **Multiplicative inverse property: $a \cdot (1/a) = 1$.** For example, $2 \cdot ½ = 1$.
- **Distributive property: $a(b + c) = a \cdot b + a \cdot c$.** For example $10(2 + 3) = 10 \cdot 2 + 10 \cdot 3 = 50$.
- **Multiplicative property of zero: $a \cdot 0 = 0$.** For example, $5 \cdot 0 = 0$.
- **Zero product property: If $a \cdot b = 0$, $a = 0$ or $b = 0$.** For example, if $x(x + 2) = 0$, then $x = 0$ or $x + 2 = 0$.

If you're trying to perform an operation that isn't on the previous list, then the operation probably isn't correct. After all, algebra has been around since 1600 BC, and if a property exists, someone has probably already discovered it. For example, it may look inviting to say that $10(2 + 3) = 10 \cdot 2 + 3 = 23$, but that's incorrect. The correct answer is $10 \cdot 2 + 10 \cdot 3 = 20 + 30 = 50$. Knowing what you *can't* do is just as important as knowing what you *can* do.

Putting Mathematical Statements in Visual Form: Fun with Graphs

Graphs are great visual tools. They're used to display what's going on in math problems, in companies, and in scientific experiments. For instance, graphs can be used to show how something (like real estate prices) changes over time. Surveys can be taken to get facts or opinions, the results of which can be displayed in a graph. Open up the newspaper on any given day and you can find a graph in there somewhere.

Hopefully that answers the question of why you need to understand how to construct graphs. Even though in real life you don't walk around with graph paper and a pencil to make the decisions you face, graphing is vital in math and in other walks of life. Regardless of the absence of graph paper, graphs indeed are everywhere.

For example, when a scientist goes out and collects data or measures things, he or she arranges the data as x and y values. Typically, the scientist is looking for some kind of general relationship between these two values to support his or her hypothesis. These values can then be graphed on a coordinate plane to show trends in data. A good scientist may show that the more you read this book, the more you understand pre-calculus! (Another scientist may show that people with longer arms have bigger feet. Boring!)

Digesting basic terms and concepts

Graphing equations is such a huge part of pre-calc, and eventually calc, so we want to review the basics of graphing before we get into the more complicated and unfamiliar graphs you see later in the book.

Although some of the graphs in pre-calc will look very familiar, some will be new — and possibly intimidating. We're here to get you familiar with these graphs so that you can study them in detail in calculus. However, the information in this chapter is mostly information that your pre-calc teacher or book will assume that you remember from Algebra II. You did pay attention then, right?

Each point on the coordinate plane on which you construct graphs — made up of the horizontal, or x-axis, and the vertical, or y-axis, creating a plane of four quadrants — is called a *coordinate pair* (x, y), which is often referred to as a *Cartesian coordinate pair.*

The name *Cartesian coordinates* comes from the French mathematician and philosopher who invented all this graphing stuff, René Descartes. Descartes worked to merge algebra and Euclidean geometry (flat geometry), and his work was influential in the development of analytic geometry, calculus, and cartography.

A *relation* is a set (meaning one or more) of ordered pairs that can be graphed on a coordinate plane. Each relation is kind of like a computer that expresses x as input and y as output. You know you're dealing with a relation when it's set in those curly-brackets (like these: { }) and has one or more points inside. For example, $R = \{(2, -1), (3, 0), (-4, 5)\}$ is a relation with three ordered pairs. Think of each point as (input, output) just like from a computer.

The *domain* of a relation is the set of all the input values from least to greatest. The domain of set R is $\{-4, 2, 3\}$.The *range* is the set of all the output values, also from least to greatest. The range of R is $\{-1, 0, 5\}$. If any value in the domain or range is repeated, you don't have to list it twice. Usually, the domain is the x variable and the range is y.

If different variables appear, such as m and n, input (domain) and output (range) usually go alphabetically, unless you're told otherwise. In this case, m would be your input/domain and n would be your output/range. But when written as a point, a relation is always (input, output).

Graphing equalities versus inequalities

When you first figured out how to graph a line on the coordinate plane, you learned to pick domain values (x) and plug them into the equation to solve for the range (y). Then you went through the process multiple times, expressed each pair as a coordinate point, and connected the dots to make a line. Some mathematicians call this the ol' *plug and chug* method.

After a while of that tedious work, somebody said to you, "Hold on! There's a shortcut." That shortcut is called *slope-intercept form* — $y = mx + b$. The variable m stands for the *slope* of the line (see the next section), and b stands for the y-intercept (or where the line crosses the y-axis). You can change equations that aren't written in slope-intercept form by solving for y. For example, graphing $2x - 3y = 12$ requires you to subtract $2x$ from both sides first to get $-3y = -2x + 12$. Then you divide every term by -3 to get $y = \frac{2x}{3} - 4$. This graph starts at -4 on the y-axis; to find the next point, you move up two and right three (using the slope). Slope is always a fraction because it's rise over run — in this case $\frac{2}{3}$.

Inequalities are used for comparisons, which are a big part of pre-calc. They show a relationship between two expressions (we're talking greater than, less than, or equal to). Graphing inequalities starts exactly the same as graphing equalities, but at the end of the graphing process (you still put the equation in slope-intercept form and graph), you have two decisions to make:

- Is the line *dashed* — $y <$ or $y >$ — or is the line *solid* — $y \leq$ or $y \geq$?
- Do you shade under the line — $y <$ or $y \leq$ — or do you shade above the line — $y >$ or $y \geq$? Simple inequalities (like $x < 3$) express all possible answers. For inequalities, you show all possible answers by shading the side of line that works in the original equation.

  For example, when graphing $y < 2x - 5$, you follow these steps:

  1. Start off at -5 on the y-axis and mark a point.
  2. Move up two and right one to find a second point.
  3. When connecting the dots, you produce a straight line that will be dashed.
  4. Shade on the bottom half of the graph to show all possible points in the solution.

Gathering information from graphs

After getting you used to coordinate points and graphing equations of lines on the coordinate plane, typical math books and teachers will begin to ask you questions about the points and lines that you've been graphing. The three main things you'll be asked to find are the distance between two points, the midpoint of the segment connecting two points, and the exact slope of a line that passes between two points. Away we go in the following sections!

Calculating distance

Knowing how to calculate distance by using the information from a graph comes in handy in pre-calc in a big way, so allow us to review a few things first. *Distance* is how far apart two objects, or two points, are. To find the distance, *d*, between the two points (x_1, y_1) and (x_2, y_2) on a coordinate plane, for example, use the following formula:

$$d = \sqrt{(x_2 - x_1)^2 + (y_2 - y_1)^2}$$

You can use this equation to find the length of the segment between two points on a coordinate plane whenever the need arises. For example, to find the distance between A(–6, 4) and B(2, 1), first identify the parts: $x_1 = -6$ and $y_1 = 4$; $x_2 = 2$ and $y_2 = 1$. Plug these values into the distance formula: $d = \sqrt{(2 - {}^-6)^2 + (1 - 4)^2}$. This simplifies to $\sqrt{73}$.

Finding the midpoint

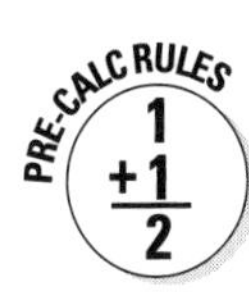

Finding the midpoint of a segment will pop up in pre-calc topics like conics (Chapter 12). To find the midpoint of the segment connecting two points, you just average their *x* values and *y* values and express the answer as an ordered pair:

$$M = \left(\frac{x_1 + x_2}{2}, \frac{y_1 + y_2}{2}\right)$$

You can use this formula to find the center of various graphs on a coordinate plane, but for now you're just finding the midpoint. You find the midpoint of the segment connecting the two points $\overline{AB}$ (see the previous section) by using the previous formula. This would give you $\left(\frac{-6+2}{2}, \frac{4+1}{2}\right)$, or (–2, 5⁄2).

Figuring a line's slope

When you graph a linear equation, slope plays a role. The *slope* of a line tells how steep the line is on the coordinate plane. When you're given two points (x_1, y_1) and (x_2, y_2) and are asked to find the slope of the line between them, you use the following formula:

$$m = \frac{y_2 - y_1}{x_2 - x_1}$$

If you use the same two points A and B from the previous sections and plug the values into the formula, the slope is $^-3/8$.

Positive slopes always move up and to the right on the plane. Negative slopes either move down and right or up and left. (Note that if you moved the slope down and left, it would be $-/-$, which is really positive.) Horizontal lines have zero slope, and vertical lines have undefined slope.

If you ever get the different types of slopes confused, remember the skier on the ski-slope:

- When he's going uphill, he's doing a lot of work (+ slope).
- When he's going downhill, the hill is doing the work for him (– slope).
- When he's standing still on flat ground, he's not doing any work at all (0 slope).
- When he hits a wall (the vertical line), he's dead and he can't ski anymore (undefined slope)!

Getting a Grip on a Graphing Calculator

We *highly* recommend that you purchase a graphing calculator for pre-calculus work. Since the invention of the graphing calculator, math classes have begun to change their scope. Some teachers feel that the majority of the work should be done using the calculator. More conservative math teachers, however, won't even let you use one. Your instructor should make his views quite apparent that first day of school. A graphing calculator does so many things for you, and even if a teacher won't allow you to use one on a test, you can almost always use one to check your work on homework problems.

There are many different types of graphing calculators, and their individual inner workings are all different. As far as which one to purchase, ask advice from someone who has already taken a pre-calc class, and then look around on the Internet for the best deal. (In our opinion, the TI-89 or TI-89 Titanium is the greatest calculator ever, as of press time — if you can figure out how to use it [we're still learning!].)

Just a hint: If you can find one with an exact/approximate mode, you'll thank us later because it will give you exact values (rather than decimal approximations), which is often what teachers are looking for.

We recommend that if, by chance, you're allowed to use a graphing calculator, you still do the work by hand. Then use your graphing calculator to check your work. This way, you won't become dependent on technology to do work for you; someday, you may not be allowed to use one (a college math placement test, for example).

Many of the more theoretical concepts in this book, and in pre-calc in general, are lost when you use your graphing calculator. All you're told is "plug in the numbers and get the answer." Sure, you get your answer, but do you really know what the calculator did to get that answer? Nope. For this purpose, this book goes back and forth between using the calculator and doing complicated problems long hand. But whether you're allowed to use the graphing calculator or not, be smart with its use. If you plan on moving on to calculus after this course, you need to know the theory and concepts behind each topic.

We can't even begin to teach you how to use your unique graphing calculator, but the good *For Dummies* folks at Wiley supply you with entire books on the use of them, depending on the type you own. We can, however, give you some general "heads up" on their use. Here's a list of hints that should help you use your graphing calculator:

- **Always double-check that the mode in your calculator is set according to the problem you're working on.** Look for a button somewhere on the calculator that says "mode." Depending on the brand of calculator, it will allow you to change things like degrees or radians, or $f(x)$ or $r(\theta)$, which we discuss in Chapter 11. For example, if you're working in degrees, you must make sure that your calculator knows that before you ask it to solve a problem. The same goes for working in radians. Some calculators have over ten different modes to choose from. Be careful!

- **Make sure you can solve for *y* before you try to construct a graph.** You can graph anything in your graphing calculator as long as you can solve for *y*. The calculators are set up to accept only equations that have been solved for *y*.

  Equations that you have to solve for *x* often aren't true functions and aren't studied in pre-calc — except conic sections, and students generally aren't allowed to use graphing calculators for this material because it's entirely based on graphing (see Chapter 12).

- **Be aware of all the shortcut menus available to you and use as many of the calculator's functions as you can.** Typically, under your calculator's graphing menu you can find shortcuts to other mathematical concepts (like changing a decimal to a fraction, finding roots of numbers, or entering matrices and then performing operations with them). Each brand of graphing calculator is unique, so read the manual. Shortcuts give you great ways to check your answers!

- **Type in an expression exactly the way it looks and the calc will do the work and simplify the expression.** All graphing calculators do order of operations for you, so you won't even have to worry about the order. Just be aware that some built-in math shortcuts automatically start with grouping parentheses.

For example, the calculator we use starts a square root off as $\sqrt{\ }($, so all information we type after that is automatically inside the square root sign until we close the parentheses. For instance, $\sqrt{\ }(4+5)$ and $\sqrt{\ }(4)+5$ represent two different calculations and, therefore, two different values (3 and 7, respectively). Some smart calculators even solve the equation for you. In the near future, you probably won't even have to take a pre-calc class; the calculator will take it for you!

Okay, now you're ready to take flight into pre-calculus. Good luck to you and enjoy the ride!

Chapter 2

Dealing with Real Numbers

In This Chapter

- Reviewing the basic elements of real numbers
- Working with equations and inequalities
- Mastering radicals and exponents

If you're taking a pre-calculus class, you've already taken Algebra I and II and survived (whew!). You may also be thinking, "I'm sure glad that's over; now I can move on to some new stuff." Although pre-calculus presents many new and wonderful ideas and techniques, these new ideas build on the solid-rock foundation of algebra. Alas, we must refresh your memory a bit and test just how sturdy your foundation is.

We assume that you have certain algebra skills down cold, but we're going to begin this book by reviewing some of the tougher ones that become the fundamentals of pre-calculus. In this chapter, we review solving inequalities, absolute value equations and inequalities, as well as radicals and rational exponents. We also introduce a new way to express solution sets: interval notation.

Solving Inequalities

By now you're familiar with equations and how to solve them. Pre-calculus teachers generally will assume that you know how to solve equations, so most courses begin with inequalities. An *inequality* is a mathematical sentence indicating that two expressions aren't equal. The following symbols express inequalities:

Less than: $<$

Less than or equal to: $\leq$

Greater than: $>$

Greater than or equal to: $\geq$

A brief how-to inequality recap

Inequalities are set up and solved the same way as equations — the inequality sign doesn't change the method of solving. In fact, to solve an inequality, you treat it exactly like an equation — with one exception.

If you multiply or divide an inequality by a negative number, you must change the inequality sign to face the opposite way.

For example, if you must solve $-4x + 1 < 13$, you do as follows:

$$-4x < 12$$

$$x > -3$$

You first subtract 1 from both sides, and then divide both sides by –4, at which point the less-than sign changes to the greater-than sign. You can check this solution by picking a number that's greater than –3 and plugging it into the original equation to make sure you get a true statement. If you do this with 0, for instance, you get $-4(0) + 1 < 13$, which is a true statement.

Switching the inequality is a step that many students forget. Look at an inequality with numbers in it, like $-2 < 10$. This statement is true. If you multiply 3 on both sides, you get $-6 < 30$, which is still true. But if you multiply –3 on both sides — and don't fix the sign — you get $6 < -30$. This statement is false, and you always want to keep the statements true. The only way for the equation to work is to switch the inequality sign to read $6 > -30$. The same rule applies if you divide $-2 < 10$ by –2 on both sides. The only way for the problem to make sense is to read $1 > -5$.

Solving equations and inequalities when absolute value is involved

If you think back to Algebra I, you'll likely remember that an absolute value equation usually has two possible solutions. Absolute value is a bit trickier to handle when you're solving inequalities. Similarly, though, inequalities have two possible solutions:

- One where the quantity inside the absolute value bars is greater than a number.
- One where the quantity inside the absolute value bars is less than a number.

In mathematical terminology, the inequality $|ax \pm b| < c$ — where a, b, and c are real numbers — always becomes two inequalities:

$$ax \pm b < c \text{ AND } ax \pm b > -c$$

The "AND" comes from the graph of the solution set on a number line, as seen in Figure 2-1a.

The inequality $|ax \pm b| > c$ becomes

$$ax \pm b > c \text{ OR } ax \pm b < -c$$

The "OR" also comes from the graph of the solution set, which you can see in Figure 2-1b.

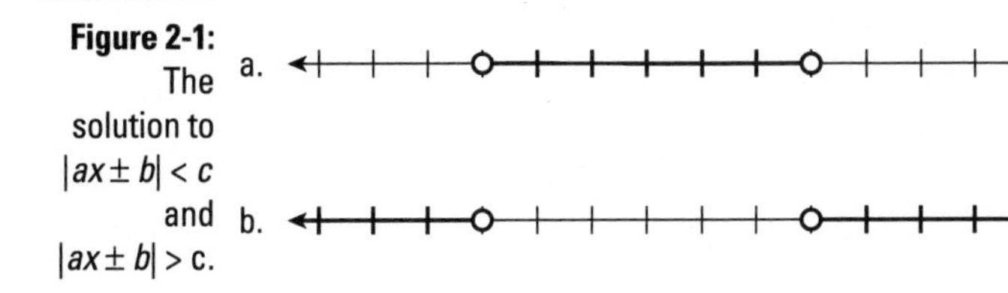

Figure 2-1: The solution to $|ax \pm b| < c$ and $|ax \pm b| > c$.

Here are two caveats to remember when dealing with absolute values:

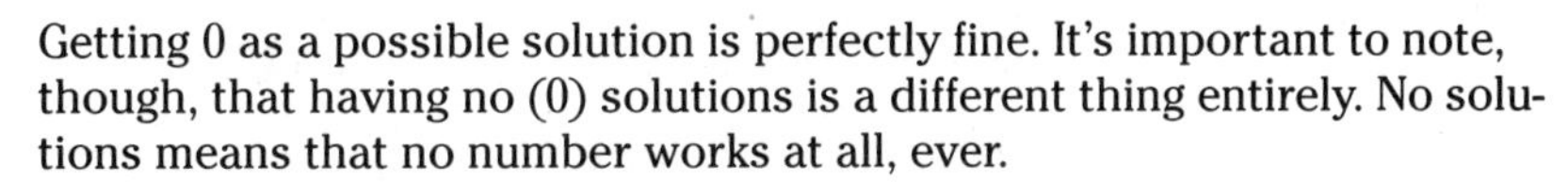

- **If the absolute value is less than (<) or less than or equal to (≤) a negative number, there's no solution.** An absolute value must always be positive (the only thing less than negative numbers is other negative numbers). For instance, the absolute value inequality $|2x - 1| < -3$ doesn't have a solution because the inequality is less than a negative number.

  Getting 0 as a possible solution is perfectly fine. It's important to note, though, that having no (0) solutions is a different thing entirely. No solutions means that no number works at all, ever.

- **If the result is greater than or equal to a negative number, the solutions are infinite.** For example, given the equation $|x - 1| > -5$, x is all real numbers. The left-hand side of this equation is an absolute value, and an absolute value always represents a positive number. Because positive numbers are always greater than negative numbers, these types of inequalities will always have a solution. Any real number that you put into this equation works.

To solve and graph an inequality with an absolute value — for instance, $2|3x - 6| < 12$ — follow these steps:

1. **Isolate the absolute value expression.**

   In this case, divide by both sides by 2 to get $|3x - 6| < 6$.

2. **Break the inequality in two.**

   This process gives you $3x - 6 < 6$ and $3x - 6 > -6$. Did you notice how the inequality sign for the second part changed? When you switch from positives to negatives in an inequality, you must change the inequality sign.

   Don't fall prey to the trap of changing the equation inside the absolute value bars. For example, $|3x - 6| < 6$ doesn't change to $3x + 6 < 6$ or $3x + 6 > -6$.

3. **Solve both inequalities.**

   The solutions to this problem are $x < 4$ and $x > 0$.

4. **Graph the solutions.**

   Create a number line and show the answers to the inequality. Figure 2-2 shows this solution.

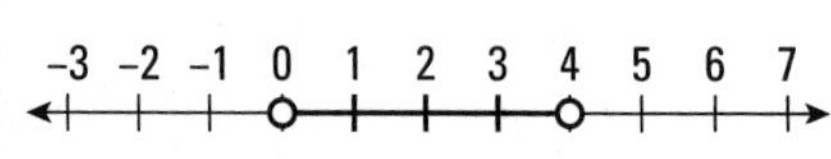

Figure 2-2: The solution to $2|3x - 6| < 12$ on a number line.

Expressing solutions for inequalities with interval notation

Now comes the time to venture into interval notation to express where a set of solutions begins and where it ends. *Interval notation* is another way to express the solution set to an inequality, and it's important because this is how you'll be expressing solution sets in calculus. Most pre-calculus books and some pre-calculus teachers now require all sets to be written in interval notation.

The easiest way to find interval notation is to draw a graph on a number line first as a visual representation of what's going on in the interval.

If the coordinate point of the number isn't included in the problem (for < or >), the interval is called an *open interval.* You show this on the graph with an open circle at the point and by using parentheses in notation. If the point is included in the solution (≤ or ≥), the interval is called a *closed interval,* which you show on the graph with a filled-in circle at the point and by using square brackets in notation.

For example, the solution set $-2 < x \leq 3$ is shown in Figure 2-3. ***Note:*** You can rewrite this solution set as an "AND" statement:

$$-2 < x \text{ AND } x \leq 3$$

In interval notation, you write this solution as

$$(-2, 3]$$

The bottom line: Both of these inequalities *have* to be true at the same time.

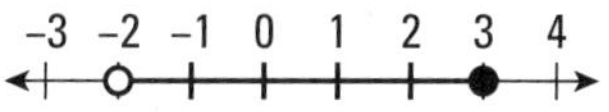

Figure 2-3: The graph of $-2 < x \leq 3$ on a number line.

-3 -2 -1 0 1 2 3 4

You can also graph "OR" statements (also known as *disjoint sets* because the solutions don't overlap). "OR" statements are two different inequalities where one or the other is true. For example, Figure 2-4 shows the graph of $x < -4$ OR $x > -2$.

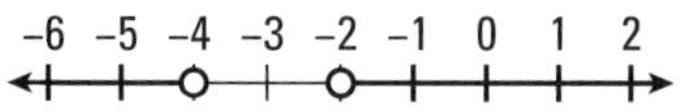

Figure 2-4: The graph of the OR statement $x < -4$ or $x > -2$.

-6 -5 -4 -3 -2 -1 0 1 2

Writing the set for Figure 2-4 in interval notation can be confusing. x can belong to two different intervals, but because the intervals don't overlap, you have to write them separately:

- The first interval is $x < -4$. This interval includes all numbers between negative infinity and –4. Because $-\infty$ isn't a real number, you use an open interval to represent it. So, in interval notation, you write this part of the set as $(-\infty, -4)$.
- The second interval is $x > -2$. This set is all numbers between –2 and positive infinity, so you write it as $(-2, \infty)$.

  You describe the whole set as $(-\infty, -4) \cup (-2, \infty)$. The symbol in between the two sets is the *union symbol* and means that the solution can belong to either interval.

When you're solving an absolute value inequality that's greater than a number, you write your solutions as "OR" statements. Take a look at the following example: $|3x - 2| > 7$. You can rewrite this inequality as $3x - 2 > 7$ or $3x - 2 < -7$. You'll have two solutions: $x > 3$ or $x < -5/3$.

In interval notation, this solution is $(-\infty, -5/3) \cup (3, \infty)$, where $\cup$ represents the union of the two disjoint sets.

Variations on Dividing and Multiplying: Working with Radicals and Exponents

Radicals and exponents (also known as *roots* and *powers*) are two common — and oftentimes frustrating — elements of basic algebra. And, of course, they follow you wherever you go in math, just like a cloud of mosquitoes follows a novice camper. The best thing you can do to prepare for calculus is to be ultra-solid on what can and can't be done when simplifying with exponents and radicals. It's good to have this knowledge so that when more challenging math problems come along, the correct answers come along also. This section gives you the solid background you need for those challenging moments.

Defining and relating radicals and exponents

Before you dig deeper into your work with radicals and exponents, make sure you remember exactly what they are and how they relate to each other:

- **A *radical* is a root of a number.** Radicals are represented by the root sign, $\sqrt{\ }$. For example, if you take the 2nd root of the number 9 (or the *square root*), you get 3 because $3 \cdot 3 = 9$. If you take the 3rd root (or the *cube root*) of 27, you get 3, because $3 \cdot 3 \cdot 3 = 27$. (In equation form, you write $\sqrt[3]{27} = 3$.)

  The square root of any number represents the principal root (the fancy term for the *positive root*) of that number. For example, $\sqrt{16}$ is 4, even though $(-4)^2$ gives you 16 as well. $-\sqrt{16}$ is –4 because it's the opposite of the principal root. When you're presented with the equation $x^2 = 16$, you have to state both solutions: $x = \pm 4$.

  Also, you can't take the square root of a negative number; however, you can take the cube root of a negative number. For example, the cube root of –8 is –2, because $(-2)^3 = -8$.

- **An *exponent* represents the power of a number.** If the exponent is a whole number — say, 2 — it means the base is multiplied by itself that many times — two times, in this case. For example, $3^2 = 3 \cdot 3 = 9$.

  Other types of exponents, including negative exponents and fractional exponents, have different meanings and are discussed in the sections that follow.

Rewriting radicals as exponents (or, creating rational exponents)

Sometimes, a different (yet equivalent) way of expressing radicals makes a solution easier to come by. For instance, it may be easier when given a problem in radical form to rewrite it by using *rational exponents* — an exponent which is a fraction. You can rewrite every radical as an exponent by using the following property — the top number in the resulting rational exponent tells you the power, and the bottom number tells you the root you're taking:

$$x^{m/n} = \sqrt[n]{x^m} = \left(\sqrt[n]{x}\right)^m$$

For example, you can rewrite $\sqrt[3]{8^2}$ or $\left(\sqrt[3]{8}\right)^2$ as $8^{2/3}$.

Fractional exponents are roots and nothing else. For example, $64^{1/3}$ doesn't mean 64 times ⅓ (that's written as $64 \cdot \frac{1}{3}$); and it doesn't mean 1 over 64 to the third power (written as 64^{-3}). In this example, you should be finding the root shown in the denominator (the cube root) and then taking it to the power in the numerator (the first power). So, the answer to $64^{1/3}$ is 4.

The order of these processes really doesn't matter. You can

1. Cube root the 8 and then square that

or

2. Square the 8 and then cube root that

Either way, the equation simplifies to 4. Depending on the original expression, though, it may be easier to take the root first and then take the power, or it may be easier to take the power first. For example $64^{(3/2)}$ is easier if you write it as $(64^{1/2})^3 = 8^3 = 512$ rather than $(64^3)^{1/2}$, because then you'd have to find the square root of 262,144.

Take a look at some steps that illustrate this process. To simplify the expression $\sqrt{x}\left(\sqrt[3]{x^2} - \sqrt[3]{x^4}\right)$, rather than work with the roots, execute the following:

1. **Rewrite the entire expression using rational exponents.**

   Now you have all the properties of exponents available to help you to simplify the expression: $x^{1/2}(x^{2/3} - x^{4/3})$.

2. **Distribute to get rid of the parentheses.**

   When you multiply monomials with the same base, you add the exponents.

   Hence, the exponent on the first term is $\frac{1}{2} + \frac{2}{3} = \frac{7}{6}$. So you get $x^{7/6} - x^{11/6}$.

3. **Because the solution is written in exponential form and not in radical form, as the original expression was, rewrite it to match the original expression.**

   This gives you $\sqrt[6]{x^7} - \sqrt[6]{x^{11}}$.

   Typically, your final answer should be in the same format as the original problem; if the original problem is in radical form, your answer should be in radical form. And if the original problem is in exponential form with rational exponents, your solution should be as well.

Getting a radical out of a denominator: Rationalizing

Another convention of mathematics is that you don't leave radicals in the denominator of an expression when you write it in its final form — called *rationalizing the denominator.* This convention makes it easy to collect like terms, and your answers will be truly simplified.

A numerator can contain a radical, just not the denominator; the final expression may look more complicated in its rational form, but that's what you have to do sometimes.

This section shows you how to get rid of pesky radicals that may show up in the denominator of a fraction. The focus is on two separate situations: expressions that contain one radical in the denominator and expressions that contain two terms in the denominator, at least one of which is a radical.

A square root

Rationalizing expressions with a square root in the denominator is easy. At the end of it all, you're just getting rid of a square root. Normally, the best way to do that in an equation is to square both sides. For example, if $\sqrt{(x-3)} = 5$, $\sqrt{(x-3)}^2 = 5^2$ or $x - 3 = 25$.

However, you can't fall for the trap of rationalizing a fraction by squaring the numerator and the denominator. If you have $\frac{2}{\sqrt{3}}$, for instance, that is *not* equivalent to $\frac{4}{3}$ by squaring the top and bottom.

Instead, follow these steps:

1. **Multiply the numerator and the denominator by the same square root.**

   Whatever you multiply to the bottom of a fraction you must multiply to the top; this way, it's really like you multiplied by one and you didn't change the fraction. Here's what it looks like: $\frac{2}{\sqrt{3}} \cdot \frac{\sqrt{3}}{\sqrt{3}}$.

2. **Multiply the tops and multiply the bottoms and simplify.**

   For this example, you get $\frac{2\sqrt{3}}{3}$.

A cube root

The process for rationalizing a cube root in the denominator is quite similar to that of rationalizaing a square root. To get rid of a cube root in the denominator of a fraction, you must cube it. If the denominator is a cube root to the first power, for example, you multiply both the numerator and the denominator by the cube root to the second power to get the cube root to the third power (in the denominator). Raising a cube root to the third power cancels the root — and you're done!

A root when the denominator is a binomial

You must rationalize the denominator of a fraction when it contains a binomial with a radical. For example, look at the following equations:

$$\frac{3}{x+\sqrt{2}}$$

$$\frac{-2}{\sqrt{x}-\sqrt{5}}$$

Getting rid of the radical in these denominators involves using the conjugate of the denominators. A *conjugate* is a binomial formed by taking the opposite of the second term of the original binomial. The conjugate of $a+\sqrt{b}$ is $a-\sqrt{b}$. The conjugate of $x+2$ is $x-2$; similarly, the conjugate of $x+\sqrt{2}$ is $x-\sqrt{2}$.

Multiplying a number by its conjugate is really the FOIL method in disguise. Remember from algebra that FOIL stands for F (first), O (outside), I (inside), and L (last). So, $(x+\sqrt{2})(x-\sqrt{2}) = x^2 - x\sqrt{2} + x\sqrt{2} - \sqrt{2^2}$. The middle two terms always cancel each other, and the radicals disappear. For this problem, you get $x^2 - 2$.

Take a look at a typical example involving rationalizing a denominator by using the conjugate. First, simplify the expression $\frac{1}{\sqrt{5}-2}$. To rationalize this denominator, you multiply the top and bottom by the conjugate of $\sqrt{5}-2$, which is $\sqrt{5}+2$. The step-by-step breakdown when you do this multiplication is

$$\frac{1}{\sqrt{5}-2}\frac{\left(\sqrt{5}+2\right)}{\left(\sqrt{5}+2\right)} = \frac{\sqrt{5}+2}{\left(\sqrt{5}\right)^2 + \cancel{2\sqrt{5}} - \cancel{2\sqrt{5}} - 4} = \frac{\sqrt{5}+2}{5-4} = \sqrt{5}+2$$

Here's a second example: Suppose you need to simplify $\frac{\sqrt{2}-\sqrt{6}}{\sqrt{10}+\sqrt{8}}$. Follow these steps:

1. **Multiply by the conjugate.**

   The conjugate of $\sqrt{10}+\sqrt{8}$ is $\sqrt{10}-\sqrt{8}$, like in $\frac{\sqrt{2}-\sqrt{6}}{\sqrt{10}+\sqrt{8}}\frac{\left(\sqrt{10}-\sqrt{8}\right)}{\left(\sqrt{10}-\sqrt{8}\right)}$.

2. **Multiply the numerators and denominators.**

   FOIL the top and the bottom. (Tricky, we know!) Here's what we did:

   $$\frac{\sqrt{20}-\sqrt{16}-\sqrt{60}+\sqrt{48}}{\left(\sqrt{10}\right)^2 - \sqrt{80} + \sqrt{80} - \left(\sqrt{8}\right)^2}.$$

3. **Simplify.**

   Both the numerator and denominator simplify first to $\frac{2\sqrt{5}-4-2\sqrt{15}+4\sqrt{3}}{10-8}$, which becomes $\frac{2\sqrt{5}-4-2\sqrt{15}+4\sqrt{3}}{2}$. This simplifies even further because the denominator divides into every term in the numerator, which gives you $\sqrt{5}-2-\sqrt{15}+2\sqrt{3}$.

Simplify any radical in your final answer — always. For example, to simplify a square root, find perfect square root factors: $\sqrt{20}=\sqrt{4\cdot 5}=2\sqrt{5}$. Also, you can add and subtract only radicals that are like terms. This means the number inside the radical and the *index* (which is what tells you whether it's a square root, a cube root, a fourth root, or whatever) are the same.

Chapter 3

The Foundation of Pre-Calc: Functions

In This Chapter

- Identifying, graphing, and translating parent functions
- Piecing together piece-wise functions
- Breaking down and graphing rational functions
- Performing different operations on functions
- Finding and verifying inverses of functions

Maps of the world identify cities as dots and use lines to represent the roads that connect them. Modern country and city maps use a grid system to help users find locations easily. If you can't find the place you're looking for, you look at an index, which gives you a letter and number. This information narrows down your search area, after which you can easily figure out how to get where you're going.

We can take this idea and use it for our own pre-calc purposes through the process of graphing. But instead of naming cities, the dots name points on the coordinate plane (which we discuss more in Chapter 1). A point on this plane relates two numbers to each other, usually in the form of input and output. The whole coordinate plane really is just a big computer, because it's based on input and output, with you as the operating system. This idea of input and output is best expressed mathematically using functions. A *function* is a set of ordered pairs where every x value gives one and only one y value (as opposed to a relation).

This chapter shows you how to perform your role as the operating system, explaining the map of the world of points and lines on the coordinate plane along the way.

Qualities of Even and Odd Functions and Their Graphs

Knowing whether a function is even or odd will help you to graph it because you will only have to graph half of the points. These types of functions are symmetrical, so whatever is on one side is exactly the same as the other side. If a function is even, the graph is symmetrical over the y-axis. If the function is odd, the graph is symmetrical about the origin.

The mathematical definition of an *even function* is $f(-x) = f(x)$ for any value of x. The simplest example of this is $f(x) = x^2$; $f(3) = 9$, and $f(-3) = 9$. Basically, opposite input, same output. Visually speaking, the graph is a mirror image across the y-axis.

The definition for an *odd function* is $f(-x) = -f(x)$ for any value of x. The opposite input gives the opposite output. These graphs will have 180-degree symmetry about the origin. If you turn the graph upside down it looks the same. For example, $f(x) = x^3$ is an odd function because $f(3) = 27$ and $f(-3) = -27$.

Dealing with Parent Functions (The Most Common) and Their Graphs

In mathematics, you'll see certain graphs over and over again. For that reason, these original, common functions are called the *parent graphs,* and they include graphs of quadratic functions, square roots, absolute value, cubics, and cube roots. In this section, we work to get you used to graphing the parent graphs so you can graduate to more in-depth graphing work.

Quadratic functions

Quadratic functions are equations in which the second power, or square, is the highest to which the unknown quantity or variable is raised. In such an equation, *either* x or y is squared, but not both. The graph of $x = y^2$ isn't a function because any x value produces two different y values — look at (4, 2) and (4, –2) for example. The equation y or $f(x) = x^2$ is a quadratic function and is the parent graph for all other quadratic functions.

The shortcut to graphing the function $f(x) = x^2$ is to start at the point (0, 0) (the *origin*) and mark the point, called the *vertex*. Note that the point (0, 0) is the vertex of the parent function only — later, when you transform graphs, the vertex will move around the coordinate plane. In calculus, this point is called a *critical point,* and some pre-calc teachers also use that terminology. Without having to get into the calc definition, it means that the point is special.

The graph of any quadratic function is called a *parabola.* All parabolas have the same basic shape (for more, see Chapter 12). To get the other points, you move horizontally from the vertex 1, up 1^2; over 2, up 2^2; over 3, up 3^2; and so on. This graphing occurs on both sides of the vertex and keeps going, but usually just a couple points on either side of the vertex gives you a good idea of what the graph will look like. Check out Figure 3-1 for an example of a quadratic function in graph form.

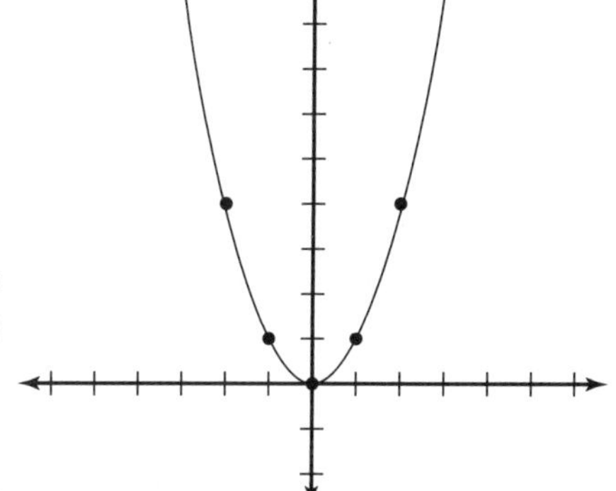

Figure 3-1: Graphing a quadratic function.

Square root functions

A *square root graph* is related to a quadratic graph (see the previous section); the quadratic graph is $f(x) = x^2$, while the square root graph is $g(x) = x^{1/2}$. The graph of a square-root function looks like a parabola that has been rotated 90° clockwise. You can also write the square root function as $g(x) = \sqrt{x}$.

However, only half of the parabola exists, for two reasons. Its parent graph exists only when x is positive (because you can't find the square root of negative numbers [and keep them real, anyway]) and when $g(x)$ is positive (because when you see $\sqrt{x}$, you're being asked to find only the principal or positive root).

This graph starts at the origin (0, 0) and then moves to the right 1 position, up $\sqrt{1}$ (1); to the right 2, up $\sqrt{2}$; to the right 3, up $\sqrt{3}$; and so on. Check out Figure 3-2 for an example of this graph.

Notice that the values you get by plotting consecutive points don't exactly give you the nicest numbers. Instead, try picking values for which you can easily find the square root. Here's how this works: Start at the origin and go right 1, up $\sqrt{1}$ (1); right 4, up $\sqrt{4}$ (2); right 9, up $\sqrt{9}$ (3); and so on.

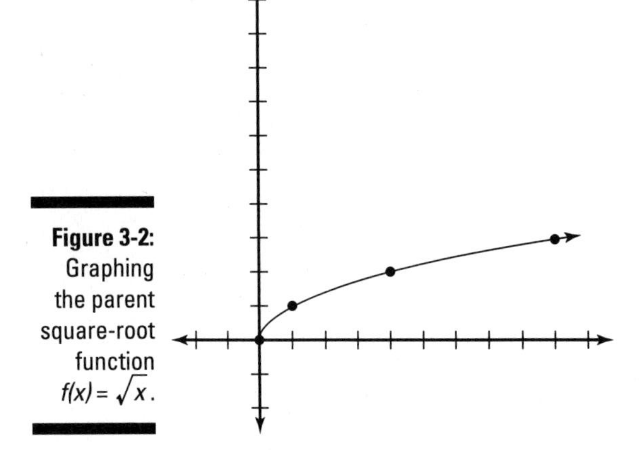

Figure 3-2: Graphing the parent square-root function $f(x) = \sqrt{x}$.

Absolute value functions

The absolute value parent graph of the function $y = |x|$ turns all inputs non-negative (0 or positive). To graph absolute value functions, you start at the origin and move in both directions along the *x*-axis and the *y*-axis from there: over 1, up 1; over 2, up 2; and on and on forever. Figure 3-3 shows this graph in action.

Figure 3-3: Staying positive with the graph of an absolute value function.

Cubic functions

In a *cubic function,* the highest degree on any variable is three — $f(x) = x^3$ is the parent function. You start graphing the cubic function parent graph at its critical point, which is also the origin (0, 0). The origin isn't, however, a critical point for every function.

From the critical point, the cubic graph moves right 1, up 1^3; right 2, up 2^3; and so on. The function x^3 is an odd function, so you rotate half of the graph 180° about the origin to get the other half. Or, you can move left –1, down $(-1)^3$; left –2, down $(-2)^3$; and so on. How you plot is based on your personal preference. Consider $g(x) = x^3$ in Figure 3-4.

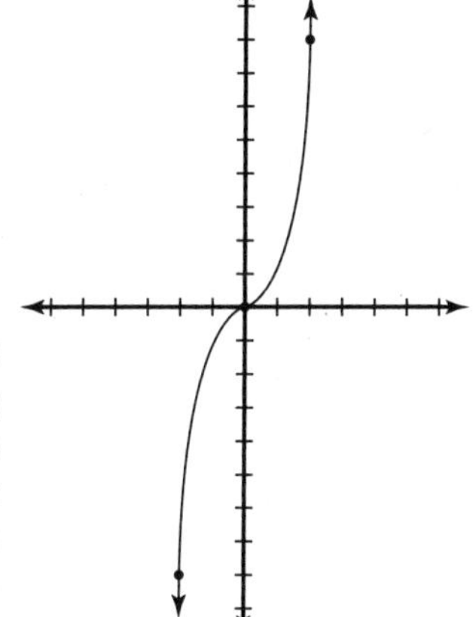

Figure 3-4: Putting the cubic parent function in graph form.

Cube root functions

Cube root functions are related to cubic functions in the same way that square root functions are related to quadratic functions: The cube root function graph is the cubic function graph rotated 90° clockwise. You write cubic functions as $f(x) = x^3$ and cube root functions as $g(x) = x^{1/3}$, or $g(x) = \sqrt[3]{x}$.

It's important to note that a cube root function is odd because it helps you graph it. The critical point of the cube root parent graph is at the origin (0, 0), as shown in Figure 3-5.

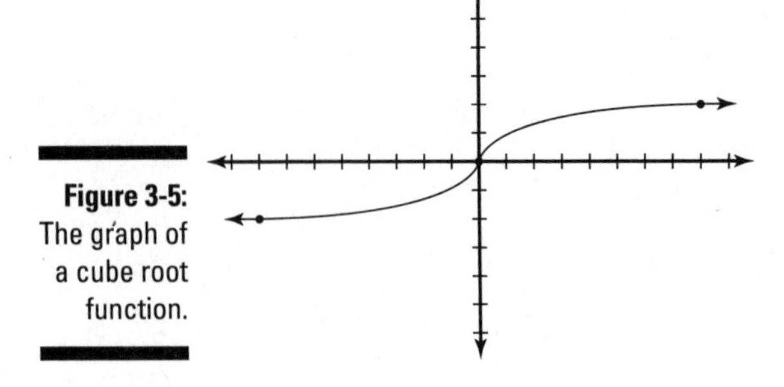

Figure 3-5: The graph of a cube root function.

Transforming the Parent Graphs

In certain situations, you need to use a parent function to get the graph of a more complicated version of the same function. For instance, you can graph each of the following by *transforming* its parent graph:

$$f(x) = -2(x + 1)^2 - 3$$

$$g(x) = \frac{1}{4}|x - 2|$$

$$h(x) = (x - 1)^4 + 2$$

As long as you have the graph of the parent function, you can transform it by using the rules we describe in this section. When using a parent function for this purpose, you can choose from different types of transformations:

- **Vertical transformations** cause the parent graph to stretch or shrink vertically.
- **Horizontal transformations** cause the parent graph to stretch or shrink horizontally.
- **Translations** cause the parent graph to shift left, right, up, or down (or a combined shift both horizontally and vertically).
- **Reflections** flip the parent graph over a horizontal or vertical line. They do just as they sound: mirror the parent graphs (unless other transformations are involved, of course).

The methods to transform quadratic functions also work for all other types of common functions, such as square roots. A function is always a function, so the rules for transforming functions always apply, no matter what type of function you're dealing with.

And if you can't remember these shortcut methods later on, you can always take the long route: picking random values for x and plugging them into the function to see what y values you get.

Vertical transformations

A number (or *coefficient*) multiplying in front of a function causes a *vertical transformation.* This term is a fancy math term for height. The coefficient always affects the height of each and every point in the graph of the function. We call the vertical transformation a *stretch* if the coefficient is greater than 1 and a *shrink* if the coefficient is between 0 and 1.

For example, the graph of $f(x) = 2x^2$ takes the graph of $f(x) = x^2$ and stretches it by a vertical factor of two. That means that each time you plot a point vertically on the graph, the value gets multiplied by two (making the graph twice as tall at each point). So, from the vertex, you move over 1, up $2 \cdot 1^2$ (2); over 2, up $2 \cdot 2^2$ (8); over 3, up $2 \cdot 3^2$ (18); and so on. Figure 3-6 shows two different graphs to illustrate vertical transformation.

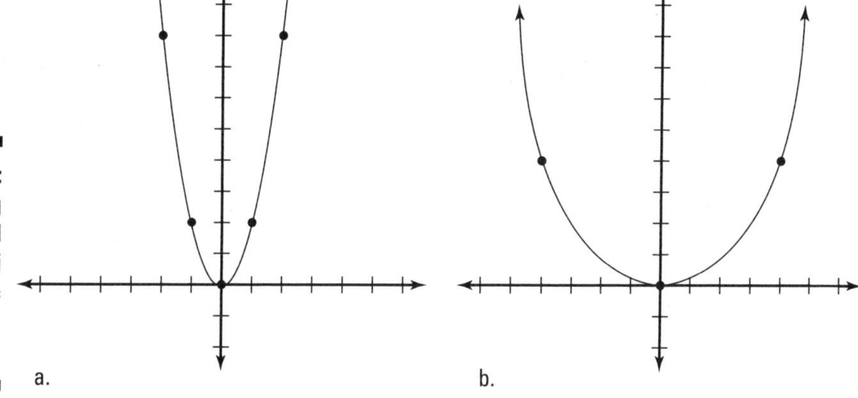

Figure 3-6: Graphing the vertical transformation of $f(x) = 2x^2$ and $g(x) = \frac{1}{4}x^2$.

The transformation rules apply to *any* function, so Figure 3-7, for instance, shows $f(x) = 4\sqrt{x}$. The 4 is a vertical stretch; it makes the graph four times as tall at every point: right 1, up $4 \cdot \sqrt{1}$ (4); right 4, up $4 \cdot \sqrt{4}$ (notice that we're using numbers that you can take the square root of easily to make graphing a simple task); and so on.

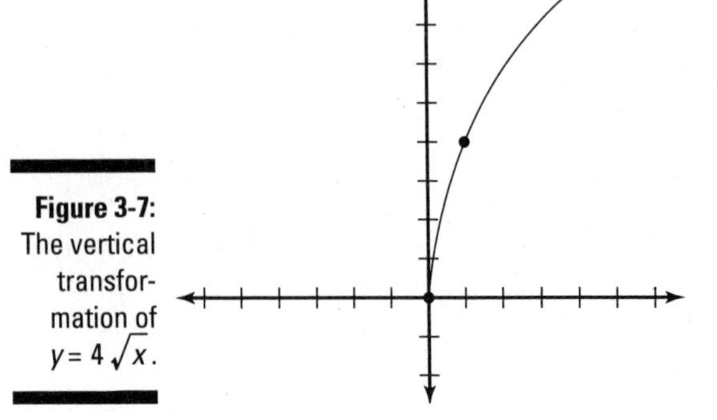

Figure 3-7: The vertical transformation of $y = 4\sqrt{x}$.

Horizontal transformations

Horizontal transformation means to stretch or shrink a graph along the *x*-axis. A number multiplying a variable inside a function affects the horizontal position of the graph — a little like the fast-forward or slow-motion button on a remote control, making the graph move faster or slower. A coefficient greater than 1 causes the function to stretch horizontally, making it appear to move faster. A coefficient between 0 and 1 makes the function appear to move slower, or a horizontal shrink.

For instance, look at the graph of $f(x) = |2x|$ (see Figure 3-8). The distance between any two consecutive values from the parent graph $|x|$ along the *x*-axis is always 1. If you set the inside of the new, transformed function equal to the distance between the *x* values, you'd get $2x = 1$. Solving the equation gives you $x = \frac{1}{2}$. This is how far you step along the *x*-axis. Beginning at the origin (0, 0), you move right ½, up $|\frac{1}{2}|$; right 1, up $|1|$; right $\frac{3}{2}$, up $|\frac{3}{2}|$; and so on.

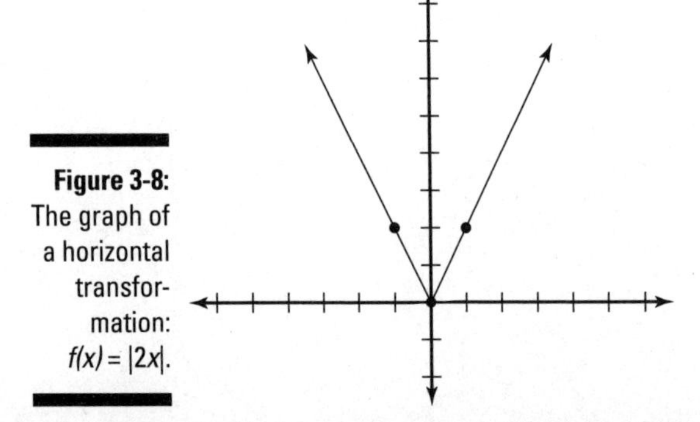

Figure 3-8: The graph of a horizontal transformation: $f(x) = |2x|$.

Translations

Moving a graph horizontally or vertically is called a *translation.* In other words, every point on the parent graph is translated left, right, up, or down. In this section, you find information on both kinds of translations: horizontal shifts and vertical shifts.

Horizontal shifts

A number adding or subtracting inside the parentheses (or other grouping device) of a function creates a *horizontal shift.* Such functions are written in the form $f(x - h)$, where h represents the horizontal shift.

The numbers in this function do the opposite of what they look like they should do. For example, if you have the equation $g(x) = (x - 3)^2$, the graph moves to the right three units; in $h(x) = (x + 2)^2$, the graph moves to the left two units.

Why does it work this way? Examine the parent function $f(x) = x^2$ and the horizontal shift $g(x) = (x - 3)^2$. When $x = 3$, $f(3) = 3^2 = 9$ and $g(3) = (3 - 3)^2 = 0^2 = 0$. The $g(x)$ function acts like the $f(x)$ function when x was 0. In other words, $f(0) = g(3)$. It's also true that $f(1) = g(4)$. Each point on the parent function gets moved to the right by three units; hence, three is the horizontal shift for $g(x)$.

Try your hand at graphing $g(x) = \sqrt{(x - 1)}$. Because – 1 is underneath the square root sign, this is a horizontal shift — the graph gets moved to the right one position. If $k(x) = \sqrt{x}$ (the parent function), you'll find that $k(0) = g(1)$, which is to the right by one. Figure 3-9 shows the graph of $g(x)$.

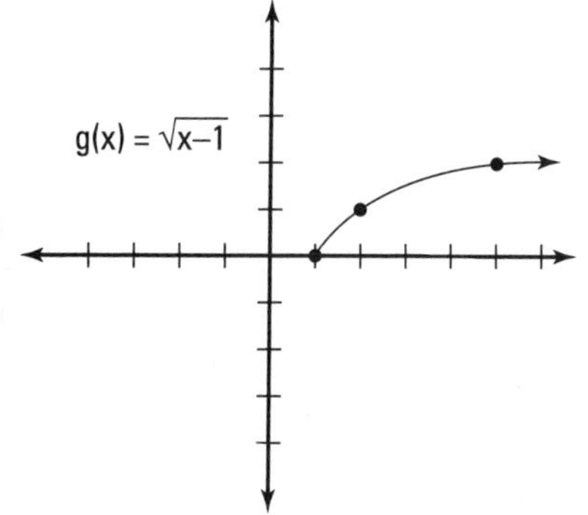

Figure 3-9: The graph of a horizontal shift: $g(x) = \sqrt{(x - 1)}$.

Vertical shifts

Adding or subtracting numbers completely separate from the function causes a *vertical shift* in the graph of the function. Consider the expression $f(x) + v$, where v represents the vertical shift. Notice that the addition of the variable exists outside the function.

Vertical shifts are less complicated than horizontal shifts (see the previous section), because reading them tells you exactly what to do. In the equation $f(x) = x^2 - 4$, you can probably guess what the graph is going to do. Right! It moves down four units, while the graph of $g(x) = x^2 + 3$ moves up three.

Note: You see no vertical stretch or shrink for either $f(x)$ or $g(x)$, because the coefficient in front of x^2 for both functions is 1. If another number multiplied with the functions, you'd have a vertical stretch or shrink.

To graph the function $h(x) = |x| - 5$, notice that the vertical shift is down five units. Figure 3-10 shows this translated graph.

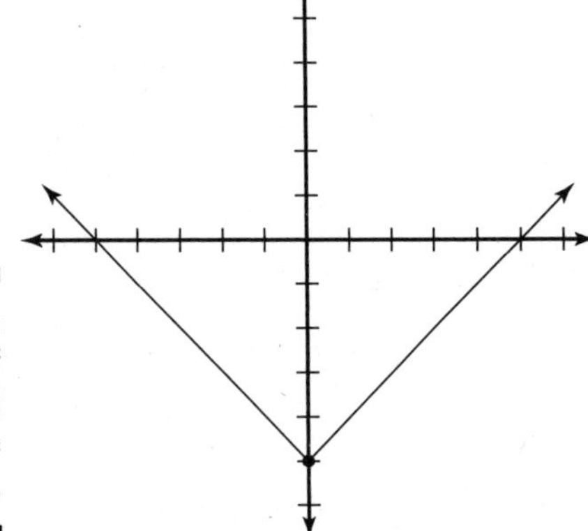

Figure 3-10: The graph of a vertical shift: $h(x) = |x| - 5$.

When translating a cubic function, the critical point moves horizontally or vertically, so the point of symmetry around which the graph is based moves as well. In the function $f(x) = x^3 - 4$ in Figure 3-11, for instance, the point of symmetry is (0, –4).

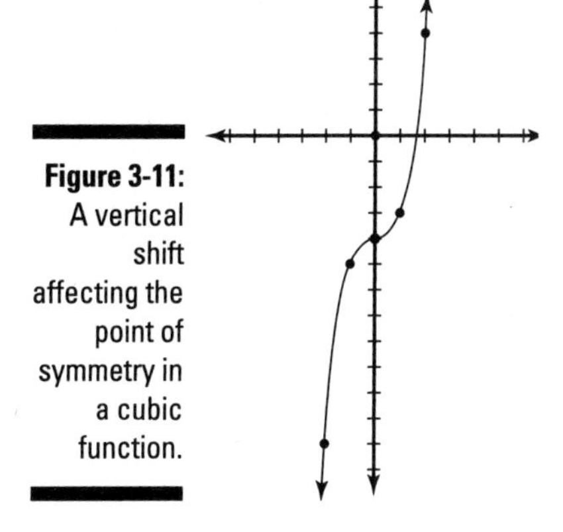

Figure 3-11: A vertical shift affecting the point of symmetry in a cubic function.

Reflections

Reflections take the parent function and provide a mirror image of it over either a horizontal or vertical line. You'll come across two types of reflections:

- **A negative number multiplies the whole function (as in $f(x) = -1\sqrt{x}$):** The negative outside the function reflects across a horizontal line, because it makes the output value negative if it was positive and positive if it was negative. Look at Figure 3-12, which shows the parent function $f(x) = x^2$ and the horizontal reflection $g(x) = -1x^2$. If you find the value of both functions at the same number in the domain, you'll get opposite values in the range. For example, if $x = 4$, $f(4) = 16$ and $g(4) = -16$.

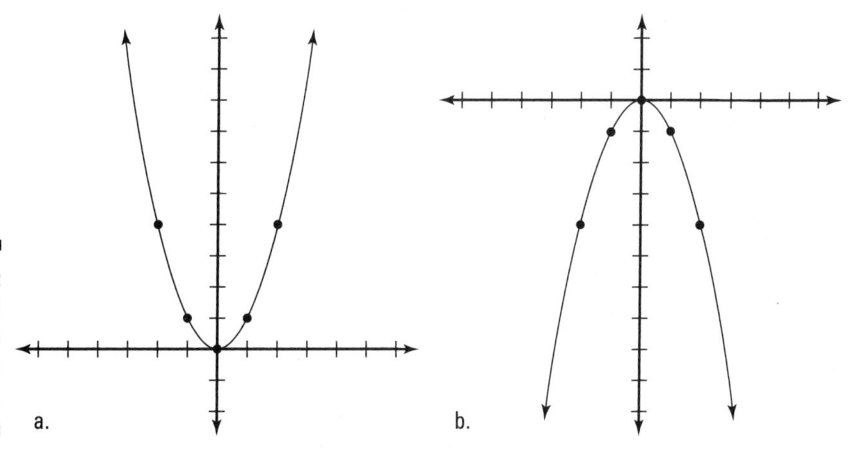

Figure 3-12: A horizontal reflection mirrors up and down.

- **A negative number multiplies only the input x (as in $g(x) = \sqrt{-x}$):** Vertical reflections work the same as horizontal reflections, except the reflection occurs across a vertical line and reflects from side to side rather than up and down. You now have a negative inside the function. For this reflection, evaluating opposite inputs in both functions will yield the same output. For example, if $f(x) = \sqrt{x}$, you can write its vertical reflection as $g(x) = \sqrt{-x}$. When $f(4) = 2$, $g(-4) = 2$ as well (check out the graph in Figure 3-13).

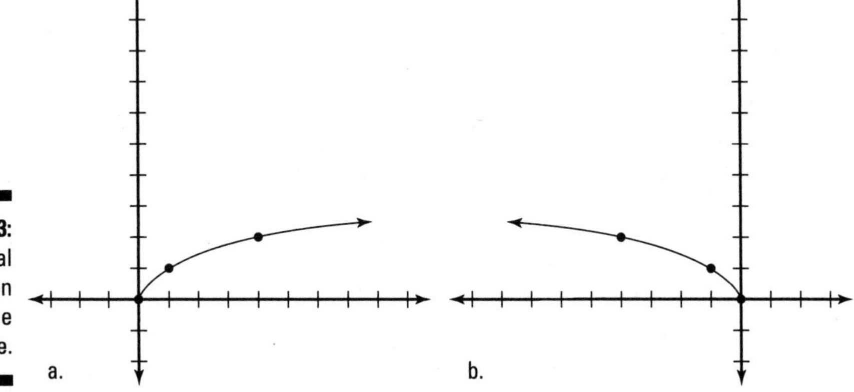

Figure 3-13: A vertical reflection mirrors side to side.

Combining various transformations (a transformation in itself!)

Certain mathematical expressions allow you to combine stretching, shrinking, translating, and reflecting a function all into one graph. An expression that shows all the transformations in one is $a \cdot f[c(x - h)] + v$, where

a is the vertical transformation.

c is the horizontal transformation.

h is the horizontal shift.

v is the vertical shift.

For instance, $f(x) = -2(x-1)^2 + 4$ moves right one and up four, stretches twice as tall, and reflects upside down. Figure 3-14 shows that

(a) is the parent graph: $k(x) = x^2$.

(b) is the horizontal shift to the right by one: $h(x) = (x-1)^2$.

(c) is the vertical shift up by four: $g(x) = (x-1)^2 + 4$.

(d) is the vertical stretch of two: $f(x) = -2(x-1)^2 + 4$. (Notice that because the value was negative, the graph was also turned upside down.)

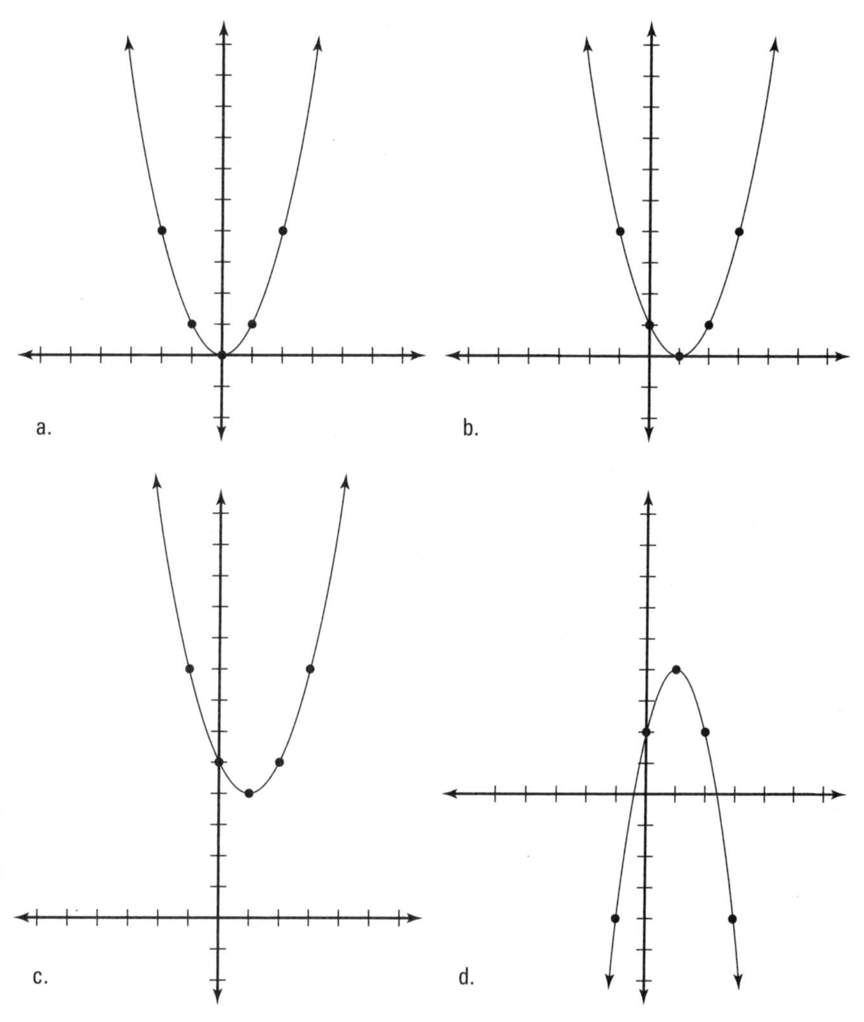

Figure 3-14: A view of multiple transformations.

Allow us to show one more transformation — and illustrate the importance of the order of the process. You graph the function $q(x) = \sqrt{(4-x)}$ with the following steps:

1. **Rewrite the function in the form $a \cdot f[c(x-h)] + v$.**

   First reorder the function so that the x comes first (in descending order). And don't forget the negative sign! Here it is: $q(x) = \sqrt{(-x+4)}$.

2. **Factor out the coefficient in front of the x.**

   You now have $q(x) = \sqrt{-1(x-4)}$.

3. **Reflect the parent graph.**

   Because the –1 is inside the square-root function, $q(x)$ is a vertical reflection of $f(x) = \sqrt{x}$.

4. **Shift the graph.**

   The factored form of $q(x)$ (from Step 2) reveals that the horizontal shift is four to the right.

Figure 3-15 shows the graph of $q(x)$.

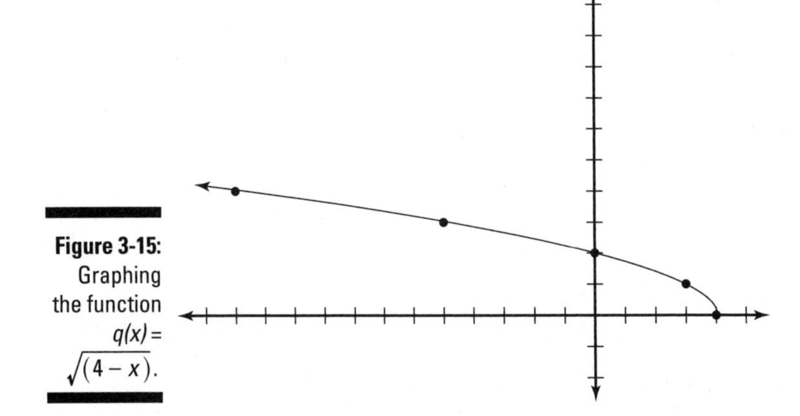

Figure 3-15: Graphing the function $q(x) = \sqrt{(4-x)}$.

Transforming functions point by point

For some problems, you may be required to transform a function given only a set of random points on the coordinate plane. Quite frankly, your textbook or teacher will be making up some new kind of function that has never existed before. Just remember that *all* functions follow the same transformation rules, not just the common functions that we've explained so far in this chapter.

For example, the graphs of $y = f(x)$ and $y = \frac{1}{2}f(x-4)-1$ are shown in Figure 3-16.

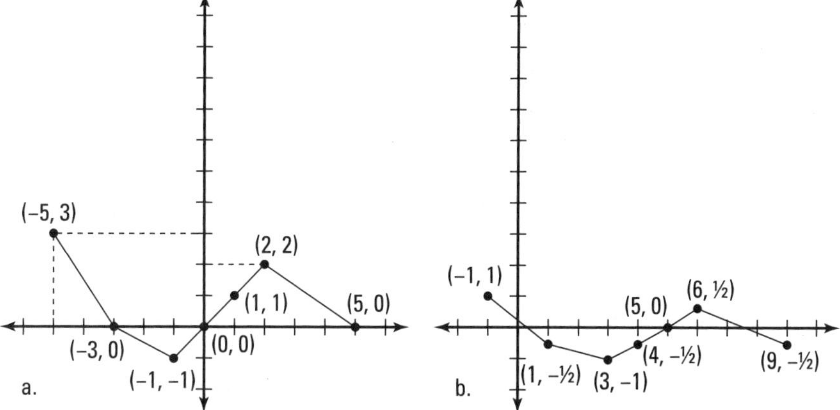

Figure 3-16: The graph of $y = f(x)$ and $y = \frac{1}{2}f(x-4)-1$.

Figure 3-16a represents the parent function (the set of random points). Figure 3-16b transforms the parent function by translating it to the right four units and down one unit, and by shrinking it by a factor of ½. The first random point on the parent function is (–5, 3); shifting it to the right by four will put you at (–1, 3), and shifting it down one will put you at (–1, 2). Because the translated height is two, you shrink the function by finding ½ of 2. You end up at the final point, which is (–1, 1).

You must repeat that process for as many points as you see on the original graph to get the transformed one.

Graphing Functions that Have More than One Rule: Piece-Wise Functions

Piece-wise functions are functions broken into pieces, depending on the input. A piece-wise function will have more than one function, but each function will be defined only on a specific interval. Basically, the output depends on the input, and the graph of the function sometimes will look like it has literally been broken into pieces.

For example, the following represents a piece-wise function:

$$f(x)=\begin{cases} x^2-1 & \text{if } x\le -2 \\ |x| & \text{if } -2<x\le 3 \\ x+8 & \text{if } x>3 \end{cases}$$

This function is broken into three pieces, depending on the domain values for each piece:

- The first piece is the quadratic function $f(x) = x^2 - 1$ and exists only on the interval $(-\infty, -2]$. As long as the input for this function is less than –2, the function exists in the first piece (the top line) only.
- The second piece is the absolute value function $f(x) = |x|$ and exists only on the interval $(-2, 3]$.
- The third piece is the linear function $f(x) = x + 8$ and exists only on the interval $(3, \infty)$.

To graph this example function, follow these steps:

1. **Lightly sketch out a quadratic function that moves down one (see the earlier section "Quadratic functions") and darken all the values to the left of $x = -2$.**

   Because of the interval of the quadratic function of the first piece, you darken all points to the left of –2. And because $x = -2$ is included (the interval is $x \le -2$), the circle at $x = -2$ is filled in.

2. **Between –2 and 3, the graph moves to the second function of the equation ($|x|$ if $-2 < x \le 3$); sketch the absolute value graph (see the earlier section "Absolute value functions"), but pay attention only to the x values between –2 and 3.**

   You don't include –2 (open circle), but the 3 is included (closed circle).

3. **For x values bigger than 3, the graph follows the third function of the equation: $x + 4$ if $x > 3$.**

   You sketch in this linear function where $b = 4$ with a slope of 1, but only to the right of $x = 3$ (that point is an open circle). The finished product is shown in Figure 3-17.

Notice that you can't draw the graph of this piece-wise function without lifting your pencil from the paper. Mathematically speaking, this is called a *discontinuous function.* You'll get to practice discontinuities with rational functions later in this chapter.

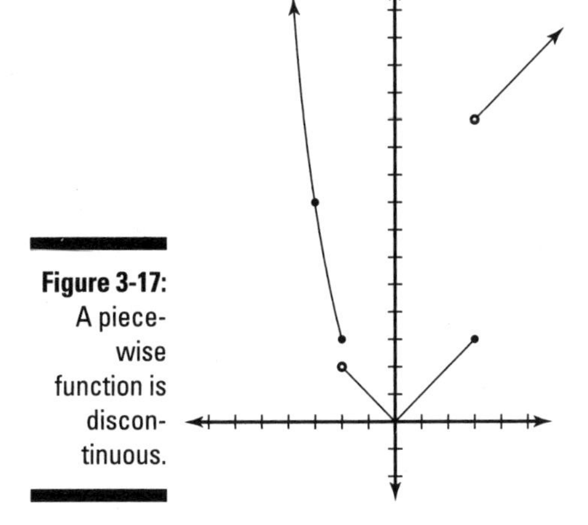

Figure 3-17: A piece-wise function is discontinuous.

Calculating Outputs for Rational Functions

In addition to the common parent functions, you'll have to graph another type of function in pre-calculus: *rational functions,* which basically are functions where the variable appears in the denominator of a fraction. (This isn't the same as the rational exponents you see in Chapter 2, though. The word "rational" means fraction; before the fraction was the exponent, and now it's the entire function.)

The mathematical definition of a *rational function* is a function that can be expressed as the quotient of two polynomials, such that $f(x) = \frac{p(x)}{q(x)}$ where the degree of $q(x)$ is greater than zero.

The variable in the denominator of a rational function could create a situation where the denominator is zero for certain numbers in the domain. Of course, division by zero is an undefined value in mathematics. Typically in a rational function, you'll find at least one value of x for which the rational function is undefined, at which point the graph will have an *asymptote* — the graph gets closer and closer to that value but never crosses it (in the case of vertical asymptotes). Knowing in advance that these values of x are undefined helps you to graph.

In the following sections, we show you the steps involved in finding the outputs of (and ultimately graphing) rational functions.

Step 1: Search for vertical asymptotes

Having the variable on the bottom of a fraction is a problem, because the denominator of a fraction can never be zero. Usually, some domain value(s) of x makes the denominator zero. The function will "jump over" this value in the graph, creating what's called a *vertical asymptote.* Graphing the vertical asymptote first shows you the number in the domain where your graph won't pass through. The graph will approach this point but never reach it. With that in mind, what value(s) for x can you *not* plug into the rational function?

The following are all rational equations:

$$f(x) = \frac{3x - 1}{x^2 + 4x - 21}$$

$$g(x) = \frac{6x + 12}{4 - 3x}$$

$$h(x) = \frac{x^2 - 9}{x + 2}$$

Now try to find the value for x in which the function is undefined. Use the following steps to find the vertical asymptote for $f(x)$ first:

1. **Set the denominator of the rational function equal to zero.**

   For $f(x)$, $x^2 + 4x - 21 = 0$.

2. **Solve this equation for x.**

   Because this equation is a quadratic (see the earlier section "Quadratic functions" and Chapter 4), try to factor it. This quadratic factors to $(x + 7)(x - 3) = 0$. Set each factor equal to zero to solve. If $x + 7 = 0$, $x = -7$. If $x - 3 = 0$, $x = 3$. Your two vertical asymptotes, therefore, are $x = -7$ and $x = 3$.

Now you can find the vertical asymptote for $g(x)$. Follow the same set of steps:

$4 - 3x = 0$. That was easy!

$x = 4/3$

Now you have your vertical asymptote for $g(x)$. Time to do it all again for $h(x)$:

$x + 2 = 0$. Easy as pie.

$x = -2$

Keep these equations for the vertical asymptotes close by because you'll need them when you graph later.

Step 2: Look for horizontal asymptotes

To find a horizontal asymptote of a rational function, you need to look at the degree of the polynomials in the numerator and the denominator. The *degree* is the highest power of the variable in the polynomial expression. Here's how you proceed:

- If the denominator has the bigger degree (like in the $f(x)$ example in the previous section), the horizontal asymptote automatically is the x-axis, or $y = 0$.
- If the numerator and denominator have an equal degree, you must divide the *leading coefficients* (the coefficients of the terms with the highest degrees) to find the horizontal asymptote.

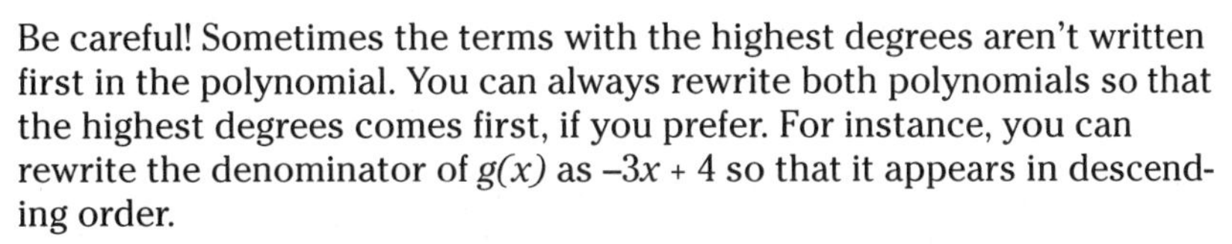

  Be careful! Sometimes the terms with the highest degrees aren't written first in the polynomial. You can always rewrite both polynomials so that the highest degrees comes first, if you prefer. For instance, you can rewrite the denominator of $g(x)$ as $-3x + 4$ so that it appears in descending order.

  The function $g(x)$ has equal degrees on top and bottom. To find the horizontal asymptote, divide the leading coefficients on the highest-degree terms: $y = 6 \div -3$, or $y = -2$. You now have your horizontal asymptote for $g(x)$. Hold onto that equation for graphing!

- If the numerator has the bigger degree of exactly one more than the denominator, the graph will have an oblique asymptote; see Step 3 for more information on how to proceed.

Step 3: Seek out oblique asymptotes

Oblique asymptotes are neither horizontal nor vertical. In fact, an oblique asymptote doesn't even have to be a straight line at all; it can be a slight curve or a really complicated curve.

To find an oblique asymptote, you have to use long division of polynomials to find the quotient. You take the denominator of the rational function and divide it into the numerator. The quotient (neglecting the remainder) gives you the equation of the line of your oblique asymptote.

We cover long division of polynomials in Chapter 4. You must understand long division of polynomials in order to complete the graph of a rational function with an oblique asymptote.

The $h(x)$ example from Step 1 has an oblique asymptote because the numerator has the higher degree in the polynomial. By using long division, you get a quotient of $x - 2$. This means the oblique asymptote follows the equation $y = x - 2$ (the quotient). Because this is a first-degree equation, you graph it by using slope-intercept form. Hold onto this oblique asymptote because graphing is coming right up!

Step 4: Locate the x- and y-intercepts

The final piece of the puzzle is to find the intercepts (where the line or curve crosses the x- and y-axes) of the rational function, if any exist:

- To find the y-intercept of an equation, set $x = 0$. (Plug in 0 wherever you see x.) The y-intercept of $f(x)$ from Step 1, for instance, is $1/21$.
- To find the x-intercept of an equation, set $y = 0$.

For any rational function, the shortcut is to set the numerator equal to zero and then solve. Sometimes when you do this, however, the equation you get is unsolvable, which means that the rational function doesn't have an x-intercept.

The x-intercept of $f(x)$ is $1/3$.

Now find the intercepts for $g(x)$ and $h(x)$ from Step 1. Doing so, you'll find

- $g(x)$ has a y-intercept at 3 and an x-intercept at -2.
- $h(x)$ has a y-intercept at $-9/2$ and x-intercepts at ± 3.

Putting the Output to Work: Graphing Rational Functions

After you calculate all the asymptotes and the x- and y-intercepts for a rational function (we take you through that process in the preceding section), you have all the information you need to start graphing the rational function. Graphing a rational function is all about the degree of the numerator and denominator. Because the numerator and the denominator are polynomials, their degrees are easy to find — just look for the highest exponent in each.

There are three types of rational functions, depending on the degree:

- The denominator has the greater degree
- The numerator and the denominator have equal degrees
- The numerator has the greater degree

The following sections describe how to graph in each case.

The denominator has the greater degree

Rational functions are really just fractions. If you look at several fractions where the numerator stays the same but the denominator gets bigger, the whole fraction gets smaller. For instance, look at $\frac{1}{2}$, $\frac{1}{20}$, $\frac{1}{200}$, and $\frac{1}{2,000}$.

In any rational function where the denominator has a greater degree as values of x get infinitely large, the fraction gets infinitely smaller until it approaches zero (this process is called a *limit;* you can see it again in Chapter 17). The following sections break down graphing this type of function.

Graphing the info you know

When the denominator has the greater degree, you begin by graphing the information that you know for $f(x)$. (See Step 1 from the section "Calculating Outputs for Rational Functions," as well as the info from the subsequent steps, for the full take on this function.) Figure 3-18 shows all these parts of the graph clearly labeled:

1. **Draw the vertical asymptote(s).**

   Whenever you graph asymptotes, be sure to use dotted lines, not solid lines, because the asymptotes aren't part of the rational function.

   For $f(x)$, you find that the vertical asymptotes are $x = -7$ and $x = 3$, so you should draw two dotted, vertical lines, one at $x = -7$ and another at $x = 3$.

2. **Draw the horizontal asymptote(s).**

   Continuing with the example, the horizontal asymptote is $y = 0$ — or the x-axis.

3. **Plot the x-intercept(s) and the y-intercept(s).**

   In the example, the y-intercept is $y = \frac{1}{21}$, and the x-intercept is $x = \frac{1}{3}$.

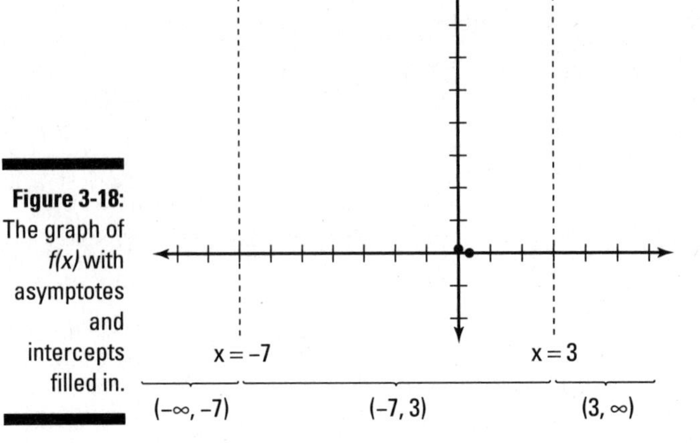

Figure 3-18: The graph of $f(x)$ with asymptotes and intercepts filled in.

Filling in the blanks by plotting outputs of test values

The vertical asymptotes divide the graph and the domain of $f(x)$ into three intervals: $(-\infty, -7)$, $(-7, 3)$, and $(3, \infty)$. For each of these three intervals, you must pick at least one test value and plug it into the original rational function; you do this to determine if the graph on that interval is above or below the horizontal asymptote (the x-axis). Follow these steps:

1. **Test a value in the first interval.**

   In the example, the first interval is $(-\infty, -7)$, so you can choose any number you want as long as it's less than –7. We'll choose $x = -8$. Now you evaluate $f(-8) = {}^{-25}/_{11}$. This negative value tells you that the function is under the horizontal asymptote on the first interval only.

2. **Test a value in the second interval.**

   If you look at the second interval $(-7, 3)$ in Figure 3-18, you'll realize that you already have two test points located in it. The y-intercept has a positive value, which tells you that the graph is above the horizontal asymptote for that part of the graph.

   Now here comes the curve ball: It stands to reason that a graph should never cross an asymptote; it should just get closer and closer to it. In this case, there's an x-intercept, which means that the graph actually crosses its own horizontal asymptote. The graph becomes negative for the rest of this interval.

Sometimes the graphs of rational functions will cross a horizontal asymptote, and sometimes they won't. In this case, where the denominator has a greater degree, and the horizontal asymptote is the x-axis, it depends on whether the function has roots or not. You can find out by setting the numerator equal to zero and solving the equation. If you find a solution, there will be a zero and the graph will cross the x-axis. If not, the graph won't cross the x-axis.

Vertical asymptotes are the only asymptotes that will *never* be crossed. A horizontal asymptote actually tells you what value the graph is approaching for infinitely large or small values of x.

3. **Test a value in the third interval.**

   For the third interval $(3, \infty)$, we use the test value of 4 (you can use any number greater than 3) to determine the location of the graph on the interval. We get $f(4) = 1$, which tells you that the graph is above the horizontal asymptote for this last interval.

Knowing a test value in each interval, you can plot the graph by starting at a test value point and moving from there toward both the horizontal and vertical asymptotes. Figure 3-19 shows the complete graph of $f(x)$.

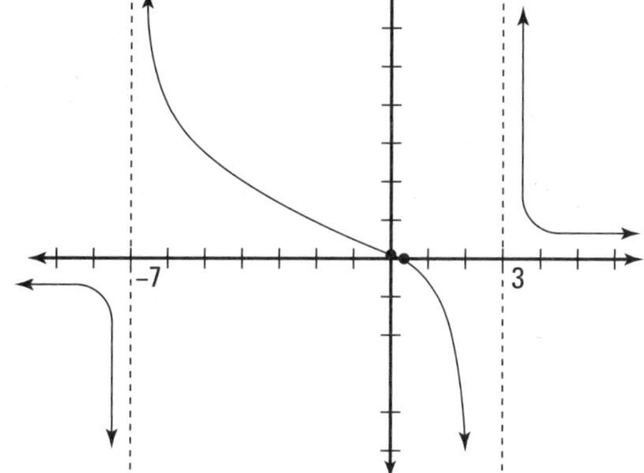

Figure 3-19: The final graph of $f(x)$.

The numerator and denominator have equal degrees

Rational functions with equal degrees in the numerator and denominator behave the way that they do because of limits (see Chapter 15). What you need to remember is that the horizontal asymptote is the quotient of the leading coefficients of the top and the bottom of the function (see the section "Step 2: Finding horizontal asymptotes" for more info).

Take a look at $g(x)$ — $\frac{6x+12}{4-3x}$ — which has equal degrees on the variables for each part of the fraction. Follow these simple steps to graph $g(x)$:

1. **Sketch the vertical asymptote(s) for $g(x)$.**

   From your work in the previous section, you find only one vertical asymptote at $x = 4/3$, which means you have only two intervals to consider: $(-\infty, 4/3)$ and $(4/3, \infty)$.

2. **Sketch the horizontal asymptote for $g(x)$.**

   You find in Step 2 from the previous section that the horizontal asymptote is $y = -2$. So, you sketch a horizontal line at that position.

3. **Plot the x- and y-intercepts for $g(x)$.**

   You find in Step 4 from the previous section that the intercepts are $x = -2$ and $y = 3$.

4. **Use test values of your choice to determine whether the graph is above or below the horizontal asymptote.**

   The two intercepts are already located on the first interval and above the horizontal asymptote, so you know that the graph on that entire interval is above the horizontal asymptote. Now, choose a test value for the second interval greater than $4/3$. We chose $x = 2$. Substituting this into the function $g(x)$ gives you -12. You know that -12 is waaaay under -2, so you know that the graph lives under the horizontal asymptote in this second interval.

Figure 3-20 shows you the complete graph of $g(x)$.

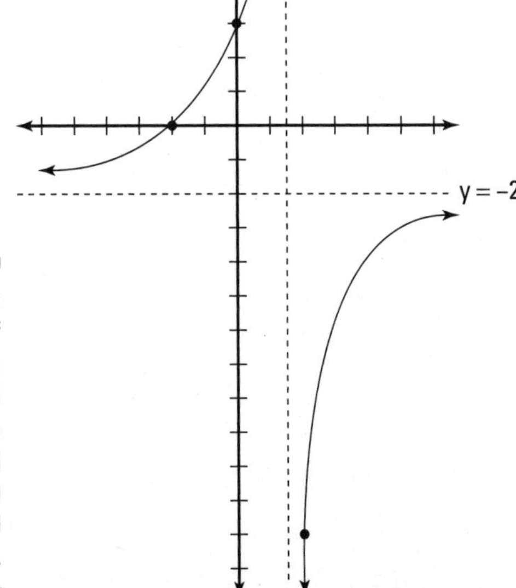

Figure 3-20: The graph of $g(x)$, which is a rational function with equal degrees on top and bottom.

The numerator has the greater degree

Rational functions where the numerator has the greater degree don't actually have horizontal asymptotes. Instead, they have oblique asymptotes, which you find by using long division (see Chapter 4).

It's time to graph *h(x)*, which is $\frac{x^2-9}{x+2}$ from Step 1 of the previous section:

1. **Sketch the vertical asymptote(s) of *h(x)*.**

   You find only one vertical asymptote for this rational function in Step 1 — $x = -2$. You find only two intervals for this graph because there's only one vertical asymptote — $(-\infty, -2)$ and $(-2, \infty)$.

2. **Sketch the oblique asymptote of *h(x)*.**

   Because the numerator of this rational function has the greater degree, the function has an oblique asymptote. Using long division, you find that the oblique asymptote follows the equation $y = x - 2$.

3. **Plot the *x*- and *y*-intercepts for *h(x)*.**

   You find that the *x*-intercepts are ± 3 and the *y*-intercept is $-9/2$.

4. **Use test values of your choice to determine whether the graph is above or below the oblique asymptote.**

   Notice that the intercepts conveniently give test points in each interval. You don't need to create your own test points, but you can if you really want to. In the first interval, the test point $(-3, 0)$, hence the graph, is located above the oblique asymptote. In the second interval, the test points $(0, -9/2)$ and $(3, 0)$, as well as the graph, are located under the oblique asymptote.

Figure 3-21 shows the complete graph of *h(x)*.

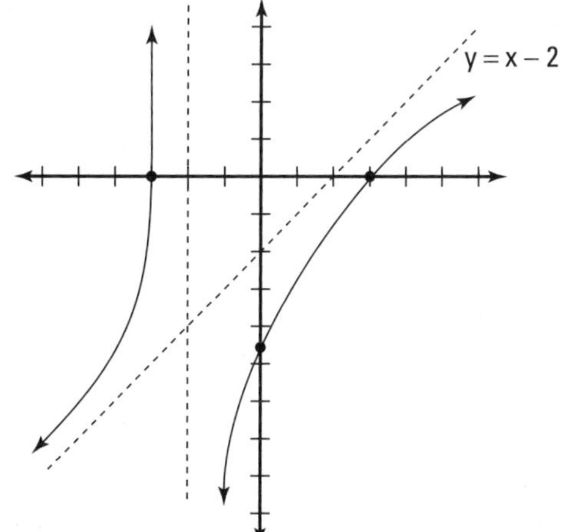

Figure 3-21: The graph of *h(x)*, which has an oblique asymptote.

No Scalpel Necessary: Operating on Functions

Yes, graphing functions is fun, but what if you want more? Well, good news; you also can operate with functions. That's right, we're here to show you how to add, subtract, multiply, or divide two or more functions.

Operating on (sometimes called *combining*) functions is pretty easy, but the graphs of new, combined functions can be hard to create, because those combined functions don't have parent functions and, therefore, no transformations of parent functions that allow you to graph easily. So, we steer clear of them in pre-calculus . . . well, except for maybe a few. If you're asked to graph a combined function, you must resort to the old plug-and-chug method (or perhaps your teacher will be nice enough to let you use your graphing calculator; see Chapter 1).

This section walks you through various operations you may be asked to perform on functions, using the following three functions throughout the examples:

$$f(x) = x^2 - 6x + 1$$

$$g(x) = 3x^2 - 10$$

$$h(x) = \sqrt{(2x - 1)}$$

Adding and subtracting

When asked to add functions, you simply combine like terms, if the functions have any. For example, $(f + g)(x)$ is asking you to add the $f(x)$ and the $g(x)$ functions:

$$(f + g)(x) = (x^2 - 6x + 1) + (3x^2 - 10) = 4x^2 - 6x - 9$$

The x^2 and $3x^2$ add to $4x^2$; $-6x$ remains because it has no like terms; and 1 and -10 add to -9.

But what do you do if you're asked to add $(g + h)(x)$? You'd get

$$(g + h)(x) = (3x^2 - 10) + (\sqrt{(2x - 1)})$$

You have no like terms to add, so you can't simplify the answer any further. You're done!

When asked to subtract functions, you distribute the negative sign throughout the second function, using the distributive property (see Chapter 1), and then treat the process like an addition problem:

$$(g - f)(x) = (3x^2 - 10) - (x^2 - 6x + 1) = (3x^2 - 10) + (-x^2 + 6x - 1) = 2x^2 + 6x - 11$$

Multiplying and dividing

Multiplying and dividing functions is a similar concept to adding and subtracting them (see the previous section). When multiplying functions, you use the distributive property over and over and then add the like terms to simplify. Dividing functions is trickier, however. We'll tackle multiplication first and save the trickster division for last. Here's the setup for multiplying $f(x)$ and $g(x)$:

$$(fg)(x) = (x^2 - 6x + 1)(3x^2 - 10)$$

Follow these steps to multiply these functions:

1. **Distribute each term of the polynomial on the left to each term of the polynomial on the right.**

   You start with $x^2(3x^2) + x^2(-10) + -6x(3x^2) + -6x(-10) + 1(3x^2) + 1(-10)$.

   You end up with $3x^4 - 10x^2 - 18x^3 + 60x + 3x^2 - 10$.

2. **Combine like terms to get the final answer to the multiplication.**

   This simple step gives you $3x^4 - 18x^3 - 7x^2 + 60x - 10$.

Operations that call for division of functions may involve factoring to cancel out terms and simplify the fraction. (If you're unfamiliar with this concept, check out Chapter 4.) If you're asked to divide $g(x)$ by $f(x)$, though, you'd write $\left(\frac{g}{f}\right)(x) = \frac{3x^2 - 10}{x^2 - 6x + 1}$. Because neither the denominator nor the numerator factor, the new, combined function is simplified and you're done.

You may be asked to find a specific value of a combined function. For example, $(f + h)(1)$ asks you to put the value of 1 into the combined function $(f + h)(x) = (x^2 - 6x + 1) + (\sqrt{(2x - 1)})$. When you plug in 1, you get

$$\begin{aligned}
&(1)^2 - 6(1) + 1 + \sqrt{2(1) - 1} \\
&= 1 - 6 + 1 + \sqrt{2 - 1} \\
&= -4 + \sqrt{1} \\
&= -4 + 1 \\
&= -3
\end{aligned}$$

Breaking down a composition of functions

A *composition* of functions is one function acting upon another. Think of it like putting one function inside of the other — $f(g(x))$, for instance, means that you plug the entire $g(x)$ function into $f(x)$. To solve such a problem, you work from the inside out:

$$(f \circ g)(x) = f(g(x)) = f(3x^2 - 10) = (3x^2 - 10)^2 - 6(3x^2 - 10) + 1$$

This process puts the $g(x)$ function into the $f(x)$ function everywhere the $f(x)$ function asks for x. This equation ultimately simplifies to $9x^4 - 78x^2 + 161$, in case you're asked to simplify the composition.

Likewise, $g(h(x)) = (g \circ h)(x) = 3\left(\sqrt{2x-1}\right)^2 - 10$, which easily simplifies to $3(2x - 1) - 10$ because the square root and square cancel each other. This equation simplifies even further to $6x - 13$.

You can't just square the square root in the combined function $h(g(x))$ without stating that the domain is restricted because its domain controls the domain of the composed function (see the following section for more on domain). Although the composition seems like it should be linear, and therefore have a domain of all real numbers, in fact it does not. If you simply don't square out the square root, the domain becomes clear: $[0.5, \infty)$. But the graph doesn't look like a square-root graph or a quadratic; it looks like a line that starts at $x = ½$. So, don't simplify the equation unless you specify that the domain is now restricted.

You may also be asked to find one value of a composed function. To find $(g \circ f)(-3)$, for instance, it helps to realize that it's like reading Hebrew: You work from right to left. In this example, you're asked to put –3 into $f(x)$, get an answer, and then plug that answer into $g(x)$. Here are these two steps in action:

$$f(-3) = (-3)^2 - 6(-3) + 1 = 28$$

$$g(28) = 3(28)^2 - 10 = 2{,}342$$

Adjusting the domain and range of combined functions (if applicable)

If you've looked over the previous sections that cover adding, subtracting, multiplying, and dividing functions, or putting one function inside of another, you may be wondering whether all these operations are messing with domain and range. Well, the answer depends on the operation performed and the original function. But yes, there *is* a possibility that the domain and range will change when you combine functions.

There are two main types of functions whose domains are *not* all real numbers:

- **Rational functions:** The denominator of a fraction can never be zero, so there will be times when rational functions are undefined.
- **Square root functions (and any root with an even index):** The *radicand* (the stuff underneath the root symbol) can't be negative. To find out how the domain is affected, set the radicand greater than or equal to zero and solve. This solution will tell you.

When you begin combining functions (like adding a polynomial and a square root, for example), it makes sense that the domain of the new combined function will also be affected. The same can be said for the range of a combined function; the new function will be based on the restriction(s) of the original functions.

The domain is affected when you combine functions with division because variables end up in the denominator of the fraction. When this happens, you need to specify the values in the domain for which the quotient of the new function is undefined. The undefined values are also called the *excluded values* for the domain. If $f(x) = x^2 - 6x + 1$ and $g(x) = 3x^2 - 10$, if you look at $\left(\frac{g}{f}\right)(x)$, this fraction has excluded values because $f(x)$ is a quadratic equation with real roots. The roots of $f(x)$ are $3 + 2\sqrt{2}$ and $3 - 2\sqrt{2}$, so these are your excluded values.

Unfortunately, we can't give you one fool-proof method for finding the domain and range of a combined function. The domain and range you find for a combined function depend on the domain and range of each of the original functions individually. The best way is to look at the functions visually, creating a graph by using the plug-and-chug method. This way, you can see the minimum and maximum of x, which is your domain, and the minimum and maximum of y, which is your range.

If you don't have the graphing option, however, you simply break down the problem and look at the individual domains and ranges first. Given two functions, $f(x)$ and $g(x)$, assume you have to find the domain of the new combined function $f(g(x))$. To do so, you need to find the domain of each individual function first. If $f(x) = \sqrt{x}$ and $g(x) = 25 - x^2$, here's how you find the domain of the composed function $f(g(x))$:

1. **Find the domain of $f(x)$.**

   Because you can't square root a negative number, the domain of f has to be all non-negative numbers. Mathematically, you write this as $x \geq 0$, or in interval notation $[0, \infty)$.

2. **Find the domain of *g(x)*.**

   Because this equation is a polynomial, its domain is all real numbers, or $(-\infty, \infty)$.

3. **Find the domain of the combined function.**

   When specifically asked to look at the composed function $f(g(x))$, note that g is inside of f. You're still dealing with a square root function, meaning that all the rules for square root functions still apply. So, the new radicand of the composed function has to be non-negative: $25 - x^2 \geq 0$. Solving this quadratic inequality gives you $x \leq 5$ and $x \geq -5$. That's the domain of the composed function: $-5 \leq x \leq 5$.

To find the range of the same composed function, you must also consider the range of both original functions first:

1. **Find the range of *f(x)*.**

   A square root function always gives non-negative answers, so its range is $y \geq 0$.

2. **Find the range of *g(x)*.**

   This function is a polynomial of even degree (specifically, a quadratic), and even-degree polynomials always have a minimum or a maximum value. The higher the degree on the polynomial, the harder it is to find the minimum or the maximum. Because this function is "just" a quadratic, you can find its min or max by locating the vertex.

   First, rewrite the function as $g(x) = -x^2 + 25$. This tells you that the function is a transformed quadratic that has been shifted up 25 and turned upside down (see the earlier section "Transforming the Parent Graphs"). Therefore, the function will never get higher than 25 in the y direction. The range is $y \leq 25$.

3. **Find the range of the composed function *f(g(x))*.**

   The function $g(x)$ reaches its maximum (25) when $x = 0$. This means the composed function also will reach its maximum at $x = 0$: $f(g(0)) = \sqrt{25 - 0^2} = 5$. Therefore, the range of the composed function has to be less than that value, or $y \leq 5$.

The graph of this combined function also depends on the range of each individual function. Because the range of $g(x)$ must be non-negative, so must be the combined function, which is written as $y \geq 0$. Therefore, the range of the combined function is $0 \leq y \leq 5$. If you graph this combined function on your graphing calculator, you'll get a half circle of radius 5 that's centered at the origin.

Flip-Flopping with Inverse Functions

Every operation in math has an inverse: addition undoes subtraction, multiplication undoes division (and vice versa for both). Because functions are just more complicated forms of operations, it's true that functions also have inverses. An *inverse function* simply undoes another function.

Perhaps the best reason to know whether functions are inverses of each other is that if you can graph the original function, you can *usually* graph the inverse as well. So that's where we begin this section. At times in pre-calc you'll be asked to show that two functions are inverses or to find the inverse of a given function, so you'll find that info later in this section as well.

If $f(x)$ is the original function, $f^{-1}(x)$ is the symbol for its inverse. This notation is used strictly to describe the inverse function and not $\frac{1}{f(x)}$. The negative is used only to represent the inverse, not the reciprocal.

Graphing an inverse

If you're asked to graph the inverse of a function, you can do it the long way and find the inverse first (see the next section), or you can remember one fact and get the graph. What's the one fact, you ask? Well, it's that a function and its inverse are reflected over the line $y = x$. This is a linear function that passes through the origin and has a slope of 1. When you're asked to draw a function and its inverse, you may choose to draw this line in as a dotted line; this way, it acts like a big mirror, and you can literally see the points of the function reflecting over the line to become the inverse function points. Reflecting over that line switches the x and the y and gives you a graphical way to find the inverse without plotting tons of points.

The best way to understand this concept is to see it in action. For instance, just trust us for now when we tell you that the functions $f(x) = 2x - 3$ and $g(x) = \frac{x+3}{2}$ are inverses of each other. To see how x and y switch places, follow these steps:

1. **Take a number (any that you want) and plug it into the first given function.**

   We picked –4. When $f(-4)$, you get –11. As a point, this is written (–4, –11).

2. **Take the value from Step 1 and plug it into the other function.**

   In this case, you need to find $g(-11)$. When you do, you get –4 back again. As a point, this is (–11, –4). Whoa!

This works with *any* number and with *any* function: The point (a, b) in the function will become the point (b, a) in its inverse. But don't let that terminology fool you. Because they're still points, you graph them the same way you've always been graphing points.

The entire domain and range swap places from a function to its inverse. For instance, knowing that just a few points from the given function $f(x) = 2x - 3$ include (–4, –11), (–2, –7), and (0, –3), you automatically know that the points on the inverse $g(x)$ will be (–11, –4), (–7, –2), and (–3, 0).

So, if you're asked to graph a function and its inverse, all you have to do is graph the function and then switch all x and y values in each point to graph the inverse. Just look at all those values switching places from the $f(x)$ function to its inverse $g(x)$ (and back again), reflected over the line $y = x$!

You can now graph the function $f(x) = 3x - 2$ and its inverse without even knowing what its inverse is. Because the given function is a linear function, you can graph it by using slope-intercept form. First, graph $y = x$. The slope-intercept form gives you at least two points: the y-intercept at (0, –2) and moving the slope once at the point (1, 1). If you move the slope again, you get (2, 4). The inverse function, therefore, moves through (–2, 0), (1, 1), and (4, 2). Both the function and its inverse are shown in Figure 3-22.

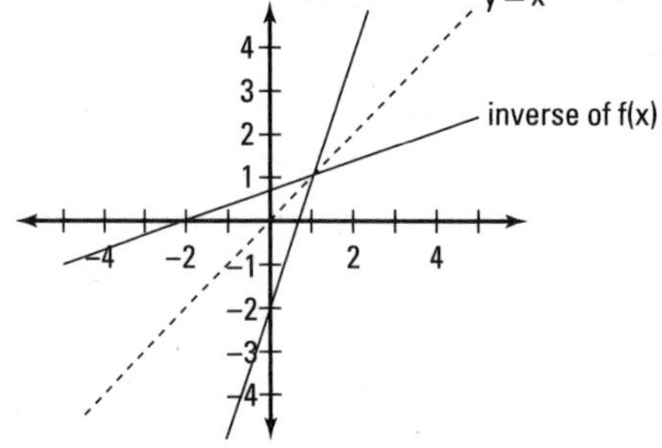

Figure 3-22: Graphing $f(x) = 3x - 2$ and its inverse $f^{-1}(x)$.

Inverting a function to find its inverse

If you're given a function and must find its inverse, first remind yourself that domain and range swap places in the functions. Literally, you exchange $f(x)$ and x in the original equation. When you make that change, you call the new $f(x)$ by its true name — $f^{-1}(x)$ — and solve for this function.

For example, to find the inverse of $f(x) = \frac{2x - 1}{3}$, follow these steps:

1. **Switch *f(x)* and *x*.**

   When you switch *f(x)* and *x*, you get $x = \frac{2f(x) - 1}{3}$. You can also change *f(x)* to *y* and then switch *x* and *y*.

2. **Change the new *f(x)* to its proper name — $f^{-1}(x)$.**

   The equation then becomes $x = \frac{2f^{-1}(x) - 1}{3}$.

3. **Solve for the inverse.**

   This involves three simple steps:

   a. Multiply both sides by 3 to get $3x = 2f^{-1}(x) - 1$.

   b. Add 1 to both sides to get $3x + 1 = 2f^{-1}(x)$.

   c. Lastly, divide both sides by 2 to get $\frac{3x+1}{2} = f^{-1}(x)$. You now have your inverse!

Verifying an inverse

At times, your textbook or teacher may ask you to verify that two given functions are actually inverses of each other. To do this, you need to show that both *f(g(x))* and *g(f(x))* = *x*.

When you're asked to find an inverse of a function (like in the previous section), it isn't a bad idea to verify on your own that what you did was correct, time permitting.

For example, to show that $f(x) = 5x - 4$ and $g(x) = \frac{x+4}{5}$ are inverses of each other, follow these steps:

1. **Show that *f(g(x))* = *x*.**

   This is a matter of plugging in all the components:

$$\begin{aligned} f(g(x)) &= \cancel{5}\left(\frac{x+4}{\cancel{5}}\right) - 4 \\ &= x + 4 - 4 \\ &= x \checkmark \end{aligned}$$

2. **Show that *g(f(x))* = *x*.**

   Again, plug in the numbers and start crossing out:

$$\begin{aligned} g(f(x)) &= \frac{5x - \cancel{4} + \cancel{4}}{5} \\ &= \frac{\cancel{5}x}{\cancel{5}} \\ &= x \checkmark \end{aligned}$$

Chapter 4

Finding and Using Roots to Graph Polynomial Functions

In This Chapter

- Exploring the factoring of quadratic equations
- Solving quadratic equations that you can't factor
- Deciphering and counting a polynomial's roots
- Employing solutions to find factors
- Plotting polynomials on the coordinate plane

Ever since those bygone days of algebra, variables have been standing in for unknowns in equations. Therefore, you should be very comfortable with using them by now. When variables and constants start multiplying, the result is called a *monomial,* which stands for "one term." Examples of monomials include -3, x^2, and $4ab^3c^2$. When you start adding and subtracting monomials, you get *polynomials,* because you create one or more terms. Usually, a monomial refers to a polynomial with one term only, a binomial refers to two terms, trinomial refers to three, while the word polynomial is reserved for four or more. Think of *polynomial* as the umbrella under which are monomial, binomial, and trinomial. Each part of a polynomial that's added or subtracted is a term; so, for example, the polynomial $2x + 3$ has two terms: $2x$ and 3.

Part of the official definition of a polynomial is that it can never have a variable in the denominator of a fraction; it can't have negative exponents; and it can't have fractional exponents.

In this chapter you'll be searching for the *solution(s)* of the given equation — the value(s) that make it true. When the given equation is equal to zero, these solutions are called *roots* or *zeros*. Textbooks and teachers will use these words interchangeably because they represent the same idea — where the graph crosses the x-axis (also called the x-intercept). We will show you how to find the roots of polynomial functions.

The Function of Degrees and Roots

The *degree* of a polynomial is closely related to its exponents, and it determines how you work with the polynomial to find the roots. To find the degree of a polynomial, you simply find the degree of each term by adding the exponents of variables. The greatest of these sums is the degree of the whole polynomial. For example, consider the expression $3x^4y^6 - 2x^4y - 5xy + 2$:

The degree of the first term is 4 + 6, or 10.

The degree of the second term is 4 + 1, or 5.

The degree of the third term is 1 + 1, or 2.

The degree of the last term is 0, because it has no variables.

Therefore, this polynomial has a degree of 10.

A *quadratic expression* is a polynomial in which the highest degree is two. One example of a quadratic polynomial is $3x^2 - 10x + 5$. The x^2 term in the polynomial is called the *quadratic term,* because it's the one that makes the whole expression quadratic. The number in front of x^2 is called the *leading coefficient* (in the example above it's the 3). The x term is called the *linear term* ($-10x$), and the number by itself is called the *constant* (5).

Without taking calculus, getting a perfectly accurate graph of a polynomial function by plotting points is nearly impossible. However, in pre-calc you can find the roots of a polynomial (if there are any), you can use them as a guide to get a more precise idea of what the graph of that polynomial looks like. You simply plug in an x-value between the two roots, which are x-intercepts, to see if the function is positive or negative in between those roots. For example, you could be asked to graph the equation $y = 3x^2 - 10x + 5$. You now know that this is a second degree polynomial, so it will have two roots and, therefore, can cross the x-axis up to two times (more on why later).

Graphing is an important concept of pre-calculus, and one which you will be asked to do numerous times. Depending on the type of function you are graphing, there are numerous strategies to get an accurate graph. For polynomials, though, start with the roots!

If you're lucky enough to own a graphing calculator *and* have a teacher who allows you to use it, you can enter any quadratic equation into the calculator's graphing utility and graph the equation. The calculator will not only identify the zeros, but also tell you the maximum and minimum values of the graph so that you can draw the best possible representation.

We're going to begin this chapter by looking at solving quadratics because the techniques required to solve them are specific: factoring, completing the square, and the quadratic formula are excellent methods to solve quadratics, but they do not work for polynomials of higher degrees. Then we'll move onto higher degree polynomials (like x^3 or x^5, for example) because the steps required to solve them are oftentimes longer and more complicated. ***Note:*** You can solve *any* polynomial equation (including quadratics) using the steps described at the end of this chapter. However, it will save you time and effort to solve quadratics using the techniques specifically reserved for them. Not to worry, though, as we will walk you through each step of solving every kind of polynomial one at a time.

Factoring a Polynomial Expression

Recall that when two or more terms are multiplied to get a product, each term is called a *factor*. You first ran into factors when multiplication was introduced (remember factor trees, prime factorization, and so on?). In mathematics, *factorization* or *factoring* is the breaking apart of a polynomial into a product of other smaller polynomials. If you choose, you could then multiply these factors together, and you should get the original polynomial (this is a great way to check yourself on your factoring skills). One set of factors, for example, of 24 is 6 and 4 because $6 \cdot 4 = 24$. When you have a polynomial, one way of solving it is to factor it into the product of two binomials.

You have multiple factoring options to choose from when solving polynomial equations:

- For a polynomial, no matter how many terms it has, always check for a *greatest common factor* (GCF) first. Literally, the greatest common factor is the biggest expression that will go into all of the terms. Using the GCF is like doing the distributive property backward (see Chapter 1).
- If the equation is a *trinomial* — it has three terms — you can use the FOIL method for multiplying binomials backward.
- If it's a binomial, look for difference of squares, difference of cubes, or sum of cubes.

Finally, after the polynomial is fully factored, you can use the zero product property to solve the equation. The following sections show each of these methods in detail.

If a polynomial doesn't factor, it's called *prime* because its only factors are 1 and itself. When you have tried all the factoring tricks in your bag (GCF, backwards FOIL, difference of squares, and so on), and the quadratic equation will not factor, then you can either complete the square or use the quadratic formula to solve the equation. The choice is yours. You could even potentially choose to *always* use either completing the square or quadratic formula (and skip the factoring) to solve an equation. Factoring can sometimes be quicker, which is why we recommend that you try it first.

Standard form for a quadratic expression (simply a quadratic equation without the equal sign) is the x^2 term, followed by the x term, followed by the constant — in other words, $ax^2 + bx + c$. If you're given a quadratic expression that isn't in standard form, rewrite it in standard form by putting the degrees in descending order. This makes factoring easier (and is sometimes even necessary to factor).

Later in this section, we show you how to solve a quadratic equation after it has been factored using what's known as the zero product property. But because that is based, sometimes, on all the following techniques, we're going to concentrate on factoring expressions only at first and not finding any roots. You'll find that most textbooks share this approach.

Always the first step: Look for a GCF

No matter how many terms a polynomial has, it is always important to check for a greatest common factor (GCF) first. If there is a GCF, it will make factoring the polynomial much easier because the number of factors of each term will be lower (because you will have factored one or more of them out!). This is especially important if the GCF includes a variable.

If you forget to factor out this GCF, you may also forget to find a solution, and that could mix you up in more ways than one! Without that solution, you could miss a root, and then you could end up with an incorrect graph for your polynomial. And then all this work would be for nothing! Well, maybe not *nothing,* but you know what we mean.

To factor the polynomial $6x^4 - 12x^3 + 4x^2$, for example, follow these steps:

1. **Break down every term into prime factors.**

   This expands the expression to $3 \cdot 2 \cdot x \cdot x \cdot x \cdot x - 2 \cdot 2 \cdot 3 \cdot x \cdot x \cdot x + 2 \cdot 2 \cdot x \cdot x$.

2. **Look for factors that appear in every single term to determine the GCF.**

   In this example, you can see one 2 and two x's in every term. We underline these in the following: $3 \cdot \underline{2 \cdot x \cdot x} \cdot x \cdot x - \underline{2} \cdot 2 \cdot 3 \cdot \underline{x \cdot x} \cdot x + 2 \cdot \underline{2 \cdot x \cdot x}$. The GCF here is $2x^2$.

3. **Factor the GCF out from every term in front of parentheses, and leave the remnants inside the parentheses.**

   You now have $2 \cdot x \cdot x(3 \cdot x \cdot x - 2 \cdot 3 \cdot x + 2)$.

4. **Multiply out to simplify each term.**

   This gives you $2x^2(3x^2 - 6x + 2)$.

5. **Distribute to make sure the GCF is correct.**

   If you multiply the $2x^2$ inside the parentheses, you get $6x^4 - 12x^3 + 4x^2$. You can now say with confidence that $2x^2$ is the GCF.

Wrap it up: The FOIL method for trinomials

After you've checked a polynomial for a GCF (regardless of whether it had one or not), try to factor again. You may find that it is easier to factor after the GCF has been factored out. The polynomial in the last section had two factors: $2x^2$ and $3x^2 - 6x + 2$. The first factor $2x^2$ is unfactorable in itself because it's a monomial. However, the second factor may be able to factor again because it's a trinomial, and if it does you'll have two more factors that are both binomials.

Most teachers show the guess-and-check method of factoring, where you write down two sets of parentheses — ( )·( ) — and literally plug in guesses for the factors to see if anything works. Maybe your first guess for this example would be $(3x - 2)(x - 1)$, but if you FOILed this out, you would get $3x^2 - 5x + 2$, and you'd have to guess again. This guess-and-check method is looooooong and tedious, at best. In fact, this particular quadratic is *prime,* so you could guess and check all day long and it would *never* factor.

If you're in pre-calculus and your teacher is using the guess-and-check method of factoring, which just isn't working for you, you've come to the right section. The following procedure, called the *FOIL method* of factoring (sometimes called the *British Method*), always works for factoring trinomials and is a very helpful tool if you can't wrap your brain around guess-and-check. When the FOIL method fails, you know for certain the given quadratic is prime.

The FOIL method of factoring calls for you to follow the steps required to FOIL binomials, only backwards. Remember that when you FOIL, you multiply the First, Outside, Inside, and Last terms together. Then you combine any like terms, which usually come from the multiplication of the Outer and Inner terms.

For example, to factor $x^2 + 3x - 10$, follow these steps:

1. **Check for the GCF first.**

   The expression $x^2 + 3x - 10$ won't have GCF when you break it down and look at it, according to the steps in the last section. The breakdown looks like this: $x \cdot x + 3 \cdot x - 2 \cdot 5$. No factors that are common to each term, so there is no GCF. That means you get to move onto the next step.

2. **Multiply the quadratic term and the constant term.**

   Be careful of the signs when you do this. In this example, the quadratic term is $1x^2$ and the constant is –10: $1 \cdot -10 = -10x^2$.

3. **Write down all the factors of the result, in pairs.**

   The factors of $-10x^2$ are

   - $-1x$ and $10x$
   - $1x$ and $-10x$
   - $-2x$ and $5x$
   - $2x$ and $-5x$

4. **From this list, find the pair that adds to produce the coefficient of the linear term.**

   You want the pair whose sum is $+3x$. For this problem, the answer is $-2x$ and $5x$ because $-2x \cdot 5x = -10x^2$ and $-2x + 5x = 3x$.

5. **Break up the linear term into two terms, using the numbers from Step 4 as the coefficients.**

   Written out, you now have $x^2 - 2x + 5x - 10$.

   TIP: It makes life easier in the long run if you always arrange the linear term with the smallest coefficient first. That's why we put the $-2x$ in front of the $+5x$.

6. **Group the four terms into two sets of two.**

   Always put a plus sign between the two sets: $(x^2 - 2x) + (5x - 10)$.

7. **Find the GCF for each set and factor it out.**

   Look at the first two terms. What do they share in common? An x. If you factor out the x, you have $x(x-2)$. Now, look at the second two terms. They share a 5. If you factor out the 5, you have $5(x-2)$. The polynomial is now written as $x(x-2)+5(x-2)$.

8. **Find the GCF of the two new terms.**

   Do you see the $(x-2)$ in both terms? We've underlined them here: $x\underline{(x-2)}+5\underline{(x-2)}$. That's a GCF because it appears in both terms (if you factor using this method, the last step should always look like this). Factor out the GCF from both terms (it's always the expression inside the parentheses) to the front; you get $(x-2)(\ )$. When you factor it out, the terms that aren't the GCF are left inside the new parentheses. In this case, you get $(x-2)(x+5)$. The $(x+5)$ is the leftover from taking away the GCF.

Sometimes the sign has to change in Step 6 in order to correctly factor out the GCF. But if you don't start off with a plus sign between the two sets, you may lose a negative sign you need to factor all the way. For example, in factoring $x^2-13x+36$, you end up in Step 5 with the following polynomial: $x^2-9x-4x+36$. When you group the terms, you get $(x^2-9x)+(-4x+36)$. Factor out the x in the first set and the 4 in the second set to get $x(x-9)+4(-x+9)$. Notice that the second set is the exact opposite of the first one? In order for you to move to the next step, the sets have to match exactly. To fix this, change the +4 in the middle to –4 and get $x(x-9)-4(x-9)$. Now that they match, you can factor again.

If you follow all the steps in the previous list, you'll have an easy time with factoring trinomials. Even when an expression has a leading coefficient besides 1, the FOIL method still works. The monkey wrench comes only if there are no factors in Step 2 that add to give you the linear coefficient. In this case, the answer is prime. For example, in $2x^2+13x+4$, when you multiply the quadratic term of $2x^2$ and the constant of 4, you get $8x^2$. However, no factors of $8x^2$ also add to be 13x, so $2x^2+13x+4$ is prime.

Recognizing and factoring special types of polynomials

The whole point of factoring is to discover the original polynomial factors that give you an end product. You spend a long time in algebra FOILing polynomials, and factoring just undoes that process. It's a little like *Jeopardy!* — you know the answer and are looking for the question.

There are special cases of FOILing binomials, which also pop back up in factoring; you should recognize them quickly so that you can save time on factoring:

- **Perfect squares:** When you FOIL a binomial times itself, the product is called a *perfect square.* $(a + b)^2$, for example, would give you the perfect square trinomial $a^2 + 2ab + b^2$. Whenever you factor a trinomial, if you end up with two factors that are the same, you express the answer as the binomial to the second power.
- **Difference of squares:** When you FOIL a binomial and its conjugate, the product is called a *difference of squares.* The product of $(a - b)(a + b)$ is $a^2 - b^2$. Factoring a difference of squares also requires its own set of steps, which we explain for you in this section.

Two other special types of factoring didn't come up when you were learning how to FOIL, because they aren't the product of two binomials:

- **Sum of cubes:** One factor is a binomial and the other is a trinomial. $(a^3 + b^3)$ can be factored to $(a + b)(a^2 - ab + b^2)$.
- **Difference of cubes:** These factor almost like a sum of cubes, except that some signs will be different in the factors: $(a^3 - b^3) = (a - b)(a^2 + ab + b^3)$.

No matter what type of problem you face, you should always check for the GCF first; however, each of the following examples won't have a GCF, so we skip over that step in the directions. In another section, you find out how to factor more than once when you can.

Seeing double with perfect squares

Because a perfect square trinomial is still a trinomial, you follow the steps in the backward FOIL method of factoring (see the previous section). However, you must account for one extra step at the very end where you express the answer as a binomial squared.

For example, to factor the polynomial $x^2 - 16x + 64$, follow these steps:

1. **Multiply the quadratic term and the constant term.**

   The product of the quadratic term x^2 and the constant 64 is $64x^2$, so that made your job easy.

2. **Write down all the factors of the result, in pairs.**

   The factors of $64x^2$ in pairs are:

   - $1x$ and $64x$
   - $-1x$ and $-64x$
   - $2x$ and $32x$

- $-2x$ and $-32x$
- $4x$ and $16x$
- $-4x$ and $-16x$
- $8x$ and $8x$
- $-8x$ and $-8x$

3. **From this list, find the pair that adds to produce the coefficient of the linear term.**

   You want to get a sum of $-16x$ in this case. The only way to do that is to use $-8x$ and $-8x$.

4. **Break up the linear term into two terms, using the terms from Step 3.**

   That now gives you $x^2 - 8x - 8x + 64$.

5. **Group the four terms into two sets of two.**

   Did you remember to include the plus sign in between the two groups to get $(x^2 - 8x) + (-8x + 64)$?

6. **Find the GCF for each set and factor it out.**

   The GCF of the first two terms is x, and the GCF of the next two terms is -8; when you factor them out, you get $x(x - 8) - 8(x - 8)$.

7. **Find the GCF of the two new terms.**

   This time the GCF is $(x - 8)$; when you factor it out, you get $(x - 8)(x - 8)$. Aha! That's a binomial times itself, which means you have one extra step.

8. **Express the resulting product as a binomial squared.**

   This step is easy: $(x - 8)^2$.

Working with differences of squares

You'll recognize a *difference of squares* because it will always be a binomial where each term is a perfect square, and there will always be a subtraction sign between them. It *always* appears as $a^2 - b^2$, or (something)^2 – (something else)^2. When you do have a difference of squares on your hands — after checking it for a GCF in both terms — you follow a simple procedure: $a^2 - b^2 = (a - b)(a + b)$.

For example, you can factor $25y^4 - 9$ with these steps:

1. **Rewrite each term as (something)^2.**

   This example becomes $(5y^2)^2 - (3)^2$, which clearly shows the difference of squares ("difference of" meaning subtraction).

2. **Factor the difference of squares $(a)^2 - (b)^2$ to $(a - b)(a + b)$.**

   Each difference of squares $(a)^2 - (b)^2$ always factors to $(a - b)(a + b)$. This example factors to $(5y^2 - 3)(5y^2 + 3)$.

Breaking down a cubic difference or sum

After you've checked to see if there's a GCF in the given polynomial and discovered it's a binomial that isn't a difference of squares, consider that it may be a sum or difference of cubes.

A *difference of cubes* sounds an awful lot like a difference of squares (see the last section), but it factors quite differently. A difference of cubes will always start off as a binomial with a subtraction sign in between, but will be written as $(\text{something})^3 - (\text{something else})^3$. To factor any difference of cubes, you use the formula $(a)^3 - (b)^3 = (a - b)(a^2 + ab + b^2)$.

A *sum of cubes* is always a binomial with a plus sign in between — the only one where that happens: $(\text{something})^3 + (\text{something else})^3$. When you recognize a sum of cubes $a^3 + b^3$, it factors as $(a + b)(a^2 - ab + b^2)$.

For example, to factor $8x^3 + 27$, you first look for the GCF. You find none, so now you use the following steps:

1. **Check to see if the expression is a difference of squares.**

   You want to consider the possibility because the expression has two terms, but you should quickly realize that it isn't because you see a plus sign between the two terms.

2. **Determine if you must use a sum or difference of cubes.**

   The plus sign tells you that it may be a sum of cubes, but it isn't foolproof. Time for some trial and error. Try to rewrite the expression as the sum of cubes; if you try $(2x)^3 + (3)^3$, you've found a winner.

3. **Break down the sum or difference of cubes by using the factoring shortcut.**

   Replace a with $2x$ and b with 3. The formula becomes $[(2x) + (3)][(2x)^2 - (2x)(3) + (3)^2]$.

4. **Simplify the factoring formula.**

   This example simplifies to $(2x + 3)(4x^2 - 6x + 9)$.

5. **Check the factored polynomial to see if it will factor again.**

   You're not done factoring until you're done. Always look at the "leftovers" to see if they'll factor again. Sometimes, the binomial term may factor again as the difference of squares. However, the trinomial factor will *never* factor again.

   In the previous example, the binomial term $2x + 3$ is a first-degree binomial (the exponent on the variable is 1) without a GCF, so it won't factor again. That means $(2x + 3)(4x^2 - 6x + 9)$ is your final answer.

Grouping to factor four or more terms

When a polynomial has four or more terms, the easiest way to factor it is to use *grouping*. In this method, you look at only two terms at a time to see if any of the previous techniques becomes apparent (you may see a GCF in two terms, or you may recognize a trinomial as a perfect square). In fact, in previous sections when we show you how to break up the linear term in a trinomial into two separate terms and then factor out the GCF twice, we're showing you a grouping tactic. The ways you can factor by using grouping far outnumber that one example, however, so we'll now show you how to group when the given polynomial *starts off* with four (or more) terms.

Sometimes, you can group a polynomial into sets with two terms each to find a GCF in each set. You should try this method first when faced with a polynomial with four or more terms. This is the most common type of grouping you'll see in a pre-calculus text.

For example, you can factor $x^3 + x^2 - x - 1$ by using grouping. Just follow these steps:

1. **Break up the polynomial into sets of two.**

   You can go with $(x^3 + x^2) + (-x - 1)$. Put the plus sign between the sets, just like when you factor trinomials.

2. **Find the GCF of each set and factor it out.**

   The square x^2 is the GCF of the first set, and -1 is the GCF of the second set. Factoring out both of them, you get $x^2(x + 1) - 1(x + 1)$.

3. **Factor again as many times as you can.**

   The two terms you've created have a GCF of $(x + 1)$. When factored out, you get $(x + 1)(x^2 - 1)$.

   However, $x^2 - 1$ is a difference of squares and factors again. In the end, you get the following factors after grouping: $(x + 1)(x + 1)(x - 1)$, or $(x + 1)^2(x - 1)$.

If the previous method doesn't work, you may have to group the polynomial some other way. Of course, after all your effort, the polynomial may end up being prime, which is okay.

For example, if you look at the polynomial $x^2 - 4xy + 4y^2 - 16$, you can group it into sets of two, and it would become $x(x - 4y) + 4(y^2 - 4)$. This, however, won't factor again. Bells and whistles should go off inside your head at this point, telling you to look again at the original. You must try grouping it in some other way. In this case, if you look at the first three terms, you'll discover a perfect square trinomial, which factors to $(x - 2y)^2 - 16$. Now you have a difference of squares, which factors again to $[(x - 2y) - 4][(x - 2y) + 4]$.

Finding the Roots of a Factored Equation

Sometimes after you've factored, the two factors may be factorable again, in which case you should, or unfactorable, in which case you can solve them only by using the quadratic formula. For example, $6x^4 - 12x^3 + 4x^2 = 0$ factors to $2x^2(3x^2 - 6x + 2) = 0$. The first term $2x^2 = 0$ is solvable, using algebra, but the second factor $3x^2 - 6x + 2 = 0$ is unfactorable and requires the quadratic formula (see the following section).

After you've factored a polynomial into its different pieces, you can set each piece equal to zero to solve for the roots with the zero product property. The *zero product property* says that if several factors are multiplying to give you zero, at least one of them has to be zero. Your job is to find all the values of x that make the polynomial equal to zero. This is much easier if the polynomial is factored, because you can set each factor equal to zero and solve for x.

Now, $x^2 + 3x - 10 = 0$ factors to $(x + 5)(x - 2)$. Moving forward is easy because each factor is linear (first degree). The term $x + 5 = 0$ gives you one solution — $x = -5$ — and $x - 2 = 0$ gives you the other solution — $x = 2$.

These each become an x-intercept on the graph of the polynomial (see the section "Graphing Polynomials").

Cracking a Quadratic Equation When It Won't Factor

When asked to solve a quadratic equation that you just can't seem to factor (or that just doesn't factor), you have to employ other ways of solving the equation. The inability to factor means that the equation has solutions that you can't find by using normal techniques. Perhaps they involve square roots of non-perfect squares; they can even be complex numbers involving imaginary numbers (see Chapter 11).

One such method is to use the *quadratic formula,* which is the formula used to solve for the variable in a quadratic equation in standard form. Another is to *complete the square,* which means to manipulate an expression to create a perfect square trinomial that you can easily factor. The following sections present these methods in detail.

Using the quadratic formula

When a quadratic equation just won't factor, you should remember your old friend from algebra, the quadratic formula, in order to solve. Given a quadratic equation in standard form $ax^2 + bx + c = 0$, $x = \dfrac{-b \pm \sqrt{b^2 - 4ac}}{2a}$.

Before you apply the formula, you should rewrite the equation in standard form (if it isn't already) and figure out the *a, b,* and *c* values.

For example, to solve $x^2 - 3x + 1 = 0$, you first say that $a = 1$, $b = -3$, and $c = 1$. The *a, b,* and *c* terms simply plug into the formula to give you the values for *x* — $x = \dfrac{-(-3) \pm \sqrt{(-3)^2 - 4(1)(1)}}{2(1)}$.

Simplify this formula one time to get $\dfrac{3 \pm \sqrt{9-4}}{2}$, and then simplify further to get your final answer, which is two *x* values (the *x*-intercepts): $x = \dfrac{3 \pm \sqrt{5}}{2}$.

Completing the square

Completing the square comes in handy when you're asked to solve an unfactorable quadratic equation and when you need to graph conic sections, which we explain in Chapter 12. For now, we recommend that you only find the roots of a quadratic using this technique when you're specifically asked to do so, because factoring a quadratic and the quadratic formula work just as well (if not better). Those methods are less complicated than completing the square (a pain in the you-know-where!).

Say your instructor calls for you to complete the square. Follow these steps to solve the equation $2x^2 - 4x + 5 = 0$ by completing the square:

1. **Divide every term by the leading coefficient; if the equation has no leading coefficient, you can skip to Step 2.**

   Be prepared to deal with fractions in this step. The equation now becomes $x^2 - 2x + 5/2 = 0$.

2. **Move the constant term to the other side of the equation by performing its inverse operation.**

   You can subtract $5/2$ from both sides to get $x^2 - 2x = -5/2$.

3. **Divide the linear coefficient by 2, square this answer, and then add that value to both sides.**

   Take –2 divided by 2 to get –1. Square this answer to get 1, and add it to both sides: $x^2 - 2x + 1 = -5/2 + 1$.

4. **Simplify the equation.**

   The equation becomes $x^2 - 2x + 1 = -3/2$.

5. **Factor the newly created quadratic equation.**

   The new equation should be a perfect square trinomial. The example equation factors to $(x - 1)(x - 1) = -3/2$, using FOIL, which means that $(x - 1)^2 = -3/2$.

6. **Get rid of the square exponent by square rooting both sides.**

   This gives you $x - 1 = \pm\sqrt{\frac{-3}{2}}$.

7. **Simplify any square roots if possible.**

   The example equation doesn't simplify, but the fraction is imaginary (see Chapter 11) and the denominator needs to be rationalized (see Chapter 2). Do the work to get $x - 1 = \frac{\pm\sqrt{-3}}{\sqrt{2}} \cdot \frac{\sqrt{2}}{\sqrt{2}} = \frac{\pm\sqrt{6}\,i}{\sqrt{2}}$.

8. **Solve for the variable by isolating it.**

   You add 1 to both sides to get $x = \frac{\pm\sqrt{6}\,i}{2} + 1$.

   Note: You may be asked to express your answer as one fraction; in this case, find the common denominator and add to get $x = \frac{\pm\sqrt{6}\,i + 2}{2}$.

Solving Unfactorable Polynomials with a Degree Higher than Two

By now you are a professional at solving second-degree polynomial equations (quadratics) and have various tools at your disposal for solving these types of problems. You may have noticed while solving quadratics that there are always two solutions to a quadratic equation. Note that sometimes both solutions are the same (this happens in perfect square trinomials). Even though you get the same solution twice, this still counts as two solutions (how many times a solution is a root is called the *multiplicity* of the solution).

When the polynomial degree is higher than two and the polynomial won't factor using any of the techniques that we discuss earlier in this chapter, it gets harder and harder to find the roots. For example, you could be asked to solve a cubic polynomial that is *not* a sum or difference of cubes or any polynomial that is fourth degree or greater that cannot be factored by grouping. The higher the degree, the more roots there are, and the harder it is to find them. To find the roots, there are many different scenarios that can guide you in the right direction. You can make very educated guesses about how many roots a polynomial has, as well as how many of them are positive or negative and how many are real or imaginary.

Counting a polynomial's total number of roots

Usually, the first step you take before solving a polynomial is to find its *degree,* which helps you determine the number of solutions you'll find later.

When you're being asked to solve a polynomial, finding its degree will be even easier as there will be only one variable in any term. Therefore, the highest exponent will always be the highest term when asked to solve. For example, $f(x) = 2x^4 - 9x^3 - 21x^2 + 88x + 48$ is a fourth-degree polynomial with up to, but no more than, four total possible solutions.

Tallying the real roots: Descartes' Rule of Signs

The terms solutions/zeros/roots are synonymous, because they all represent where the graph of the polynomial intersects the x-axis. The roots that are found when the graph meets with the x-axis are called *real roots;* you can see them and deal with them as real numbers in the real world. Also, because they cross the x-axis, some roots may be *negative roots* (which means they intersect the negative x-axis), and some may be *positive roots* (which intersect the positive x-axis).

If you know how many total roots you have (see the last section), you can use a pretty cool theorem called *Descartes' Rule of Signs* to count how many roots are real numbers (both positive *and* negative) and how many are imaginary (see Chapter 11). You see, the same man who pretty much invented graphing, Descartes, also came up with a way to figure out how many times a polynomial crosses the x-axis — in other words, how many roots it has. All you have to be able to do is count!

Descartes' Rule of Signs calls for you to look at the polynomial, written in descending order, and count how many times the sign changes from term to term. This value represents the maximum number of positive roots in the polynomial. For example, in the polynomial $f(x) = 2x^4 - 9x^3 - 21x^2 + 88x + 48$, you see two changes in sign (don't forget to count the first term!) — from the first term to the second and from the third term to the fourth. That means this equation can have up to two positive solutions. Descartes' Rule of Signs also says the number of positive roots is equal to changes in sign of $f(x)$, or is less than that by an even number (so you keep subtracting 2 until you get either 1 or 0). Therefore, the previous $f(x)$ may have 2 or 0 positive roots.

The rule then calls for you to find $f(-x)$ and count again. But, because negative numbers raised to even powers are positive and negative numbers raised to odd powers are negative, this change affects only terms with odd powers. This step is the same as changing each term with an odd degree to its opposite sign and counting again, which will give you the maximum number of negative roots. The example equation becomes $f(-x) = 2x^4 + 9x^3 - 21x^2 - 88x + 48$, which changes sign twice. There can be, at most, two negative roots. However, similar to the rule for positive roots, the number of negative roots is equal to the changes in sign for $f(-x)$, or must be less than that by an even number. Therefore, this example can have either 2 or 0 negative roots.

Accounting for imaginary roots: The Fundamental Theorem of Algebra

Imaginary roots appear in a quadratic equation when the discriminant of the quadratic equation is negative. Recall from Algebra II that the *discriminant* is the part of the quadratic formula under the square root sign: $b^2 - 4ac$. If this value is negative, you can't actually take the square root, and the answers will not be real. In other words, there is no solution; therefore, the graph won't cross the x-axis.

When dealing with the quadratic formula, there are always two solutions, because the ± sign means you're both adding and subtracting and getting two completely different answers. When the number underneath the square root sign in the quadratic formula is negative, the answers are called *complex conjugates.* One is $a + bi$ and the other is $a - bi$. These numbers have both real (the a) and imaginary (the bi) parts.

The Fundamental Theorem of Algebra says that every polynomial function has at least one root in the complex number system. This concept is one you may remember from Algebra II. (For reference, read the parts on imaginary and complex numbers in Chapter 11 first.)

The highest degree of a polynomial gives you the highest possible number of *complex* roots for the polynomial. Between this fact and Descartes' Rule of Signs, you can figure out how many pure imaginary roots (no real part at all, unlike complex numbers) a polynomial has. Pair up every possible number of positive real roots with every possible number of negative real roots (see the previous section); the remaining number of roots for each situation represents the number of pure imaginary roots.

Continuing with the example $f(x) = 2x^4 - 9x^3 - 21x^2 + 88x + 48$ from the previous section, the polynomial has a degree of 4, with 2 or 0 positive real roots, and 2 or 0 negative real roots. Pair up the possible situations:

- If there are 2 positive and 2 negative real roots, that leaves 0 pure imaginary roots.
- If there are 2 positive and 0 negative real roots, that leaves 2 pure imaginary roots.
- If there are 0 positive and 2 negative real roots, that leaves 2 pure imaginary roots.
- If there are 0 positive and 0 negative real roots, that leaves 4 pure imaginary roots.

Allow us to arrange this information in a chart to make things easier to picture:

Positive real roots	*Negative real roots*	*Pure imaginary roots*
2	2	0
2	0	2
0	2	2
0	0	4

Complex numbers are written in the form $a + bi$ and have both a real and an imaginary part, which is why every polynomial has at least one root in the complex number system (see Chapter 11). Real and imaginary numbers are both included in the complex number system. Real numbers have no imaginary part, and pure imaginary numbers have no real part. For example, if $x = 7$ is one root of the polynomial, this root is considered both real and complex because it can be rewritten as $x = 7 + 0i$ (the imaginary part is 0).

The Fundamental Theorem of Algebra gives the total number of complex roots (say there are 7); Descartes' Rule of Signs tells you how many possible real roots there are and how many are positive and how many are negative (say there are, at most, 2 positive roots but only 1 negative root). Assume you've found them all, using the techniques we discuss throughout this section; they are $x = 1$, $x = 7$, and $x = -2$. It gets confusing because these are real, but they're also complex because they can all be rewritten, as in the previous example.

You see, the first two columns in the chart find the pure real roots and classify them as positive or negative. The third column is actually finding, specifically, the non-real numbers: both the pure imaginary and the pure complex.

Guessing and checking the real roots

After you've worked through the last section, you can determine exactly how many roots (and what type of roots) there are. Now, the *Rational Root Theorem* is another method you can use to narrow down the search for roots of polynomials. Descartes' Rule of Signs only narrows down the real roots into positive and negative. The Rational Root Theorem says that it's possible that some real roots are rational (they can be expressed as a fraction). It also helps you create a list of the *possible* rational roots of any polynomial.

The problem? Not every root is rational, because some are irrational. It's even possible for a polynomial to have *only* irrational roots. But this theorem is always the next place to start in your search for roots; it will at least give you a diving-off point. Besides, the problems you'll be presented with in precalc will more than likely have at least one rational root, so the information in this section will greatly improve your odds of finding more!

Follow these general steps to ensure that you've found every root:

1. **Use the Rational Root Theorem to list all possible rational roots.**
2. **Pick one root from the list in Step 1 and use long or synthetic division to find out if it is, in fact, a root.**
   a. If the root doesn't work, try another guess.
   b. If the root works, proceed to Step 3.
3. **Using the depressed polynomial (the one you get after doing the synthetic division in Step 2b), test the root that worked to see if it works again.**
   a. If it works, repeat Step 3 *again.*
   b. If it doesn't work, return to Step 2 where you must try a different root from the list in Step 1 and use synthetic division to check.
4. **List all the roots you find that work; there should be as many roots as the degree of the polynomial.**

   Don't stop until you've found them all. It's entirely possible that some will be real and some will be imaginary.

Finding possible real roots with the Rational Root Theorem

PRE-CALC RULES $\frac{1+1}{2}$

The *Rational Root Theorem* says that if you take all the factors of the constant term in a polynomial and divide by all the factors of the leading coefficient, you produce a list of all the possible rational roots of the polynomial. However, keep in mind that you're finding only the *rational* ones, and sometimes the roots of a polynomial are irrational. Some of your roots could also be imaginary, but it's best to save those until the end of your search.

For example, consider the equation $f(x) = 2x^4 - 9x^3 - 21x^2 + 88x + 48$. The constant term is 48, and its factors are as follows:

$\pm 1, \pm 2, \pm 3, \pm 4, \pm 6, \pm 8, \pm 12, \pm 16, \pm 24, \pm 48$

The leading coefficient is 2, and its factors are as follows:

± 1 and ± 2

So, the list of possible real roots includes the following:

$\pm 1/1, \pm 2/1, \pm 3/1, \pm 4/1, \pm 6/1, \pm 8/1, \pm 12/1, \pm 16/1, \pm 24/1, \pm 48/1, \pm 1/2, \pm 2/2, \pm 3/2, \pm 4/2, \pm 6/2, \pm 8/2,$

$\pm 12/2, \pm 16/2, \pm 24/2,$ and $\pm 48/2$

Thankfully, these all simplify to $\pm 1/2, \pm 1, \pm 3/2, \pm 2, \pm 3, \pm 4, \pm 6, \pm 8, \pm 12, \pm 16, \pm 24,$ and ± 48.

Testing roots by dividing polynomials

Dividing polynomials follows the same algorithm as long division with real numbers. The polynomial you're dividing by is called the *divisor.* The polynomial being divided is called the *dividend.* The answer is called the *quotient,* and the leftover polynomial is called the *remainder.*

One way, other than synthetic division, that you can test possible roots from the Rational Root Theorem is to use long division of polynomials and hope that when you divide you get a remainder of 0. For example, when you have your list of possible rational roots (as found in the last section), pick one and assume that it's a root. If $x = c$ is a root, $x - c$ is a factor. So, if you pick $x = 2$ as your guess for the root, $x - 2$ should be a factor. We explain in this section how to use long division to test if $x - 2$ is actually a factor and, therefore, $x = 2$ is a root.

Dividing polynomials to get a specific answer isn't something you do everyday, but the idea of a function or expression that is written as the quotient of two polynomials is important for pre-calculus. If you divide a polynomial by another and get a remainder of 0, the divisor is a factor, which in turn gives a root. The following sections review two methods of checking your real roots: long division and synthetic division.

In math lingo, the division algorithm states the following: If $f(x)$ and $d(x)$ are polynomials such that $d(x)$ isn't equal to 0, and the degree of $d(x)$ is not larger than the degree of $f(x)$, there are unique polynomials $q(x)$ and $r(x)$ such that $f(x) = d(x) \cdot q(x) + r(x)$. In plain English, this means that dividend = divisor · quotient + remainder. You can always check your results by remembering this information.

Long division

You can use long division to find out if your possible rational roots are actual roots or not. We don't recommend doing this, but you can do it. Instead, we suggest you use synthetic division, which we cover later. But you may be asked to do long division. The directions of a problem will specifically ask you to find the quotient using long division in this case, or perhaps you won't be able to use synthetic division, leaving you with no choice but to use long division. We show you how in the following steps and try to figure out a root at the same time.

Remember the mnemonic device Dirty Monkeys Smell Bad when doing long division to check your roots. Make sure all terms in the polynomial are listed in descending order, and that every degree is represented. In other words, if x^2 is missing, put in a placeholder of $0x^2$ and then do the division. This is just to make the division process easier.

To divide two polynomials, follow these steps:

1. **Divide.**

   Divide the leading term of the dividend by the leading term of the divisor. Write this quotient directly above the term you just divided into.

2. **Multiply.**

   Multiply the quotient term from Step 1 by the entire divisor. Write this polynomial under the dividend so that like terms are lined up.

3. **Subtract.**

   Subtract the whole line you just wrote from the dividend.

   You can change all the signs and add if it makes you feel more comfortable. This way, you won't forget signs.

4. **Bring down the next term.**

   Do exactly what this says; bring down the next term in the dividend.

5. **Repeat Steps 1–4 over and over until the remainder polynomial has a degree that's less than the dividend's.**

The following list explains how to divide $2x^4 - 9x^3 - 21x^2 + 88x + 48$ by $x - 2$. Each step corresponds with the numbered step in the illustration in Figure 4-1. (Note that in the earlier section on Descartes' Rule of Signs, you find that this particular example may have positive roots, so it's efficient to try a positive number here. If Descartes' Rule of Signs had said that no positive roots existed, you wouldn't test any positives!)

1. **Divide:** What do you have to multiply x in the divisor by to make it become $2x^4$ in the dividend? The quotient, $2x^3$, goes above the $2x^4$ term.
2. **Multiply:** Multiply this quotient times the divisor and write it under the dividend.
3. **Subtract:** Subtract this line from the dividend — $(2x^4 - 9x^3) - (2x^4 - 4x^3) = -5x^3$. If you've done the job right, the subtraction of the first terms will always produce 0.
4. **Bring down:** Bring down the other terms of the dividend.
5. **Divide:** What do you have to multiply x by to make it $-5x^3$? Put the answer, $-5x^2$, above the $-21x^2$.
6. **Multiply:** Multiply the $-5x^2$ times the $x - 2$ to get $-5x^3 + 10x^2$. Write this under the remainder with the degrees lined up.
7. **Subtract:** You now have $(-5x^3 - 21x^2) - (-5x^3 + 10x^2) = -31x^2$.
8. **Bring down:** The $+88x$ takes its place.
9. **Divide:** What to multiply this with to make x become $-31x^2$? The quotient $-31x$ goes above $-21x^2$.
10. **Multiply:** The value $-31x$ times $(x - 2)$ is $-31x^2 + 62x$; write this under the remainder.
11. **Subtract:** You now have $(-31x^2 + 88x) - (-31x^2 + 62x)$, which is $26x$.
12. **Bring down:** The $+48$ comes down.
13. **Divide:** The term $26x$ divided by x is 26. This answer goes on top.
14. **Multiply:** The constant 26 times $(x - 2)$ is $26x - 52$.
15. **Subtract:** You subtract $(26x + 48) - (26x - 52)$ to get 100.
16. **Stop:** The remainder 100 finally has a degree that's less than the divisor of $x - 2$.

Wow . . . now you know why they call it *long* division. You went through all that to find out that $x - 2$ isn't a factor of the polynomial, which means that $x = 2$ isn't a root.

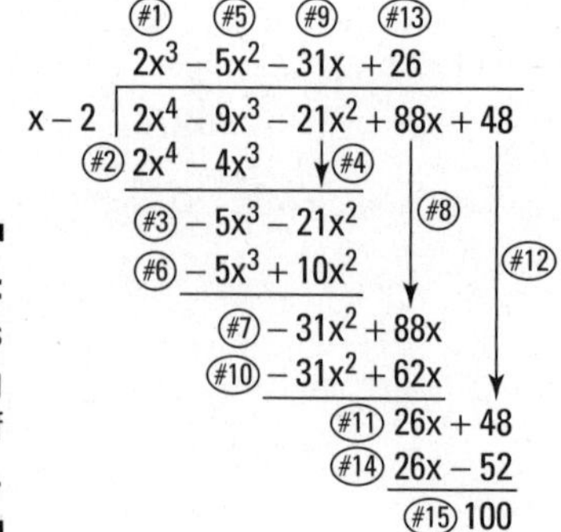

Figure 4-1: The process of long division of polynomials.

If you divide by c and the remainder is 0, this implies that the linear expression $(x - c)$ is a factor and that c is a root. A remainder other than 0 implies that $(x - c)$ isn't a factor and that c isn't a root.

Synthetic division

Want some good news? A shortcut exists for long division, and that shortcut is synthetic division. This is a special case of division when the divisor is a linear factor the form $x + c$, where c is a constant.

The bad news, however, is that the shortcut only works if the divisor $(x + c)$ is a first-degree binomial with a leading coefficient of 1 (you can always make it 1 by dividing everything by the leading coefficient first). The *great* news — yep, more news — is that you can always use synthetic division to figure out if a possible root is actually a root.

In the previous example, you eliminated $x = 2$ by using long division, so you know not to start there. We choose to do synthetic division in Figure 4-2 for $x = 4$ to show you how it works.

4	2	-9	-21	88	48
	↓	8	-4	-100	-48
	2	-1	-25	-12	0

Figure 4-2: Synthetic division is a shortcut for long division when testing possible roots.

The 4 on the outside in Figure 4-2 is the root you're testing. The numbers on the inside are the coefficients of the polynomial. Here's the synthetic process, step by step:

1. The 2 below the line just drops down from the line above.
2. Multiply $4 \cdot 2$ to get 8 and write that under the next term, –9.
3. Add $-9 + 8$ to get –1.
4. Multiply $4 \cdot -1$ to get –4, and write that under the –21.
5. Add $-21 + -4$ to get –25.
6. Multiply $4 \cdot -25$ to get –100, and write that under 88.
7. Add $88 + -100$ to get –12.
8. Multiply $4 \cdot -12$ to get –48, and write that under 48.
9. Add $48 + -48$ to get 0.

See that? All you do is multiply and add, which is why synthetic division is the shortcut. The last number, 0, is your remainder. Because you get a remainder of 0, $x = 4$ is a root.

The other numbers are the coefficients of the quotient, in order from the greatest degree to the least; however, your answer is always one degree lower than the original. So, the quotient in the previous example is $2x^3 - x^2 - 25x - 12$.

Automatically, whenever a root works, you should always test it again in the answer quotient to see if it's a double root, using the same process. A *double root* occurs when a factor has a multiplicity of two. A double root is one example of multiplicity (as we describe earlier in the "Accounting for imaginary roots: The Fundamental Theorem of Algebra" section). We test $x = 4$ again in Figure 4-3.

Figure 4-3: Testing an answer root again, just in case it's a double root.

4	2	-1	-25	-12
	↓	8	28	12
	2	7	3	0

Whaddya know, you get a remainder of 0 again, so $x = 4$ is a double root. (In math terms, you say that $x = 4$ is a root with *multiplicity of two*.) You have to check it again, though, to see if it has a higher multiplicity. When you synthetically divide $x = 4$ one more time, it doesn't work. Figure 4-4 illustrates this failure. Because the remainder isn't 0, $x = 4$ isn't a root again.

4	2	7	3
	↓	8	60
	2	15	63

Figure 4-4: Testing the root again shows that it's only a double root as far as the multiplicity goes.

Always work off the newest quotient when using synthetic division. This way, the degree will get lower and lower until you end up with a quadratic. At that point, you can solve the quadratic by using any of the techniques we talk about earlier in this chapter: factoring, completing the square, or the quadratic formula. (Some math teachers require you to use the quadratic formula; what else is the formula for?)

Before you tested $x = 4$ for a final time, the polynomial (called a *depressed polynomial*) was down to a quadratic: $2x^2 + 7x + 3$. If you factor this expression, you get $(2x + 1)(x + 3)$. This gives you two more roots of $-\frac{1}{2}$ and -3. To sum it all up, you've found $x = 4$ (multiplicity two), $x = -\frac{1}{2}$, and $x = -3$. You found four complex roots — two of them are negative real numbers, and two of them are positive real numbers.

The *remainder theorem* says that the remainder you get when you divide a polynomial by a binomial is the same as the result you get from plugging the number into the polynomial. For example, when you used long division to divide by $x - 2$, you were testing to see if $x = 2$ is a root. You could've used synthetic division to do this, because you still get a remainder of 100. And if you plug 2 into $f(x) = 2x^4 - 9x^3 - 21x^2 + 88x + 48$, you also get 100.

For really hard polynomials, it's much easier to do the synthetic division to determine the roots than it is to substitute the number. For example, if you try to plug 8 into the previous polynomial, you have to figure out first what $2(8)^4 - 9(8)^3 - 21(8)^2 + 88(8) + 48$ is. This only leads to bigger (and uglier) numbers, while in the synthetic division, all you do is multiply and add — no more exponents!

Put It in Reverse: Using Solutions to Find Factors

The *factor theorem* states that you can go back and forth between the roots of a polynomial and the factors of a polynomial. If you know one, in other words, you know the other. At times, your teacher or your textbook will ask you to factor a polynomial with a degree higher than two. If you can find its roots, you can find its factors. We show you how in this section.

In symbols, the factor theorem states that if $x - c$ is a factor of the polynomial $f(x)$, then $f(c) = 0$. The variable c is a zero or a root or a solution — whatever you want to call it (the terms all mean the same thing).

In the previous sections of this chapter, you employ many different techniques to find the roots of the polynomial $f(x) = 2x^4 - 9x^3 - 21x^2 + 88x + 48$. You find that they are $x = -\frac{1}{2}$, $x = -3$, and $x = 4$ (multiplicity two). How do you use those roots to find the factors of the polynomial?

The factor theorem states that if $x = c$ is a root, $(x - c)$ is a factor. For example, look at the following roots:

- If $x = -\frac{1}{2}$, $(x - (-\frac{1}{2}))$ is your factor, which is the same thing as $(x + \frac{1}{2})$.
- If $x = -3$ is a root, $(x - (-3))$ is a factor, which is also $(x + 3)$.
- If $x = 4$ is a root, $(x - 4)$ is a factor with multiplicity two.

You can now factor $f(x) = 2x^4 - 9x^3 - 21x^2 + 88x + 48$ to get $f(x) = (x + \frac{1}{2})(x + 3)(x - 4)^2$.

Graphing Polynomials

The hard graphing work is over after you've found the zeros of a polynomial function (using the techniques we present earlier in this chapter). Finding the zeros is very important to graphing the polynomial, because they give you a general template for what your graph should look like. Remember that zeros are x-intercepts, and knowing where the graph crosses the x-axis is half the battle. The other half is knowing what the graph does in between these points. This section shows you how to figure that out.

When all the roots are real numbers

We use many different techniques in this chapter to find the zeros for the example polynomial $f(x) = 2x^4 - 9x^3 - 21x^2 + 88x + 48$. The time has come to put this work to use to graph the polynomial. Follow these steps to start graphing like a pro:

1. **Plot the critical points on the coordinate plane.**

   Mark the zeros that you've found starting in the section "Solving Unfactorable Polynomials with a Degree Higher than Two": $x = -3$, $x = {}^{-1}\!/_2$ and $x = 4$ are all zeros.

   Now plot the *y*-intercept of the polynomial. The *y*-intercept is *always* the constant term of the polynomial — in this case, $y = 48$. If no constant term is written, the *y*-intercept is 0 (because it's understood).

2. **Determine which way the ends of the graph will point.**

   You can use a handy test called the *leading coefficient test,* which helps you figure out how the polynomial begins and ends. The degree and leading coefficient of a polynomial always explain the end behavior of its graph (see the section "The Function of Degrees and Roots" for more on finding degree):

   - If the degree of the polynomial is even and the leading coefficient is positive, both ends of the graph point up.
   - If the degree is even and the leading coefficient is negative, both ends of the graph point down.
   - If the degree is odd and the leading coefficient is positive, the left side of the graph points down and the right side points up.
   - If the degree is odd and the leading coefficient is negative, the left side of the graph points up and the right side points down.

   Figure 4-5 displays this concept in correct mathematical terms.

   The function $f(x) = 2x^4 - 9x^3 - 21x^2 + 88x + 48$ is even in degree and has a positive leading coefficient, so both ends of its graph point up (they go to positive infinity).

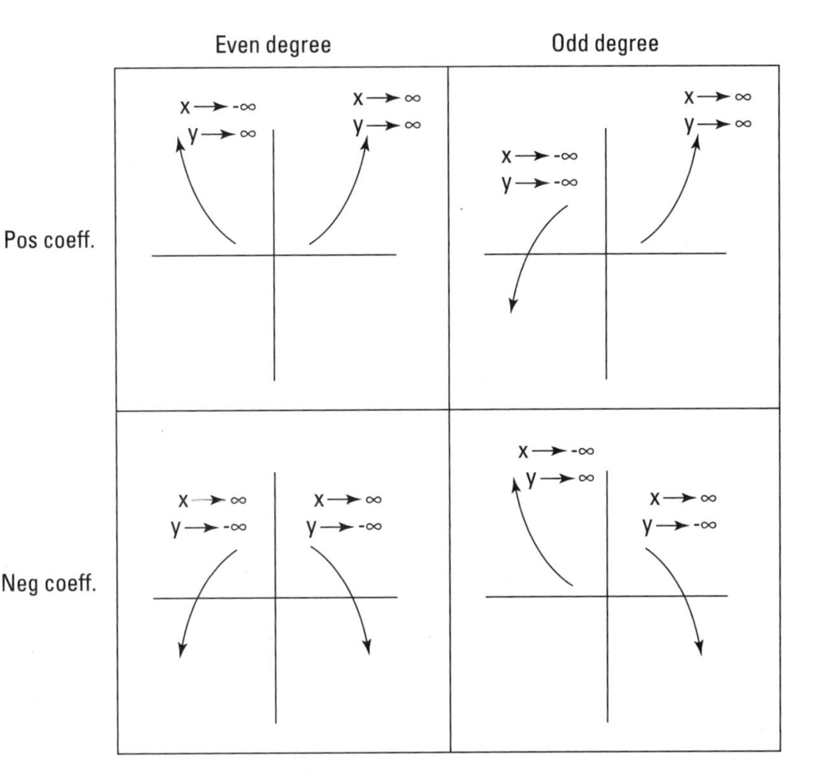

Figure 4-5: An illustration of the leading coefficient test.

3. **Figure out what happens between the critical points by picking any value to the left and right of each intercept and plugging it into the function.**

   You can either simplify each one or just figure out whether the end result is positive or negative. For now, you don't really care about the exact look of the graph. (In calculus, you learn how to find additional values that lead to the most accurate graph you can get.)

   A graphing calculator gives a very accurate picture of the graph. Calculus allows you to find the relative max and min exactly, using an algebraic process, but you can easily use the calculator to find them. You can use your graphing calculator to check your work and make sure the graph you've created looks like the one the calculator gives you.

   Using the zeros for the function, set up a table to help you figure out whether the graph is above or below the *x*-axis between the zeros. See Table 4-1 for our table example.

Table 4-1 Using the Roots to Frame the Graph

Interval	*Test Value (x)*	*Result [f(x)]*
$(-\infty, -3)$	-4	Positive (above the x-axis)
$(-3, -\frac{1}{2})$	-2	Negative (below the x-axis)
$(-\frac{1}{2}, 4)$	0	Positive (above the x-axis)
$(4, \infty)$	5	Positive (above the x-axis)

The first interval $(-\infty, -3)$ and the last interval $(4, \infty)$ both confirm the leading coefficient test from Step 2 — this graph points up (to positive infinity) in both directions.

4. **Plot the graph.**

   Now that you know where the graph crosses the x-axis, how the graph begins and ends, and whether the graph is positive (above the x-axis) or negative (below the x-axis), you can sketch out the graph of the function. Typically, in pre-calc, this is all the information you want or need when graphing. Calculus does show you how to get several other critical points that create an even better graph. If you'd like, you can always pick more points in the intervals and graph them to get a better idea of what the graph looks like. Figure 4-6 shows the completed graph.

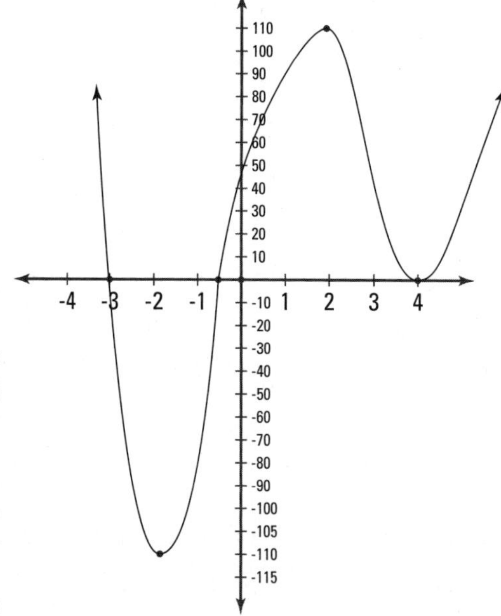

Figure 4-6: Graphing the polynomial $2x^4 - 9x^3 - 21x^2 + 88x + 48$.

Did you notice that the double root (with multiplicity two) causes the graph to "bounce" on the x-axis instead of actually crossing it? This is true for any root with even multiplicity. For any polynomial, if the root has an odd multiplicity at root c, the graph of the function crosses the x-axis at $x = c$. If the root has an even multiplicity at root c, the graph meets but doesn't cross the x-axis at $x = c$.

When some (or all) of the roots are imaginary numbers: Combining all techniques

At times in pre-calc and in calculus, a polynomial function will have non-real roots in addition to real roots (and some of the more complicated functions will have *all* imaginary roots). When you must find both, start off by finding the real roots, using all the techniques we describe earlier in this chapter (such as synthetic division). Then you'll be left with a depressed quadratic polynomial to solve that's unsolvable using real number answers. No fear! You just have to use the quadratic formula, through which you'll end up with a negative number under the square root sign. Therefore, you express the answer as a complex number (for more, see Chapter 11).

For instance, the polynomial $g(x) = x^4 + x^3 - 3x^2 + 7x - 6$ has non-real roots. Follow these steps to find *all* the roots for this (or any) polynomial; each step involves a major section of this chapter:

1. **Classify the real roots as positive and negative by using Descartes' Rule of Signs.**

   Three changes of sign in the $g(x)$ function reveals you could have three or one positive real root. One change in sign in the $g(-x)$ function reveals that you have one negative real root.

2. **Find how many roots are possibly imaginary by using the Fundamental Theorem of Algebra.**

   The theorem reveals that, in this case, there are up to four complex roots. Combining this fact with Descartes' Rule of Signs gives you several possibilities:

   a. One real positive root and one real negative root means that two roots aren't real.

   b. Three real positive roots and one real negative root means that all roots are real.

3. **List the possible rational roots, using the Rational Root Theorem.**

   The possible rational roots include ±1, ±2, ±3, and ±6.

4. **Determine the rational roots (if any), using synthetic division.**

   Utilizing the rules of synthetic division, you find that $x = 1$ is a root and that $x = -3$ is another root. These are the only real roots.

5. **Use the quadratic formula to solve the depressed polynomial.**

   Having found all the real roots of the polynomial, you're left with the depressed polynomial $x^2 - x + 2$. Because this is a quadratic, you can use the quadratic formula to solve for the last two roots. In this case, $x = \frac{1 \pm \sqrt{7}i}{2}$.

6. **Graph the results.**

   The leading coefficient test (see the previous section) reveals that the graph points up in both directions. The intervals include the following:

   - $(-\infty, -3)$ is positive
   - $(-3, 1)$ is negative
   - $(1, \infty)$ is positive

   Figure 4-7 shows the graph of this function.

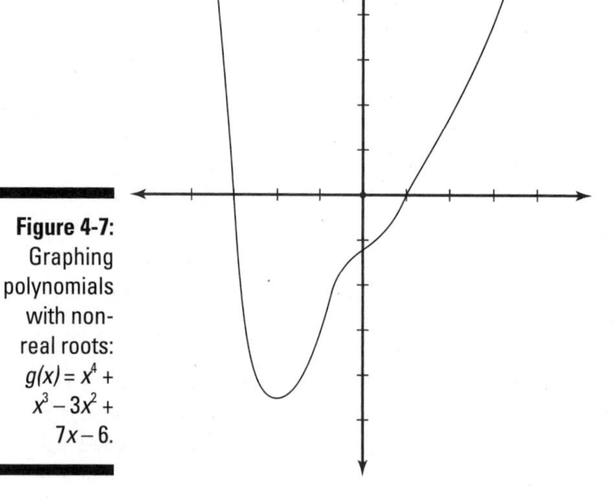

Figure 4-7: Graphing polynomials with non-real roots: $g(x) = x^4 + x^3 - 3x^2 + 7x - 6$.

Chapter 5

Powering Up with Exponential and Logarithmic Functions

In This Chapter

- Simplifying, solving, and graphing exponential functions
- Checking all the ins and outs of logarithms
- Working through equations with exponents and logs
- Conquering some growth and decay example problems

If someone presented you with the choice of taking \$1 million right now or taking one penny, with the stipulation that this amount would double every day for 30 days, which would you choose? Most people would take the million without even thinking about it, and it would surely surprise them that the other plan is the better offer. Take a look: On the first day, you have only a penny, and you feel like you've been duped. By the tenth day, you have \$5.12, and you're still feeling like you took the wrong path. By the 20th day, you have \$5,242.88, and you may be feeling better. And by the last day, you have \$5,368,709.12! Actually, on the 28th day, you'd reach over a million — \$1,342,177.28, to be exact — and that amount would continue to double until the end. As you can see, doubling something (in this case, your money) makes it get big pretty fast. This is the basic idea behind an exponential function; bet we have your attention now, don't we?

In this chapter, we cover two unique types of functions from pre-calc: the exponential and the logarithm. These functions can be graphed, solved, or simplified just like any other function we discuss in this book. We cover all the new rules you need to work with these functions; they may take some getting used to, but we break them down into the simplest terms.

That's great and all, you may be saying, but when will I ever use this complex stuff? (No one in their right mind would offer you the money, anyway.) Well, this chapter's info on exponential and logarithmic functions will come in handy when working with numbers that will grow or shrink (usually with respect to time). Populations usually grow (get larger), while the monetary value of objects usually shrinks (gets smaller). You can describe these ideas with exponential functions. In the real world, you can also figure out compounded interest, carbon dating, inflation, and so much more!

Exploring Exponential Functions

An *exponential function* is a function where a variable exists in the exponent. In math terms, you write $f(x) = b^x$, where b is the base. If you've read Chapter 2, you know all about exponents and their place in math. So, what's the difference between exponents and exponential functions? We're glad you asked!

Prior to now, the variable was always the base — as in $g(x) = x^2$, for example. The exponent always stayed the same. In an exponential function, however, the variable is the exponent and the base stays the same — as in the function $f(x) = b^x$.

The concepts of exponential growth and exponential decay play an important role in biology. Bacteria and viruses especially love to grow exponentially. If one cell of a cold virus gets into your body and then doubles every hour, at the end of one day you'll have 8,388,608 of the little bugs moving around inside your body (no wonder you have to take so many pills to kick the cold). So, the next time you get a cold, just remember to thank (or curse) your old friend, the exponential function.

In this section, we dig deeper to uncover what an exponential function really is and how you can use one to describe the growth or decay of anything that gets bigger or smaller.

Searching the ins and outs of an exponential function

Exponential functions follow all the rules of functions, which we discuss in Chapter 3. But because they also make up their own unique family, they have their own subset of rules. The following list outlines some basic rules that apply to exponential functions:

- **The parent exponential function $f(x) = b^x$ always has a horizontal asymptote at $y = 0$.** You can't raise a positive number greater than 1 to any power and get 0 (it also will never become negative). For more on asymptotes, refer to Chapter 3.

- **The domain of any exponential function is $(-\infty, \infty)$.** This is true because you can raise a positive number to any power. However, because all exponential functions have horizontal asymptotes, the ranges must reflect this. All parent exponential functions have ranges greater than 0, or $(0, \infty)$.
- **The order of operations still governs how you act on the function.** When the idea of a vertical transformation (see Chapter 3) applies to an exponential function, most people take the order of operations and throw it out the window. Avoid this mistake. For example, $y = 2 \cdot 3^x$ doesn't become $y = 6^x$. You can't multiply before you deal with the exponent.
- **You can't have a base that's negative.** For example, $y = (-2)^x$ isn't an equation you have to worry about graphing in pre-calc. If you're asked to graph $y = -2^x$, don't fret. You read this as "the opposite of 2 to the *x*," which means that (remember the order of operations) you raise 2 to the power first and then multiply by –1, or $y = -1 \cdot 2^x$. This simple change flips the graph upside down and changes its range to $(-\infty, 0)$.
- **Negative exponents take the reciprocal of the number to the positive power.** For instance, $y = 2^{-3}$ doesn't equal $(-2)^3$ or -2^3. Raising any number to a negative power takes the reciprocal of the number to the positive power: $2^{-3} = \frac{1}{2^3}$, or $\frac{1}{8}$.
- **When you multiply monomials with exponents, you add the exponents.** For instance, $x^3 \cdot x^2$ doesn't equal x^6. If you break it down, the function is easier to see: $x \cdot x \cdot x \cdot x \cdot x$, which is the same as x^5 (you add the exponents as the shortcut).
- **When you have multiple factors inside parentheses raised to a power, you raise every single term to that power.** For instance, $(4x^3y^5)^2$ isn't $4x^3y^{10}$; it's $16x^6y^{10}$.
- **When graphing an exponential function, remember that base numbers greater than 1 always get bigger (or *rise*) as they move to the right; as they move to the left, they always approach 0 but never actually get there.** For example, $f(x) = 2^x$ is an exponential function, as is $g(x) = \left(\frac{1}{3}\right)^x$. Table 5-1 shows the *x* and *y* values of these exponential functions. These parent functions illustrate that a base greater than 1 gets bigger — an example of exponential growth — while a fraction (between 0 and 1) gets smaller — an example of exponential decay. As long as the exponent is positive *x*, this will always be the case.
- **Base numbers that are fractions between 0 and 1 always rise to the left and approach 0 to the right.** This holds true until you start to transform the parent graphs, which we get to in the next section.

Table 5-1 The Values of x in Two Exponential Functions

x	$f(x) = 2^x$	$g(x) = \left(\frac{1}{3}\right)^x$
–3	1/8	27
–2	1/4	9
–1	1/2	3
0	1	1
1	2	1/3
2	4	1/9
3	8	1/27

Graphing and transforming an exponential function

Graphing an exponential function is helpful when you want to visually analyze the function. Doing so allows you to really see the growth or decay of what you're dealing with. The basic parent graph of any exponential function is $f(x) = b^x$, where b is the base. Figure 5-1a, for instance, shows the graph of $f(x) = 2^x$, and Figure 5-1b shows $g(x) = \left(\frac{1}{3}\right)^x$. Using the x and y values from Table 5-1, you simply plot the coordinates to get the graphs.

The parent graph of any exponential function crosses the y-axis at (0, 1), because anything raised to the 0 power is always 1. Some teachers refer to this as the *key point* because it's shared among all exponential parent functions.

Because an exponential function is simply a function, you can transform the parent graph, using all the same rules from Chapter 3. Keep on reading to find out how.

You can transform the parent graph of an exponential function in the same way as any other function:

$y = a \cdot \text{base}^{x-h} + v$, where

a is the vertical transformation

h is the horizontal shift

v is the vertical shift

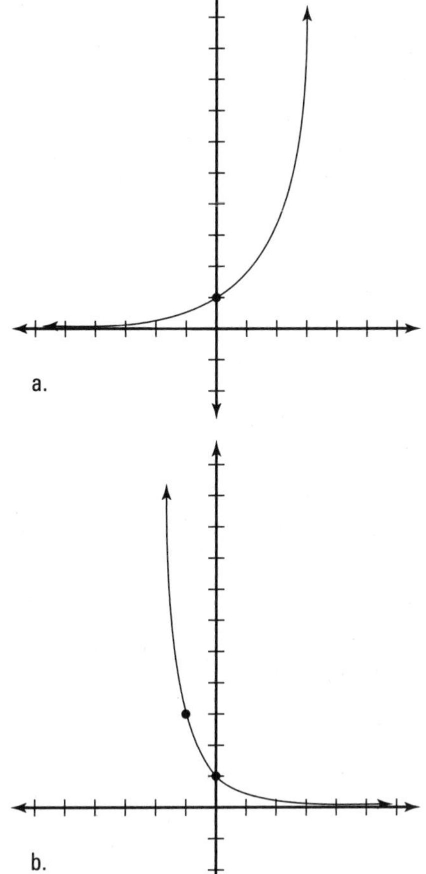

Figure 5-1: The graphs of the exponential functions $f(x) = 2^x$ and $g(x) = \left(\frac{1}{3}\right)^x$.

For example, you can graph $h(x) = 2^{(x+3)} + 1$ by transforming the parent graph of $f(x) = 2^x$. Based on the previous equation, $h(x)$ has been shifted to the left three ($h = -3$) and up one ($v = 1$). Figure 5-2 shows each of these as steps: Figure 5-2a is the horizontal transformation, showing the parent function $y = 2^x$ as a dotted line, and Figure 5-2b is the vertical transformation.

Moving an exponential function up or down moves the horizontal asymptote. The function in Figure 5-2 has a horizontal asymptote at $y = 1$ (for more info on horizontal asymptotes, see Chapter 3). This also shifts the range up 1 to $(1, \infty)$.

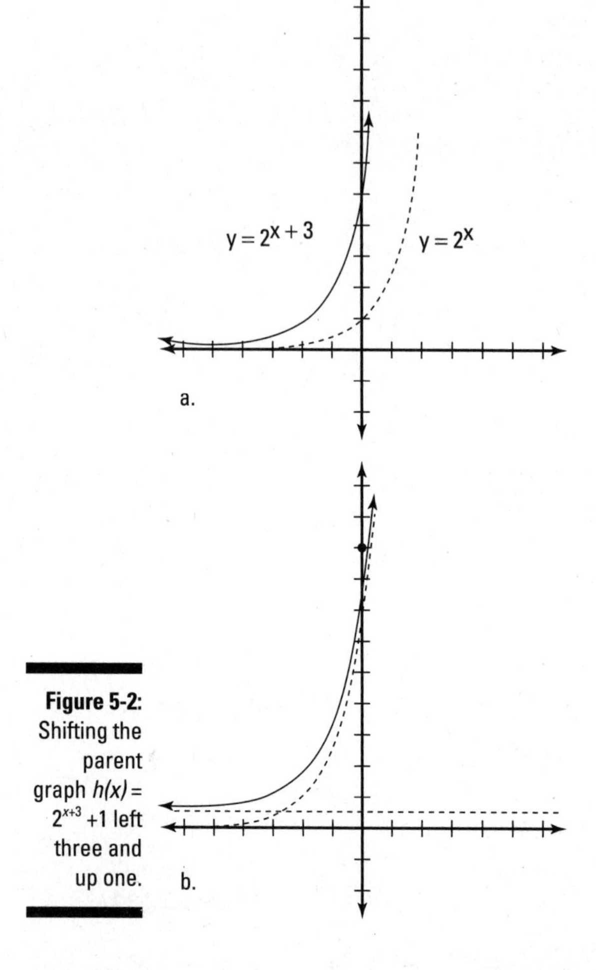

Figure 5-2: Shifting the parent graph $h(x) = 2^{x+3} + 1$ left three and up one.

Logarithms: Investigating the Inverse of Exponential Functions

Almost every function has an inverse. (Chapter 3 discusses what an inverse function is and how to find one.) But this question once stumped mathematicians: What could possibly be the inverse for an exponential function? They couldn't find one, so they invented one! They defined the inverse of an exponential function to be a *logarithm* (or *log*).

A logarithm *is* an exponent, plain and simple. Recall, for example, that $4^2 = 16$; 4 is called the *base* and 2 is called the *exponent.* The logarithm is $\log_4 16 = 2$, where 2 is called the logarithm of 16 with base 4. In math, a logarithm is written $\log_b y = x$. The b is the base of the log, y is the number you're taking the log of, and x is the logarithm. So really, the logarithm and exponential forms are saying the same thing in different ways.

Getting a better handle on logarithms

If an exponential function reads $b^x = y$, its inverse, or *logarithm,* is $\log_b y = x$. Notice that the logarithm is the exponent. Figure 5-3 presents a diagram that may help you remember how to change an exponential function to a log and vice versa.

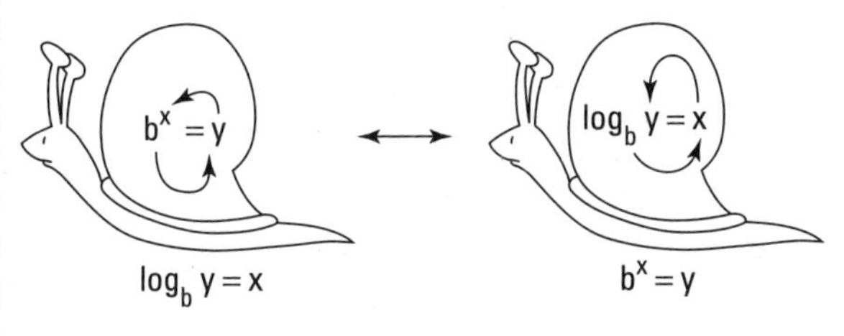

Figure 5-3: The snail rule helps you remember how to change exponentials and logs.

Two types of logarithms are special. They're considered special because you don't have to write their base (unlike any other kind of log) — it's simply understood:

- **Common logarithms:** Because our entire number system is in base 10, $\log y$ (without a base written) is always meant as log base 10. For example, $10^3 = 1{,}000$, so $\log 1{,}000 = 3$. This occurrence is called a *common logarithm* because it happens so frequently.
- **Natural logarithms:** A logarithm with base e is called a *natural logarithm.* The symbol for a natural log is *ln.* Here's an example equation: $\log_e y = \ln y$.

Managing the properties and identities of logs

You need to know several properties of logs in order to solve equations that contain them. Each of these properties applies to any base, including the common and natural logs (see the previous section):

- **$\log_b 1 = 0$**

  If you change back to an exponential function, $b^0 = 1$ no matter what the base is. So, it makes sense that $\log_b 1 = 0$.

- **$\log_b x$ exists only when $x > 0$**

  The domain $(-\infty, \infty)$ and range $(0, \infty)$ of the original exponential parent function switch places in any inverse function. Therefore, any logarithm parent function has the domain of $(0, \infty)$ and range of $(-\infty, \infty)$.

- **$\log_b b^x = x$**

  You can change this property into an exponential by using the snail rule: $b^x = b^x$. (Refer to Figure 5-3 for an illustration.) No matter what value you put in for b, this equation always works. $\log_b b = 1$ no matter what the base is (because it's really just $\log_b b^1$).

  The fact that you can use any base you want in this equation illustrates how this property works for common and natural logs: $\log 10^x = x$ and $\ln e^x = x$.

- **$b^{\log_b x} = x$**

  You can change this equation back to a log to confirm that it works: $\log_b x = \log_b x$.

- **$\log_b x + \log_b y = \log_b(x \cdot y)$**

  According to this rule, called the *product rule,* $\log_4 10 + \log_4 2 = \log_4(10 \cdot 2) = \log_4 20$.

- **$\log_b x - \log_b y = \log_b\left(\frac{x}{y}\right)$**

  According to this rule, called the *quotient rule,* $\log 4 - \log(x - 3) = \log\left(\frac{4}{x-3}\right)$.

- **$\log_b x^y = y \cdot \log_b x$**

  According to this rule, called the *power rule,* the $\log_3 x^4 = 4 \cdot \log_3 x$.

It's important to keep the properties of logs straight so you don't get confused and make a critical mistake. The following list highlights many of the mistakes that people make when it comes to working with logs:

- **Misusing the product rule:** $\log_b x + \log_b y \neq \log_b(x + y)$; this equals $\log_b(x \cdot y)$. You can't add two logs inside of one. Similarly, $\log_b x \cdot \log_b y \neq \log_b(x \cdot y)$.
- **Misusing the quotient rule:** $\log_b x - \log_b y \neq \log_b(x - y)$; this equals $\log_b\left(\frac{x}{y}\right)$. Also, $\frac{\log_b x}{\log_b y} \neq \log_b\left(\frac{x}{y}\right)$. This error messes up the change-of-base formula (see the following section).
- **Misusing the power rule:** $\log_b(xy^p) \neq p \cdot \log_b(xy)$ because the power is on the second variable only. If the formula was written as $\log_b(xy)^p$, it would equal $p \cdot \log_b(xy)$.

  Note: Watch what those exponents are doing. You should split up the multiplication from $\log_b(xy^p)$ first by using the product rule: $\log_b x + \log_b y^p$. Only then can you apply the power rule to get $\log_b x + p \cdot \log_b y$.

Changing a log's base (when the log isn't natural or common)

Calculators usually come equipped with only common log or natural log buttons, so you must know what to do when a log has a base your calculator can't recognize, such as $\log_5 2$; the base is 5 in this case. In these situations, you must use the *change of base formula* to change the base to either base 10 or base *e* (the decision depends on your personal preference) in order to use the buttons that your calculator does have.

The change of base formula is as follows:

$$\log_m n = \frac{\log_b n}{\log_b m}, \text{ where } m \text{ and } n \text{ are real numbers}$$

You can make the new base anything you want by using the change of base formula (5, 30, or even 3,000), but remember that your goal is to be able to utilize your calculator by using either base 10 or base *e* to simplify the process. For instance, if you decide that you want to use the common log in the change of base formula, you find that $\log_3 5 = \frac{\log 5}{\log 3} = 1.465$. However, if you're a fan of natural logs, you can go this route: $\log_3 5 = \frac{\ln 3}{\ln 5}$, which is still 1.465.

Calculating a number when you know its log: Inverse logs

If you know the logarithm of a number but need to find out what the original number actually was, you must use an *inverse logarithm,* which is also known as an *antilogarithm.* If $\log_b y = x$, y is the antilogarithm. An inverse logarithm undoes a log (makes it go away) so that you can solve certain log equations. For example, if you know that $\log x = 0.699$, you have to change it back to an exponential (take the inverse log) to solve it: $10^{\log x} = 10^{0.699}$, which simplifies to $x = 10^{0.699}$, so $x \approx 5$.

You can do this process with natural logs as well. If $\ln x = 1.099$, for instance, then $e^{\ln x} = e^{1.099}$, or $x = e^{1.099}$, so $x \approx 3$.

The base you use in an antilogarithm depends on the base of the given log. For example, if you're asked to solve the equation $\log_5 x = 3$, you must use base 5 on both sides to get $5^{\log_5 x} = 5^3$, which simplifies to $x = 5^3$, or $x = 125$.

Graphing logs

Want some good news, free of charge? Graphing logs is a snap! You can change any log into an exponential, so this is your first step. You then graph the exponential (or its inverse), remembering the rules for transforming (see Chapter 3), and then use the fact that exponentials and logs are inverses to get the graph of the log. The following sections explain these steps for both parent functions and transformed logs.

A parent function

Exponential functions each have a parent function that depends on the base; logarithmic functions also have parent functions for each different base. The parent function for any log is written $f(x) = \log_b x$. For example, $g(x) = \log_4 x$ is a different family than $h(x) = \log_8 x$. Here we graph the common log: $f(x) = \log x$.

1. **Change the log to an exponential.**

   Because $f(x)$ and y represent the same thing mathematically, and because it's easier to deal with y in this case, you can rewrite the equation as $y = \log x$. The exponential equation of this log is $10^y = x$.

2. **Find the inverse function by switching *x* and *y*.**

   As you discover in Chapter 3, you find the inverse function $10^x = y$.

3. **Graph the inverse function.**

   Because you're now graphing an exponential function, you can plug and chug a few x values to find y values and get points. The graph of $10^x = y$ gets really big, really fast. You can see its graph in Figure 5-4.

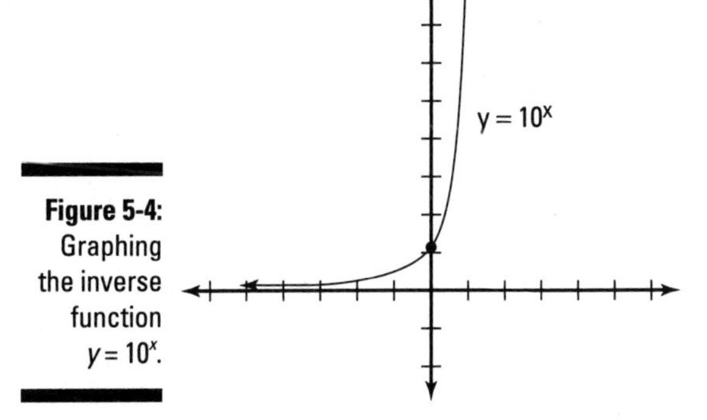

Figure 5-4: Graphing the inverse function $y = 10^x$.

4. **Reflect every point on the inverse function graph over the line $y = x$.**

   Figure 5-5 illustrates this last step, which yields the parent log's graph.

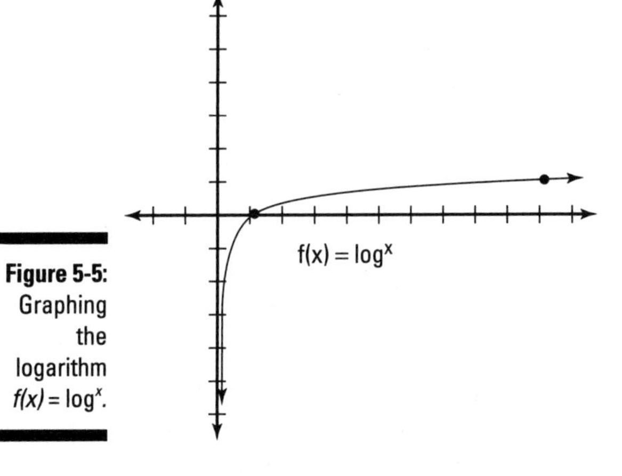

Figure 5-5: Graphing the logarithm $f(x) = \log^x$.

A transformed log

All transformed logs can be written as $f(x) = a \cdot \log_b(x - h) + v$, where

a is the vertical stretch or shrink.

h is the horizontal shift.

v is the vertical shift.

So, if you can find the graph of the parent function $\log_b x$, you can transform it. However, we find that most of our students still prefer to change the log function to an exponential one and then graph. So, the following steps show you how to do just that when graphing $f(x) = \log_3(x - 1) + 2$:

1. **Get the logarithm by itself.**

   First, rewrite the equation as $y = \log_3(x - 1) + 2$. Then subtract 2 from both sides to get $y - 2 = \log_3(x - 1)$.

2. **Change the log to an exponential and find the inverse function.**

   If $y - 2 = \log_3(x - 1)$ is the logarithmic function, $3^{y-2} = x - 1$ is the exponential; the inverse function is $3^{x-2} = y - 1$ because x and y switch places in the inverse.

3. **Solve for the variable not in the exponential of the inverse.**

   To solve for y in this case, add 1 to both sides to get $3^{x-2} + 1 = y$.

4. **Graph the exponential function.**

   The parent graph of $y = 3^x$ transforms right two $(x - 2)$ and up one $(+ 1)$, as shown in Figure 5-6. Its horizontal asymptote is at $y = 1$ (for more on graphing exponentials, refer to Chapter 3).

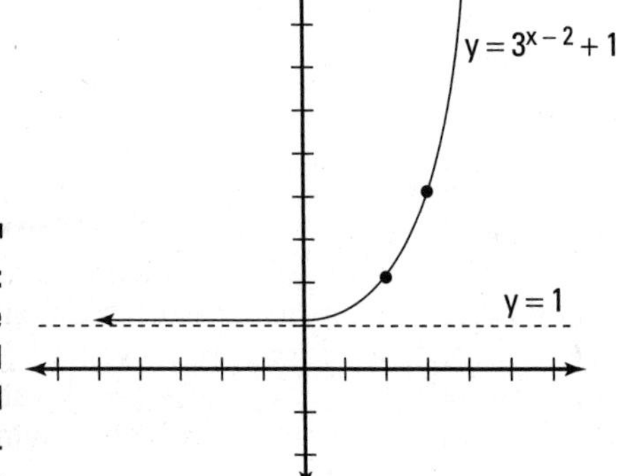

Figure 5-6: The transformed exponential function.

5. **Swap the domain and range values to get the inverse function.**

   Switch every x and y value in each point to get the graph of the inverse function. Figure 5-7 shows the graph of the logarithm.

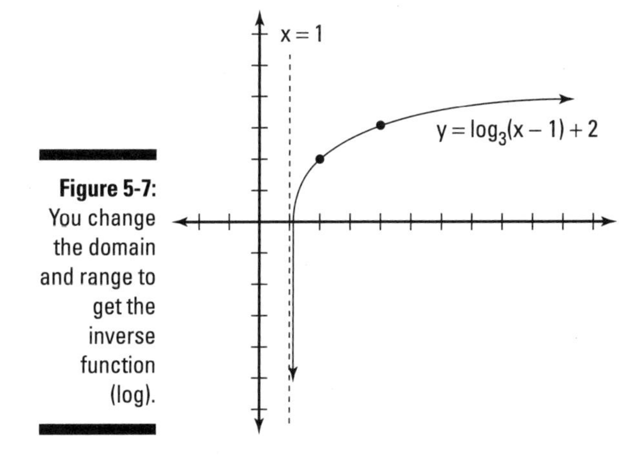

Figure 5-7: You change the domain and range to get the inverse function (log).

Did you notice that the asymptote for the log changed as well? You now have a vertical asymptote at $x = 1$. The parent function for any log will have a vertical asymptote at $x = 0$. The function $f(x) = \log_3(x - 1) + 2$ is shifted to the right one and up two from its parent function $p(x) = \log_3 x$ (using transformation rules; see Chapter 3), so the vertical asymptote is now $x = 1$. Another reason this works is that the exponential and the log are inverses; any asymptote, as well as the translation shifts, will reflect over the line $y = x$ as well.

Solving Equations with Exponents and Logs

We're sure that at some point, your instructor (or boss, perhaps) will ask you to solve an equation with an exponent or a logarithm in it. Have no fear, *Pre-Calculus For Dummies* is here! You must remember one simple rule, and it's all about the base. If you can make the base on one side the same as the base on the other, you can use the properties of exponents or logs (see the corresponding sections earlier in this chapter) to simplify the equation. You now have it made in the shade, because this makes ultimately solving the problem a heck of a lot easier!

In the following sections, you discover how to solve exponential equations with the same base. You also find out how to deal with exponential equations that don't have the same base. And to round things out, we end with the process of solving logarithmic equations.

Stepping through the process of exponential equation solving

The type of exponential equation you're asked to solve determines the steps you'll take to solve it. The following sections break down the types of equations you'll see, and we give our advice on how to solve them.

The basics: Solving an equation with a variable on one side

The basic type of exponential equation is when the variable is on only one side, and each side can be written using the same base. For example, if you're asked to solve $4^{x-2} = 64$, you follow these steps:

1. **Rewrite both sides of the equation so that the bases match.**

   You know that $64 = 4 \cdot 4 \cdot 4$, which is the same as 4^3. So, you can say $4^{x-2} = 4^3$.

2. **Drop the base on both sides and just look at the exponents.**

   When the bases are equal, the exponents *have* to be equal. This gives you the equation $x - 2 = 3$.

3. **Solve the equation.**

   This example has the solution $x = 5$.

Getting fancy: Solving when variables appear on both sides

If you must solve an equation with variables on both sides, you have to do a little more work (sorry!). For example, to solve $2^{x-5} = 8^{x-3}$, follow these steps:

1. **Rewrite all exponential equations so that they have the same base.**

   This gives you $2^{x-5} = (2^3)^{x-3}$.

2. **Use the properties of exponents to simplify.**

   A power to a power signifies that you multiply the exponents. This process gives you $2^{x-5} = 2^{3x-9}$. Distributing the exponent inside the parentheses is how you get $3(x - 3) = 3x - 9$.

3. **Drop the base on both sides.**

   The result is $x - 5 = 3x - 9$.

4. **Solve the equation.**

   Subtract x from both sides to get $-5 = 2x - 9$. Add 9 to each side to get $4 = 2x$. Lastly, divide both sides by 2 to get $2 = x$.

Solving when you can't simplify: Taking the log of both sides

Sometimes you just can't express both sides as powers of the same base. When this is the case, you can make the exponent go away by taking the log of both sides. For example, suppose you're asked to solve $4^{3x-1} = 11$. No integer with the power of 4 gives you 11, so you have to use the following technique:

1. **Take the log of both sides.**

   You can take any log you want, but remember that you actually need to solve the equation with this log, so we suggest sticking with common or natural logs only (see "Getting a better handle on logarithms" earlier in this chapter for more info).

   Using the common log on both sides gives you $\log 4^{3x-1} = \log 11$.

2. **Use the power rule to drop down the exponent.**

   This gives you $(3x - 1)\log 4 = \log 11$.

3. **Divide the log away to isolate the variable.**

   You get $3x - 1 = \frac{\log 11}{\log 4}$.

4. **Solve for the variable.**

   Taking the logs gives you $3x - 1 \approx 1.73$. That means $3x \approx 2.73$, or $x \approx 0.91$.

In the previous problem, you have to use the power rule on only one side of the equation, because the variable appeared on only one side. When you have to use the power rule on both sides, the equations can get a little messy. But with persistence, you can figure it out. For example, to solve $5^{2-x} = 3^{3x+2}$, follow these steps:

1. **Take the log of both sides.**

   As with the previous problem, we suggest you use either a common log or a natural log. If you use a natural log, you get $\ln 5^{2-x} = \ln 3^{3x+2}$.

2. **Use the power rule to drop down both exponents.**

   Don't forget to include your parentheses! You get $(2 - x)\ln 5 = (3x + 2)\ln 3$.

3. **Distribute the logs over the inside of the parentheses.**

   This gives you $2\ln 5 - x\ln 5 = 3x\ln 3 + 2\ln 3$.

4. **Isolate the variables on one side and move everything else to the other by adding or subtracting.**

   You now have $2\ln 5 - 2\ln 3 = 3x\ln 3 + x\ln 5$.

5. **Factor out the x variable from all the appropriate terms.**

   That leaves you with $2\ln5 - 2\ln3 = x(3\ln3 + \ln5)$.

6. **Divide the quantity in parentheses from both sides to solve for x.**

   $x = \dfrac{2\ln5 - 2\ln3}{3\ln3 + \ln5}$

Taking steps to solve logarithm equations

The first thing you need to know before solving equations with logs in them is that there are four types of log equations:

- **Type 1:** The variable you need to solve for is inside the log, with one log on one side of the equation and a constant on the other. If the variable is inside the log, turn it into an exponential equation (which is all about the base, of course). For example, to solve $\log_3 x = -4$, change it to the exponential equation $3^{-4} = x$, or $1/81 = x$.

- **Type 2:** The variable you need to solve for is the base. If the base is what you're looking for, you still change the equation to an exponential equation. If $\log_x 16 = 2$, for instance, change it to $x^2 = 16$, in which x equals ±4.

  Because logs don't have negative bases, you throw the negative one out the window and say $x = 4$ only.

- **Type 3:** The variable you need to solve for is inside the log, but the equation has more than one log and a constant. Using the rules we present in "Managing the properties and identities of logs," you can solve equations with more than one log. To solve $\log_2(x - 1) + \log_2 3 = 5$, for instance, first combine the two logs that are adding into one log by using the product rule: $\log_2[(x - 1) \cdot 3] = 5$. Turn this into $2^5 = (x - 1) \cdot 3$ to solve it. The solution is $x = 35/3$.

- **Type 4:** The variable you need to solve for is inside the log, and all the terms in the equation involve logs. If all the terms in a problem are logs, they have to have the same base in order for you to solve the equation. You can combine all the logs so that you have one log on the left and one log on the right, and then you can drop the log from both sides. For example, to solve $\log_3(x - 1) - \log_3(x + 4) = \log_3 5$, first apply the quotient rule to get $\log_3 \dfrac{x-1}{x+4} = \log_3 5$. You can drop the log base 3 from both sides to get $\dfrac{x-1}{x+4} = 5$, which you can solve easily by using algebra techniques. When solved, you get $x = -21/4$.

The number inside a log can never be negative. Plugging this answer back into part of the original equation gives you $\log_3(^{-21}\!/_4 - 1)$, which is $\log_3(^{-25}\!/_4)$. You don't even have to look at the rest of the equation. The solution to this equation, therefore, is actually the empty set: $\emptyset$, or no solution.

Always plug your answer to a logarithm equation back into the equation to make sure you get a positive number inside the log (not 0 or a negative number).

Surviving Exponential Word Problems

You can use exponential equations in many real-world applications: to predict populations in people or bacteria, to estimate financial values, and even to solve mysteries! Almost every pre-calc textbook includes one section specifically dedicated to exponential word problems, so we thought we'd do the same in this chapter. If the object grows continuously, then the base of the exponential function is e.

Exponential word problems come in many different varieties, but they all follow one simple formula:

$B(t) = Pe^{rt}$, where

P stands for the initial value of the function — usually referred to as the number of objects whenever $t = 0$.

t is the time (measured in many different units, so be careful!).

$B(t)$ is the value of how many people, bacteria, money, and so on you have after time t.

r is a constant that describes the rate at which the population is changing. If r is positive, it's called the *growth constant.* If r is negative, it's called the *decay constant.*

e is the base of the natural logarithm, used for continuous growth or decay.

Take a look at the following sample word problem, which this formula enables you to solve:

Exponential growth exists in your kitchen on daily basis in the form of bacteria. Everything grows in your kitchen really fast (sometimes doubling or tripling its growth) for each time interval. Whenever you see text talking about bacteria, you know automatically that it grows exponentially.

Suppose that you leave your leftover breakfast on the kitchen counter when leaving for work. Assume that 5 bacteria are present on the breakfast at 8:00 a.m., and 50 bacteria are present at 10:00 a.m. You can use $B(t) = Pe^{rt}$ to find out how long it will take for the population of bacteria to grow to 1 million if the growth is continuous.

You need to solve two parts of this problem: First, you need to know the rate at which the bacteria are growing, and then you can use that rate to find the time at which the population of bacteria will reach 1 million. Here are the steps for solving this word problem:

1. **Calculate the time that elapsed between the initial reading and the reading at time *t*.**

   Two hours elapsed between 8:00 a.m. and 10:00 a.m.

2. **Identify the population at time *t*, the initial population, and the time and plug these values into the formula.**

   You get $50 = 5 \cdot e^{r \cdot 2}$.

3. **Divide the initial population on both sides to isolate the exponential.**

   You get $10 = e^{r \cdot 2}$.

4. **Take the appropriate logarithm of both sides, depending on the base.**

   In the case of continuous growth, the base will always be *e:* $\ln 10 = \ln e^{r \cdot 2}$.

5. **Using the power rule (see the section "Managing the properties and identities of logs"), simplify the equation.**

   You get $\ln 10 = r \cdot 2 \cdot \ln e$, which simplifies to $\ln 10 = r \cdot 2$.

6. **Divide by the time to find the rate; use your calculator to find the decimal approximation.**

   You get $\frac{\ln 10}{2} = r$, or $r \approx 1.1513$. This rate means that the population is growing by more than 115 percent.

7. **Plug *r* back into the original equation and leave *t* as the variable.**

   This gives you $B(t) = 5 \cdot e^{1.1513(t)}$.

8. **Plug the final amount in B(*t*) and solve for *t*, leaving the initial population the same.**

   You now have $1{,}000{,}000 = 5e^{1.1513(t)}$.

9. **Divide by the initial population to isolate the exponential.**

   You get $200{,}000 = e^{1.1513(t)}$.

10. **Take the log (or ln) of both sides.**

    This gives you ln 200,000 = 1.1513(t).

    We recommend not simplifying ln 200,000 at all, but rather plugging it into your calculator all as one step. This gives less rounding error in the final answer.

11. **Divide by the rate on both sides.**

    Finally, you come up with 10.61 hours = t.

Phew, that was quite a workout! One million bacteria in a little more than ten hours is a good reason to refrigerate promptly.

Part II
The Essentials of Trigonometry

"Nothing personal, but I'm going to run several trigonometric formulas just to prove the identity of this casserole."

In this part . . .

Trigonometry is a subject that most students cover briefly in geometry. However, pre-calculus expands on those ideas. We begin with a review of angles and how to build the valuable tool known as the unit circle, based on previous knowledge of the special triangles. We examine right triangles and the basic trig functions built in them. Graphing trig functions may or may not be a review, depending on the Algebra II course you've taken, so we take the time to establish how to graph the parent graph of the six basic trig functions and then explain how to transform those graphs to get to the more complicated ones.

This part covers the often feared and seldom understood formulas and identities for trig functions. We break each identity down into more-manageable pieces. We show you how to simplify trig expressions and then how to solve trig equations for an unknown variable using formulas and identities. Lastly, we cover how to solve triangles that aren't right triangles using the Law of Sines and the Law of Cosines.

Chapter 6

Angling In on the Unit Circle

In This Chapter

- Discovering alternate trig function definitions
- Placing angle measurements on a unit circle
- Calculating trig functions on the unit circle

In this chapter, we move on to right triangles drawn on the coordinate plane (x- and y-axes). Moving right triangles onto the coordinate plane brings out many more interesting concepts (such as evaluating trig functions and solving trig equations) that we'll look at in this chapter, all through a very handy tool known as the unit circle.

The unit circle is extremely important in the real world and in mathematics; for instance, you're at its mercy whenever you fly in an airplane. Pilots use the unit circle, along with vectors, to fly airplanes in the correct direction and over the correct distance. Imagine what a disaster it would be to have a pilot trying to land a plane a bit to the left of the runway!

In this chapter, we work on building the unit circle together as we review the basics of angles in radians and triangles. With that information, you can place the triangles onto the unit circle (which is also located in the coordinate plane) to solve the problems we present at the end of this chapter. (We build on these ideas even further as we move into graphing trig functions in Chapter 7.)

Introducing Radians: The Basic Pre-Calc Measurement

When you study geometry, you measure every angle in degrees, based on a portion of a 360° circle around the coordinate plane. As it turns out, the number 360 was picked to represent the degrees in a circle only for convenience.

What's the convenience of the number 360, you ask? Well, you can divide a circle into many different, equal parts by using the number 360, because it is divisible by 2, 3, 4, 5, 6, 8, 9, 10, 12, 15, 18, 20, 24, 30, 36, 40, 45 . . . and these are just the numbers less than 50! Basically, the number 360 is pretty darn flexible for performing calculations.

The truth that not many people realize, though, is that degrees weren't the first form of angle measurement — radians were. The word *radian* is based on the same root word as radius, which is the building block of a circle. An angle measurement of 360°, or a complete circle, is equal to 2π radians, which breaks down in the same way that degrees do.

In pre-calculus, you draw every angle with its vertex at the origin of the coordinate plane (0, 0), and you place one side on the positive x-axis (this is called the *initial side* of the angle, and it will always be in this location). The other side of the angle extends from the origin to anywhere on the coordinate plane (this is called the *terminal side*). An angle whose initial side lies on the positive x-axis is said to be in *standard position.*

If you move from the initial side to the terminal side in a counterclockwise direction, the angle has a *positive measure.* If you move from the initial side to the terminal side in a clockwise direction, you say that this angle has a *negative measure.*

A positive/negative discussion of angles brings up another related and important point: *co-terminal angles.* Co-terminal angles are angles that have different measures, but their terminal sides lie in the same spot. These angles can be found by adding or subtracting 360° (or 2π radians) from an angle as many times as you want; there are infinitely many of them, which will become quite handy in future chapters!

Trig Ratios: Taking Right Triangles a Step Further

Reach back into your brain for a second and recall that a *ratio* is the comparison of two things. If a pre-calc class has 20 guys and 14 girls, the ratio of guys to girls is $^{20}/_{14}$, and because that's a fraction, the ratio reduces to $^{10}/_{7}$. Ratios are important in many areas of life. For instance, if you have 20 people at a cookout with only 10 burgers, it's time to worry!

Because you spend a ton of time in pre-calc working with trigonometric functions, you need to understand ratios. In this section, we look at three very important ratios in right triangles — sine, cosine, and tangent — as well as three not-so-vital but still important ratios — cosecant, secant, and cotangent. These ratios are all *functions,* where an angle is the input — so some textbooks use the words *trig function* and *trig ratio* interchangeably. Each function looks at an angle of a right triangle, known or unknown, and then uses the definition of its specific ratio to help you find missing information in the triangle quickly and easily. And to round out the section, we show you how to use inverse trig functions to solve for an unknown angle in a right triangle.

You can remember the trig functions and their definitions by using the mnemonic device SOHCAHTOA, which stands for

- **S**ine = **O**pposite over **H**ypotenuse
- **C**osine = **A**djacent over **H**ypotenuse
- **T**angent = **O**pposite over **A**djacent

Making a sine

The *sine* of an angle is defined as the ratio of the opposite leg to the hypotenuse. In symbols, you write $\sin\theta$. Here's what the ratio looks like:

$$\sin\theta = \frac{opp}{hyp}$$

In order to find the sine of an angle, you must know the lengths of the opposite side and the hypotenuse. You will always be given the lengths of two sides, but if the two sides aren't the ones you need to find a certain ratio, you can use the Pythagorean Theorem to find the missing one. For example, to find the sine of angle F (sinF) in Figure 6-1

1. **Identify the hypotenuse.**

   Where's the right angle? It's $\angle R$, so side *r*, across from it, is the hypotenuse. You can label it "hyp."

2. **Locate the opposite side.**

   Look at the angle in question, which is $\angle F$ here. Which side is across from it? Side *f* is the opposite leg. You can label it "opp."

3. **Label the adjacent side.**

   The only side that's left, side *k*, has to be the adjacent leg. You can label it "adj."

4. **Locate the two sides that you use in the trig ratio.**

   Since you are finding the sine of $\angle F$, you need the opposite side and the hypotenuse. For this triangle, $(\text{leg})^2 + (\text{leg})^2 = (\text{hypotenuse})^2$ becomes $f^2 + k^2 = r^2$. Plug in what you know to get $f^2 + 7^2 = 14^2$. When you solve this for *f*, you get $f = 7\sqrt{3}$.

5. Find the sine.

With the information from Step 4, you can find that the sine of ∠F is $\sin F = \frac{7\sqrt{3}}{14}$, which reduces to $\frac{\sqrt{3}}{2}$.

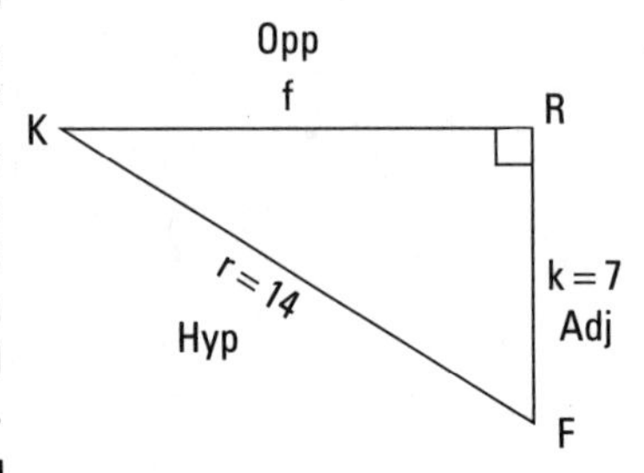

Figure 6-1: △KRF illustrates how you can find sine with minimal information.

Looking for a cosine

The *cosine* of an angle, or cosθ, is defined as the ratio of the adjacent leg to the hypotenuse, or

$$\cos\theta = \frac{adj}{hyp}$$

Consider this example: A ladder leans against a building, creating an angle of 75° with the ground. The base of the ladder is 3 feet away from the building. How long is the ladder? Did your heart just sink when you realized we're giving you a . . . *word problem?* No problem! Just follow these steps to solve; here, you're looking for the length of the ladder:

1. Draw a picture so you can see a familiar shape.

Figure 6-2 represents the ladder leaning against the building.

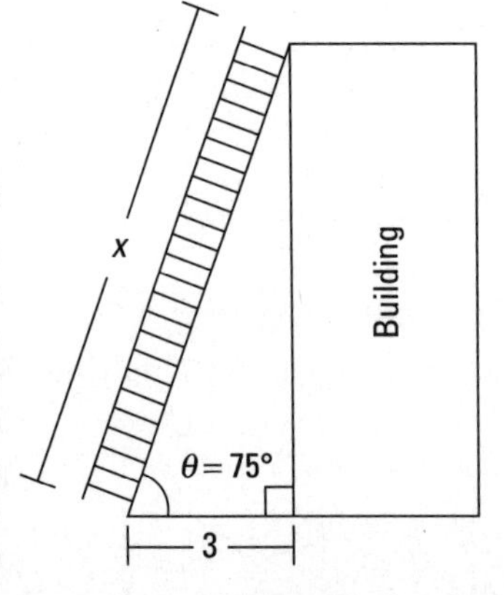

Figure 6-2: One ladder plus one building equals one cosine problem.

The right angle goes between the building and the ground, because otherwise the building would be crooked and fall down. Because you know where the right angle is, you know that the hypotenuse is the ladder itself. The given angle is down on the ground, which means the opposite leg is the distance on the building from where the ladder touches to the ground. The third side, the adjacent leg, is the distance the ladder rests from the building.

2. **Set up a trigonometry equation, using the information from the picture.**

   You have to use the cosine ratio because it's the ratio of the adjacent leg to the hypotenuse, and you know that the adjacent side is 3 feet; you're looking for the length of the ladder, or the hypotenuse. You have $\cos 75° = \frac{3}{x}$. The building has nothing to do with this problem right now, other than it's what's holding up the ladder.

   Why do you use 75° in the cosine function? Because you actually know how big the angle is; you don't have to use θ to represent an unknown angle.

3. **Solve for the unknown variable.**

   Multiply the unknown x to both sides to get $x \cdot \cos 75° = 3$. The $\cos 75°$ is just a number. When you plug it into your calculator, you get a decimal answer (make sure you set your calculator to degree mode before attempting to do this problem). Now divide both sides by $\cos 75°$ to isolate x; you get $x = \frac{3}{\cos 75°}$. This produces the answer $x \approx 11.6$, which means the ladder is about 11.6 feet long.

Going on a tangent

The *tangent* of an angle, or $\tan\theta$, is the ratio of the opposite leg to the adjacent leg. This is what it looks like in equation form:

$$\tan\theta = \frac{opp}{adj}$$

Imagine for a moment that you're an engineer. You're working with a 39-foot tower with a wire attached to the top of it. The wire needs to attach to the ground and make an angle of 80° with the ground to keep the tower from moving. Your task is to figure out how far from the base of the tower the wire should attach to the ground. Follow these steps:

1. **Draw a diagram that represents the given information.**

   Figure 6-3 shows the wire, the tower, and the known information.

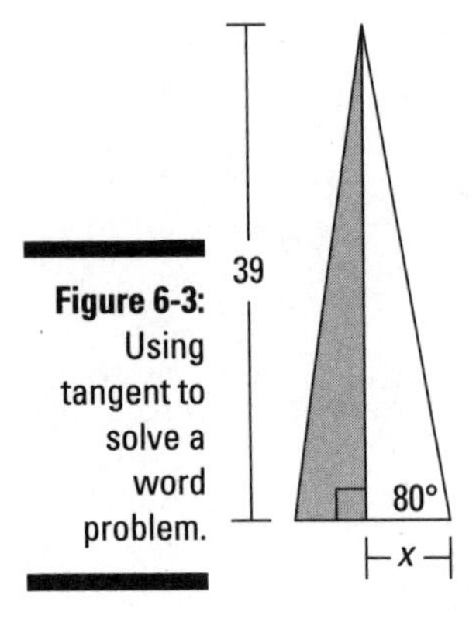

Figure 6-3: Using tangent to solve a word problem.

2. **Set up a trigonometry equation, using the information from the picture.**

   For this problem, you must set up the trig equation that features tangent, because the opposite side is the length of the tower, the hypotenuse is the wire, and the adjacent side is what you need to find. You get $\tan 80° = \frac{39}{x}$.

3. **Solve for the unknown.**

   Multiply both sides by the unknown x to get $x \cdot \tan 80° = 39$. Divide both sides by the $\tan 80°$ to get $x = \frac{39}{\tan 80°}$. Simplify to get $x = 6.88$. The wire attaches to the ground 6.88 feet from the base of the tower to form the 80° angle.

Discovering the flip side: Reciprocal trig functions

Three additional trig ratios — secant, cosecant, and cotangent — are called *reciprocal functions* because they're the reciprocals of sine, cosine, and tangent. These three functions open up three more ways in which you can solve equations in pre-calculus. The following list breaks down these functions and how you use them:

- ***Cosecant,* or csc*θ*, is the reciprocal of sine.** The reciprocal of a is $\frac{1}{a}$, so $\csc\theta = \frac{1}{\sin\theta}$. And if $\sin\theta = \frac{opp}{hyp}$, $\csc\theta = \frac{1}{\frac{opp}{hyp}}$, or $\frac{hyp}{opp}$. In other words, cosecant is the ratio of the hypotenuse to the opposite leg.
- ***Secant,* or sec*θ*, is the reciprocal of cosine.** Based on a formula similar to cosecant, $\sec\theta = \frac{1}{\cos\theta} = \frac{1}{\frac{adj}{hyp}} = \frac{hyp}{adj}$. Secant, in other words, is the ratio of the hypotenuse to the adjacent leg.

A common mistake is to think that secant is the reciprocal of sine and that cosecant is the reciprocal of cosine, but this isn't the case, as the previous bullets illustrate.

- ***Cotangent*, or cotθ, is the reciprocal of tangent.** (How's that for obvious?) You have the hang of this if you've looked at the previous bullets: $\cot\theta = \frac{1}{\tan\theta} = \frac{1}{\frac{opp}{adj}} = \frac{adj}{opp}$.

  Tanθ may also be written as $\frac{\sin\theta}{\cos\theta}$. Therefore, cot$\theta$ may be written as the reciprocal — in other words, $\cot\theta = \frac{1}{\tan\theta} = \frac{adj}{opp} = \frac{\cos\theta}{\sin\theta}$.

Secant, cosecant, and cotangent are all reciprocals, but you'll find no button for them on your calculator. You must use their reciprocals — sine, cosine, and tangent. Don't get confused and use the $\sin^{-1}$, $\cos^{-1}$, and $\tan^{-1}$ buttons, either. Those buttons are for inverse trig functions, which we describe in the following section.

Working in reverse: Inverse trig functions

Almost every function has an inverse. An *inverse function* basically undoes a function. The trigonometry functions sine, cosine, and tangent all have inverses, and they're often called arcsin, arccos, and arctan.

In trig functions, theta (θ) is the input, and the output is the ratio of the sides of a triangle. If you're given the ratio of the sides and need to find an angle, you must use the inverse trig function:

Inverse sine (arcsin): $\theta = \sin^{-1}\left(\frac{opp}{hyp}\right)$

Inverse cosine (arccos): $\theta = \cos^{-1}\left(\frac{adj}{hyp}\right)$

Inverse tangent (arctan): $\theta = \tan^{-1}\left(\frac{opp}{adj}\right)$

Here's what an inverse trig function looks like in action. To find the angle θ in degrees in a right triangle if the $\tan\theta = 1.7$, follow these steps:

1. **Isolate the trig function on one side and move everything else to the other.**

   This is done already. Tangent is on the left and the decimal 1.7 is on the right: $\tan\theta = 1.7$.

2. **Isolate the variable.**

   You're given the ratio for the trig function and have to find the angle. To work backward and figure out the angle, use some algebra. You have to undo the tangent function, which means using the inverse tangent function on both sides: $\tan^{-1}(\tan\theta) = \tan^{-1}(1.7)$. This equation simplifies to $\theta = \tan^{-1}(1.7)$.

3. **Solve the simplified equation.**

   $\theta = \tan^{-1}(1.7)$ gives you $\theta = 59.53°$.

Read the problem carefully so you know whether the angle you're looking for should be expressed in degrees or radians. Set your calculator to the correct mode.

Understanding How Trig Ratios Work on the Coordinate Plane

The unit circle that we build in this chapter lies on the coordinate plane — that same plane you've been graphing on since algebra. The *unit circle* is a very small circle centered at the origin (0, 0). The radius of the unit circle is 1, which is why it's called the unit circle. To carry out the work of this chapter, all the angles specified need to be drawn on the coordinate plane, so we have to redefine the SOHCAHTOA ratios from earlier to make sense of them on the coordinate plane and the unit circle.

Essentially, all you do is look at the trig ratios in terms of x and y values rather than opposite, adjacent, and hypotenuse. Redefining these ratios to fit the coordinate plane (sometimes called the *point-in-the-plane* definition) makes it easier for you to visualize the differences. Some of the angles, for instance, will be larger than 180°, but using the new definitions will allow you to make a right triangle by using a point and the x-axis. You then use the new ratios to find missing sides of right triangles and/or trig function values of angles.

When a point (x, y) exists on a coordinate plane, you can calculate all the trig functions for the point by following these steps:

1. **Locate the point on the coordinate plane and connect it to the origin, using a straight line.**

   Say, for example, that you're asked to evaluate all six trig functions of the point in the plane (–4, –6). The line segment moving from this point to the origin is your hypotenuse and is now called the radius r (see Figure 6-4).

2. **Draw a perpendicular line connecting the given point to the x-axis, creating a right triangle.**

   The legs of the right triangle are –4 and –6. Don't let the negative signs scare you; the lengths of the sides are still 4 and 6. The negative signs just reveal the location of that point on the coordinate plane.

3. **Find the length of the hypotenuse *r* by using the distance formula, or the Pythagorean Theorem.**

   The distance you want to find is the length of *r* from Step 1. Using the distance formula between (x, y) and the origin $(0, 0)$, you get $r = \sqrt{x^2 + y^2}$.

   This equation implies the principal or positive root only, so the hypotenuse for these point-in-the-plane triangles is always positive.

   For our example, you get $\sqrt{52}$, which simplifies to $2\sqrt{13}$. Check out what our triangle looks like in Figure 6-4.

4. **Evaluate the trig function values, using their alternate definitions.**

   With the labels from Figure 6-4, you get the following formulas:

   - $\sin\theta = y/r \rightarrow \csc\theta = r/y$
   - $\cos\theta = x/r \rightarrow \sec\theta = r/x$
   - $\tan\theta = y/x \rightarrow \cot\theta = x/y$

   Substitute the numbers from our example to pinpoint the trig values:

   - $\sin\theta = \frac{-6}{2\sqrt{13}}$. Simplify first to $\frac{-3}{\sqrt{13}}$, and then rationalize the denominator to get $\frac{-3\sqrt{13}}{13}$.
   - $\cos\theta = \frac{-4}{2\sqrt{13}}$. Simplify first to get $\frac{-2}{\sqrt{13}}$, and then rationalize to get $\frac{-2\sqrt{13}}{13}$.
   - $\tan\theta = \frac{-6}{-4}$. This simplifies to $\frac{3}{2}$.
   - $\cot\theta = \frac{2}{3}$.
   - $\sec\theta = \frac{-\sqrt{13}}{2}$.
   - $\csc\theta = \frac{-\sqrt{13}}{3}$.

   Notice that the rules of trig functions and their reciprocals still apply. For example, if you know $\sin\theta$, you automatically know $\csc\theta$ because they're reciprocals.

When the point you're given is a point on one of the axes, you can still find all the trig function values. For instance, if the point is on the *x*-axis, the cosine and the radius have the same absolute value (because the cosine can be negative but the radius can't). If the point is on the positive *x*-axis, the cosine is 1 and the sine is 0; if the point is on the negative *x*-axis, the cosine is –1. Similarly, if the point is on the *y*-axis, the sine value and the radius are of the same absolute value; the sine will be either 1 or –1, and the cosine will always be 0.

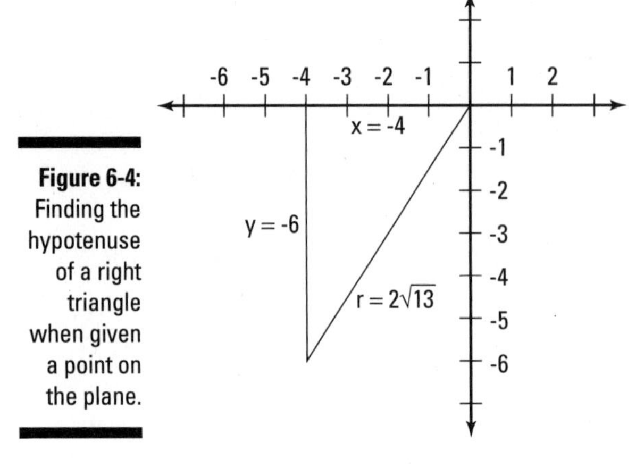

Figure 6-4: Finding the hypotenuse of a right triangle when given a point on the plane.

Getting a Good Grasp on the Unit Circle

The unit circle is a vital part of the study of trigonometry. To visualize the tool, imagine a circle drawn on the coordinate plane, centered at the origin. The trig functions sine, cosine, and tangent rely heavily on the shortcuts that you can figure out by using the unit circle. It may be a new idea, but don't be intimidated; the circle is simply built off concepts from geometry. In the following sections, we dissect the unit circle and use special right triangles to build it, which will be imperative to your studies in pre-calc.

Familiarizing yourself with the most common angles

In pre-calculus, you often need to draw an angle on the coordinate plane in order to make certain calculations. But we don't recommend that you memorize where all the angles' terminal sides lie in the unit circle because that's a waste of your time. Of course, because the 30° angle, the 45° angle, and the 60° angle are so common in pre-calc problems, it isn't a bad idea to remember exactly where their terminal sides lie. This information gives you a good foundation for figuring out where the rest of the unit-circle angles lie. These angles will help you find the trig function values for those special (or more common) angles on the unit circle. And, in Chapter 7, these special values help you graph the trig functions.

Figure 6-5 shows the 30°, 45°, and 60° angles. (For the full unit circle picture, refer to the Cheat Sheet at the front of this book.)

Instead of memorizing the locale of all the major angles on the unit circle, let the quadrants be your guide. Remember that each quadrant contains 90° (or $\pi/2$ radians), and that the measures increase when you move counterclockwise around the vertex. With that information and a little bit of math, you can figure out the location of the angle you need.

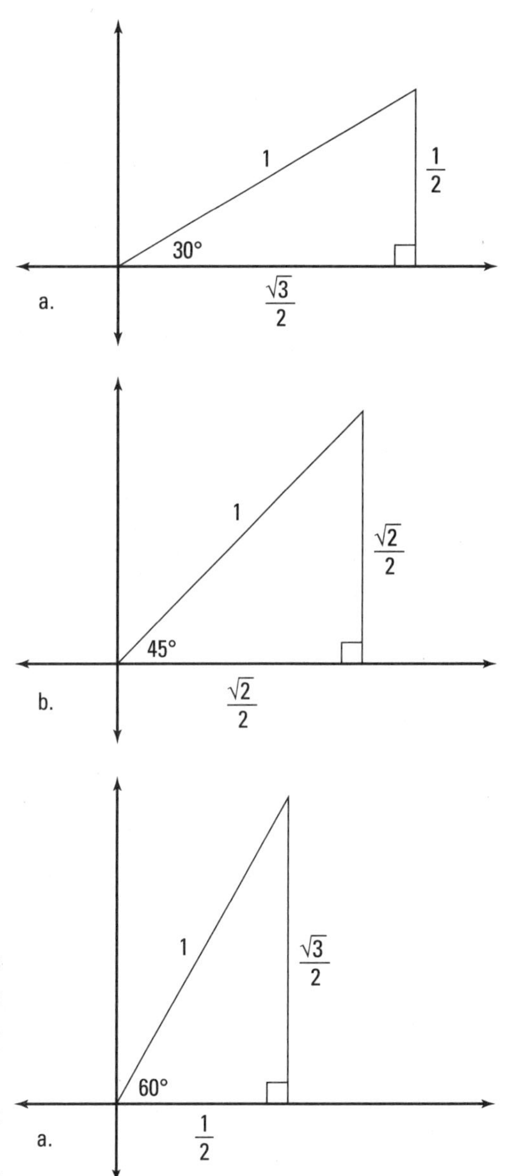

Figure 6-5: The 30°, 45°, and 60° angles on the unit circle.

Drawing uncommon angles

Many times in your journey through trig — actually, all the time — drawing a figure will help you solve a given problem (refer to any example problem in this chapter for an illustration of how to use a picture to help). Trig always starts with the basics of drawing angles so that by the time you get to the problems, the actual drawing stuff is second nature.

Always draw a picture for any trig problem. We know many students who've raised their grades by following this very simple advice. Drawing a picture makes the information given to you more visual and enables to you to picture what's going on.

We help you draw these angles by sketching their terminal sides in the correct places. Then, by drawing a vertical line up or down to the x-axis, you can make right triangles that fit in the unit circle, with smaller angles that are more familiar to you.

What do you do if you're asked to draw an angle that has a measure greater than 360°? Or a negative measure? How about both? We bet your head is spinning! No worries; this section gives you the steps.

For example, suppose that you need to draw a –570° angle. Here's what you do:

1. **Find a co-terminal angle by adding 360°.**

   Adding 360° to –570° gives you –210°.

2. **If the angle is still negative, keep adding 360° until you get a positive angle in standard position.**

   Adding 360° to –210° gives you 150°. This angle is 30° less than 180° (much closer to the 180° line than 90°).

3. **Draw the angle you create in Step 2.**

   You need to draw a –570° angle, so be careful which way your arrow points and how many times you travel around the unit circle before you stop at the terminal side.

   This angle starts at 0 on the x-axis and moves in a clockwise direction, because you're finding a negative angle. Figure 6-6 shows what the finished angle looks like.

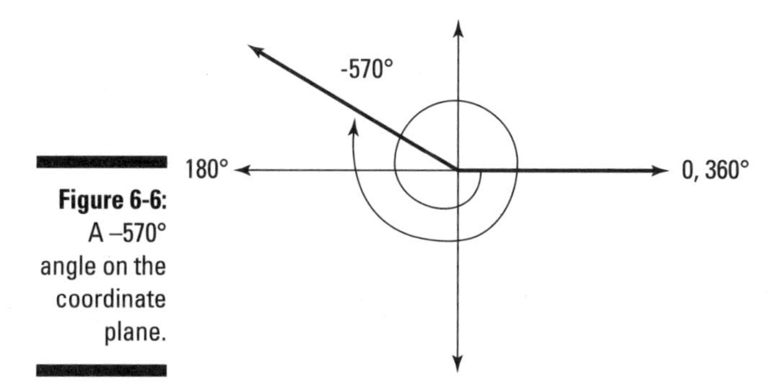

Figure 6-6: A –570° angle on the coordinate plane.

Digesting Special Triangle Ratios

You'll see two triangles over and over again in trigonometry; we call them the 45er and the old 30-60. In fact, you'll see them so often that most mathematicians recommend that you just memorize them. Relax! We show you how in this section. (And, yes, these are two triangles you've seen before in geometry.)

The 45er: 45°-45°-90° triangles

All 45°-45°-90° triangles have sides that are in a unique ratio. The two legs are the exact same length, and the hypotenuse is that length times $\sqrt{2}$. Figure 6-7 shows the ratio. (If you look at the 45er triangle in radians, you have $\frac{\pi}{4}-\frac{\pi}{4}-\frac{\pi}{2}$. Either way, it's still the same ratio.)

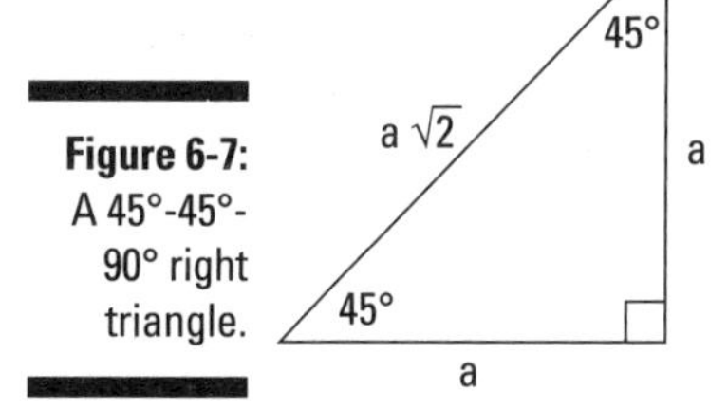

Figure 6-7: A 45°-45°-90° right triangle.

Why is this triangle important? Because any time you're given one side of a 45er triangle, you can figure out the other two sides. There are two basic cases for completing calculations with this type of triangle:

- **Type 1: You're given one leg.**

  Because you know both legs are equal, you know the length of both the legs. You can find the hypotenuse by multiplying this length by $\sqrt{2}$.

- **Type 2: You're given the hypotenuse.**

  Divide the hypotenuse by $\sqrt{2}$ to find the legs (which are equal).

Here's an example calculation: The diagonal in a square is 16 cm long. How long is each side of the square? Draw it out first. Figure 6-8 shows the square.

The diagonal of a square divides the angles into 45° pieces, so you have the hypotenuse of a 45er triangle. To find the legs, divide the hypotenuse by $\sqrt{2}$. When you do, you get $\frac{16}{\sqrt{2}}$. You must now rationalize the denominator to get $\frac{16}{\sqrt{2}} \cdot \frac{\sqrt{2}}{\sqrt{2}} = \frac{16\sqrt{2}}{2} = 8\sqrt{2}$ — the measure of each side of the square.

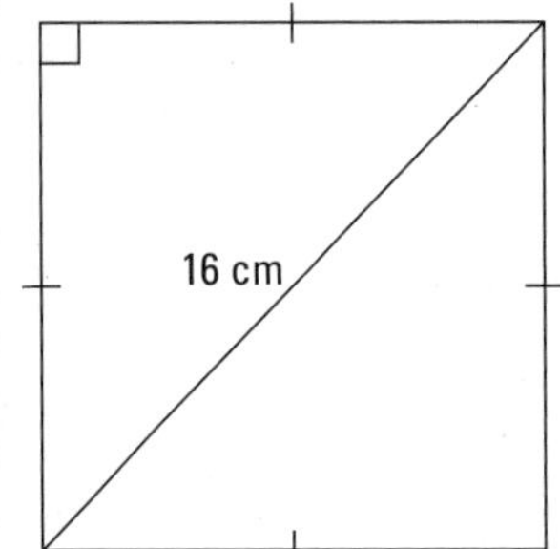

Figure 6-8: A square with a diagonal (which doubles as the hypotenuse) of 16 cm.

The old 30-60: 30°-60°-90° triangles

All 30°-60°-90° triangles have sides with the same basic ratio (if you look at the 30°-60°-90° triangle in radians, you see $\frac{\pi}{6} - \frac{\pi}{3} - \frac{\pi}{2}$):

- The shortest leg is across from the 30° angle.
- The length of the hypotenuse is always two times the length of the shortest leg.
- You can find the long leg by multiplying the short leg by $\sqrt{3}$. (Don't be scared, it's just a square root sign; it won't hurt you!)

Note: The hypotenuse is the longest side in a right triangle, which is different from the long leg. The long leg is the leg opposite the 60° angle.

Figure 6-9 illustrates the ratio of the sides for the 30-60 triangle.

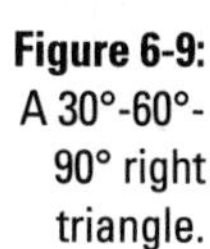

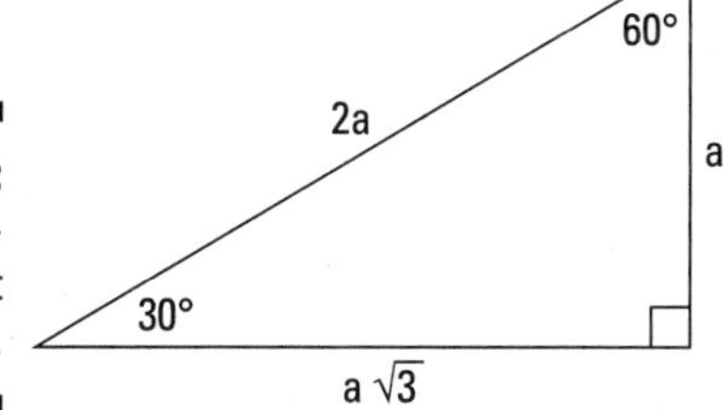

Figure 6-9: A 30°-60°-90° right triangle.

If you know one side of a 30-60 triangle, you can find the other two by using shortcuts. Here are the three situations you come across when doing these calculations:

- **Type 1: You know the short leg (the side across from the 30° angle).**

  Double its length to find the hypotenuse. You can multiply the short side by $\sqrt{3}$ to find the long leg.

- **Type 2: You know the hypotenuse.**

  Divide the hypotenuse by 2 to find the short side. Multiply this answer by $\sqrt{3}$ to find the long leg.

- **Type 3: You know the long leg (the side across from the 60° angle).**

  Divide this side by $\sqrt{3}$ to find the short side. Double that figure to find the hypotenuse.

In the triangle TRI in Figure 6-10, the hypotenuse is 14 inches long; how long are the other sides?

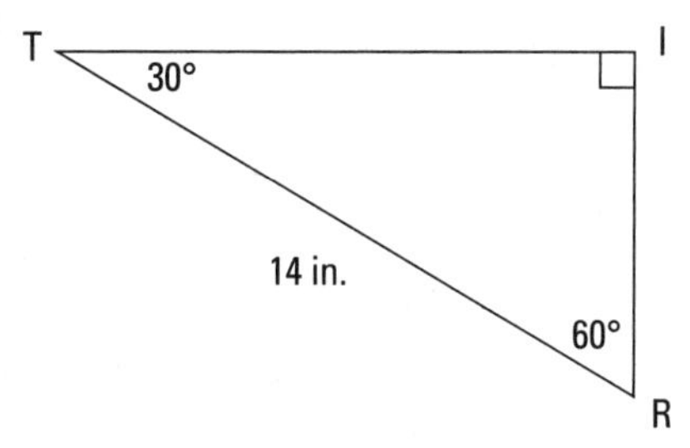

Figure 6-10: Finding the other sides of a 30-60 when you know the hypotenuse.

Because you have the hypotenuse TR = 14, you can divide by 2 to get the short side: RI = 7. Now you multiply this by $\sqrt{3}$ to get the long side: IT = $7\sqrt{3}$.

The Fusion of Triangles and the Unit Circle: Working Together for Good

Rejoice in the fusion of right triangles, common angles (see the previous section), and the unit circle, because they come together for the greater good of pre-calculus! The special right triangles that we reviewed play an important role in finding specific trig function values that you can see on the unit circle. Specifically, if you know the measure of an angle, you can make a special right triangle that will fit onto the unit circle. Using this triangle, you can evaluate all kinds of trig functions without a calculator!

All congruent angles (angles with the same measure) have the same values for the different trig functions. Some noncongruent angles also have identical values for certain trig functions; you can use a reference angle to find out the measures for these angles.

Brush up on the special right triangles before attempting to evaluate the complicated functions in this section. Although many of the values look identical, looks can be deceiving. The numbers may be the same, but the signs and locations of these numbers change as you move around the unit circle.

Placing the major angles correctly, sans protractor

In this section, we take the unit circle angles (see the Cheat Sheet for a picture of the unit circle) and the special right triangles and put them together to create neat little package: the full unit circle. We create special triangles on the unit circle one at a time, because they're all points on the coordinate plane.

Regardless of how long the sides are that make up a particular angle in a triangle, the trig function values for that specific angle will always be the same. Therefore, mathematicians "shrank" all the sides of right triangles so that they'd all fit into the unit circle.

The hypotenuse of every triangle in a unit circle is always 1, and the calculations that involve the triangles are much easier to compute. Because of the unit circle, you can draw *any* angle with *any* measurement, and all right triangles with the same reference angle are the same size.

Starting in quadrant I: Calculate the points to plot

Look at an angle marked 30° in the unit circle (see Figure 6-11) and follow these steps to build a triangle out of it — similar to the steps from the section "Understanding How Trig Ratios Work on the Coordinate Plane":

1. **Draw the angle and connect it to the origin, using a straight line.**

   The terminal side of a 30° angle should be in the first quadrant, and the size of the angle should be rather small. In fact, it should be one-third of the way between 0° and 90°.

2. **Draw a perpendicular line connecting the point where the ray stops to the *x*-axis, creating a right triangle.**

   The triangle's hypotenuse is the radius of the unit circle; one of its legs is on the *x*-axis; and the other leg is parallel to the *y*-axis. You can see what this 30°-60°-90° triangle looks like in Figure 6-11.

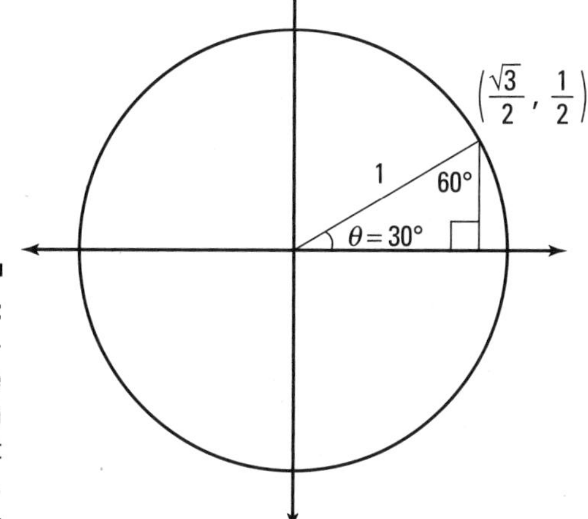

Figure 6-11: A 30°-60°-90° triangle drawn on the unit circle.

3. **Find the length of the hypotenuse.**

   The radius of the unit circle is always 1, which means the hypotenuse of the triangle is also 1.

4. **Find the lengths of the other sides.**

   To find the other two sides, you use the techniques we discuss in the 30°-60°-90° triangle section. Find the short leg first by dividing by 2, which gives you ½. To find the long leg, multiply this by $\sqrt{3}$ to get $\frac{\sqrt{3}}{2}$.

5. **Identify the point on the unit circle.**

   The unit circle is on the coordinate plane, centered at the origin. So, each of the points on the unit circle has unique coordinates. You can now name the point at 30° on the circle $\left(\frac{\sqrt{3}}{2}, \frac{1}{2}\right)$.

After going through the previous steps, you can easily find the points of other angles on the unit circle as well. For instance,

- Look at the point on the circle marked 45° (see the Cheat Sheet at the front of the book). You can draw a triangle from it, using Steps 1 and 2. Its hypotenuse is still 1, the radius of the unit circle. To find the length of the legs of a 45°-45°-90° triangle, you divide the hypotenuse by $\sqrt{2}$. You then rationalize the denominator to get $\frac{\sqrt{2}}{2}$. You can now name this point on the circle $\left(\frac{\sqrt{2}}{2}, \frac{\sqrt{2}}{2}\right)$.
- Moving counterclockwise to the 60° angle, you can create a triangle with Steps 1 and 2. If you look closely, you'll realize that the 30° angle is at the top, so the short side is the side on the x-axis. That makes the point at 60° $\left(\frac{1}{2}, \frac{\sqrt{3}}{2}\right)$, due to the radius of 1 (divide 1 by 2 and multiply ½ by $\sqrt{3}$).

Moving along to the other quadrants

Quadrants II–IV in the coordinate plane are just mirror images of the first quadrant (see the previous section). However, the signs are different because the points on the unit circle are on different locations of the plane:

- In quadrant I, both x and y values are positive.
- In quadrant II, x is negative and y is positive.
- In quadrant III, both x and y are negative.
- In quadrant IV, x is positive and y is negative.

The good news is that you never have to memorize the whole unit circle. You can simply apply the basics of what you know about right triangles and the unit circle! Figure 6-12 shows the whole pizza pie of the unit circle.

Retrieving trig-function values on the unit circle

You may be wondering why we spewed all this information about triangles, angles, and unit circles in the previous sections of this chapter. You need to be comfortable with the unit circle and the special triangles in it so that you can evaluate trig functions quickly and with ease, which you do in this section. You don't want to waste precious moments during an exam constructing the entire unit circle just to evaluate a couple angles. And the more comfortable you are with trig ratios and the unit circle as a whole, the less likely you'll be to make an error with a negative sign or get trig values mixed up.

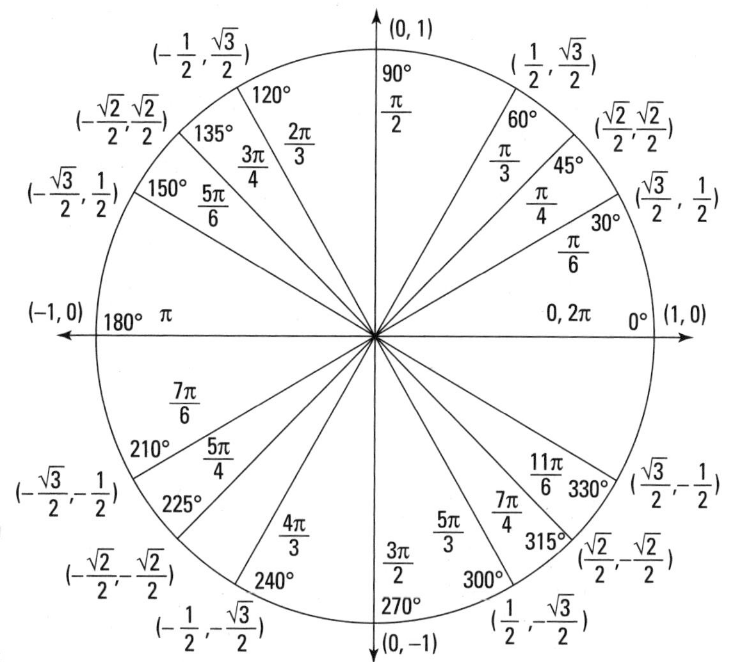

Figure 6-12: The whole unit circle.

Finding values for the six trig functions

In pre-calc, you'll need to evaluate the six trig functions — sine, cosine, tangent, cosecant, secant, and cotangent — for a single value on the unit circle. For each angle on the unit circle, there are three other angles with similar trig function values. The only difference is that the signs of these values will be opposite, depending on which quadrant the angle is in. Sometimes, the angle won't be on the unit circle, and you'll have to use your calculator.

If you don't have the unit circle at your disposal (if you're taking a test, for instance), you can draw a picture and find the values you need the long way, which we explain in this section. (However, as you progress, you'll discover a shortcut you can use to simplify the process, which we explain in the following section.)

The point-in-the-plane definition of cosine in a right triangle is $\cos\theta = x/r$. Because the hypotenuse r is always 1 in the unit circle, the x value *is* the cosine value. And if you remember the alternate definition of sine — $\sin\theta = y/r$ — you'll realize that the y value is the sine value. That means any point, anywhere on the unit circle is always $(\cos\theta, \sin\theta)$. Talk about putting all the pieces together!

Alphabetically, x comes before y and c comes before s (cosine comes before sine, in other words). This should help you remember which one is which.

Tangent, cotangent, secant, and cosecant require a little more effort than the sine and cosine do. For many angles on the unit circle, evaluating these functions requires some careful work with fractions and square roots. Remember to always rationalize the denominator for any fraction in your final answer. Also, remember that any number divided by 0 is undefined. The tangent and secant functions, for instance, are undefined when the cosine value is 0. Similarly, the cotangent and cosecant values are undefined when the sine value is 0.

Time for an example. To evaluate the six trigonometric functions of 225° on the unit circle, follow these steps.

1. **Draw the picture.**

   When you're asked to find the trig function of an angle, you don't have to draw out a unit circle every time. Instead, use your smarts to figure out the picture. For this example, 225° is 45° more than 180°. Draw out a 45°-45°-90° triangle in the third quadrant only (see the earlier section "Placing the major angles correctly, sans protractor").

2. **Fill in the lengths of the legs and the hypotenuse.**

   Use the rules of the 45er triangle. The coordinate of the point at 225° is $\left(\frac{-\sqrt{2}}{2}, \frac{-\sqrt{2}}{2}\right)$. Figure 6-13 shows the triangle, as well as all the information to evaluate the six trig functions.

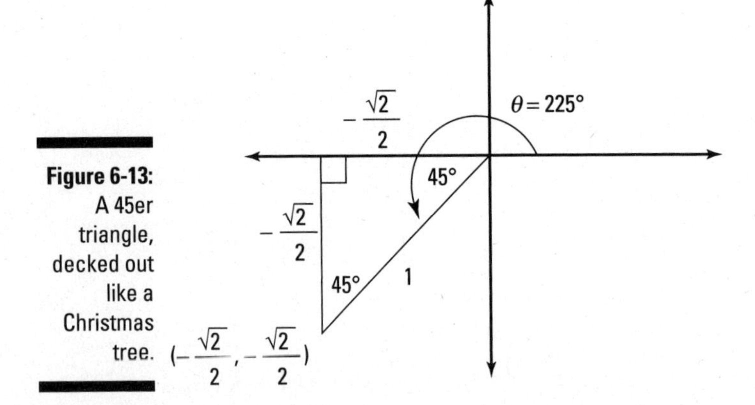

Figure 6-13: A 45er triangle, decked out like a Christmas tree.

Be careful! Use what you know about the positive and negative axes on the coordinate plane to help you. Because the triangle is in the third quadrant, both the *x* and *y* values should be negative.

3. **Find the sine of the angle.**

   The sine of an angle is the *y* value, or the vertical line that extends from the point on the unit circle to the *x*-axis. For 225°, the *y* value is $\frac{-\sqrt{2}}{2}$, so $\sin(225°) = \frac{-\sqrt{2}}{2}$.

4. **Find the cosine of the angle.**

   The cosine value is the *x* value, so it must be $\frac{-\sqrt{2}}{2}$.

5. **Find the tangent of the angle.**

   To find the tangent of an angle on the unit circle, you use the tangent's alternate definition: $\tan\theta = y/x$. Another way of looking at it is that $\tan\theta = \frac{\sin\theta}{\cos\theta}$, because in the unit circle, the *y* value is the sine and the *x* value is the cosine.

   So, if you know the sine and cosine of any angle, you also know the tangent. (Thanks, unit circle!) Because the sine and the cosine of 225° are both $\frac{-\sqrt{2}}{2}$, divide the sine by the cosine to get the tangent of 225°, which is 1.

6. **Find the cosecant of the angle.**

   The cosecant of any angle is 1 over $\sin\theta$, or r/y, using the point-in-the-plane definition. Use the fact that $\sin(225°) = \frac{-\sqrt{2}}{2}$ from Step 3 and divide 1 by $\frac{-\sqrt{2}}{2}$:

   $$\frac{1}{\left(-\frac{\sqrt{2}}{2}\right)} = 1 \cdot \frac{-2}{\sqrt{2}} = \frac{-2}{\sqrt{2}}$$

   Don't forget to rationalize the denominator: $\csc\theta = -\sqrt{2}$.

7. **Find the secant of the angle.**

   The secant of any angle is 1 over $\cos\theta$. Because the cosine of 225° is also $\frac{-\sqrt{2}}{2}$, found in Step 4, the secant of 225° is $-\sqrt{2}$.

8. **Find the cotangent of the angle.**

   The cotangent of an angle is 1 over $\tan\theta$. From Step 5, $\tan(225°) = 1$. So, $\cot(225°) = 1/1 = 1$. Easy as pie!

The tangent is always the slope of the radius *r.* This gives you a nice check for your work. Because the radius of the unit circle (the hypotenuse of the triangle) in the previous problem slants up, it has a positive slope, as does the tangent value.

The shortcut: Finding trig values of the 30°, 45°, and 60° families

Good news! We have a shortcut that can help you avoid the work of the previous section. You'll have to do less memorizing when you realize that certain special angles (and, therefore, their special triangles) on the unit circle always follow the same ratio of the sides. All you have to do is use the quadrants of the coordinate plane to figure out the signs. Solving trig function problems on the unit circle will be a blast after this section!

Perhaps you already figured out the shortcut by looking at Figure 6-12. If not, here are the families on the unit circle (for *any* family, the hypotenuse *r* is always 1):

- **The first family is the $\pi/6$ family (multiples of 30°).** Any angle with the same denominator of 6 has these qualities:
  - The longer leg is the *x* leg: $\frac{\sqrt{3}}{2}$
  - The shorter leg is the *y* leg: ½
- **The second family is the $\pi/3$ family (multiples of 60°).** Any angle with the same denominator of 3 has these qualities:
  - The shorter leg is the *x* leg: ½
  - The longer leg is the *y* leg: $\frac{\sqrt{3}}{2}$
- **The last family is the $\pi/4$ family (multiples of 45°).** Any angle with the same denominator of 4 has this quality: The two legs are equal in length — $\frac{\sqrt{2}}{2}$.

Finding the reference angle to solve for angles on the unit circle

A simple trig equation has a trig function on one side and a value on the other. The easiest trig equations to work are the ones where the value is a unit circle value, because the solutions will come from the two special right triangles. However, in this section we start to express the solutions to trig equations in radians rather than degrees, simply for the sake of consistency. (Don't worry; although the units used to measure angles will be different, the lengths of the sides still remain the same.) Radians show clear relationships between each of the families on the unit circle (see the previous section), and they're helpful when finding a reference angle to solve for the solutions. (Radians are also the units used in graphing trig functions, which we cover in Chapter 7.)

Before pre-calc, you were asked to solve such algebraic equations as $3x^2 - 1 = 26$. You learned to isolate the variable, using inverse operations. Now you'll be asked to do the same thing with trig functions in an attempt to find the value of the variable, which is now the angle that makes the equation true. After you find an angle that makes the equation true, you use that as the *reference angle* to find other angles on the unit circle that will also work in the equation. Usually you can find two, but you may find none, one, or more than two.

You can use your knowledge of trig functions to make an educated guess about how many solutions an equation can have. If the sine or cosine values are greater than 1 or less than –1, for instance, there are no solutions to the equation. If either the sine or cosine is exactly 1 or –1, there is only one solution. However, if the angles fall into specific quadrants, you'll find two solutions. (Later, in Chapter 9, you see that there can be many solutions.)

Using a reference angle to find the solution angle(s)

θ' (theta prime) is the name given to the reference angle, and θ is the actual solution to the equation, so you can find solutions by using the following quadrant rules, as seen in Figure 6-14:

QI: $\theta = \theta'$ because the reference angle and the solution angle are the same

QII: $\theta = \pi - \theta'$ because it falls short of π by however much the reference angle is

QIII: $\theta = \pi + \theta'$ because the angle is greater than π

QIV: $\theta = 2\pi - \theta'$ because it falls short of a full circle by however much the reference angle is

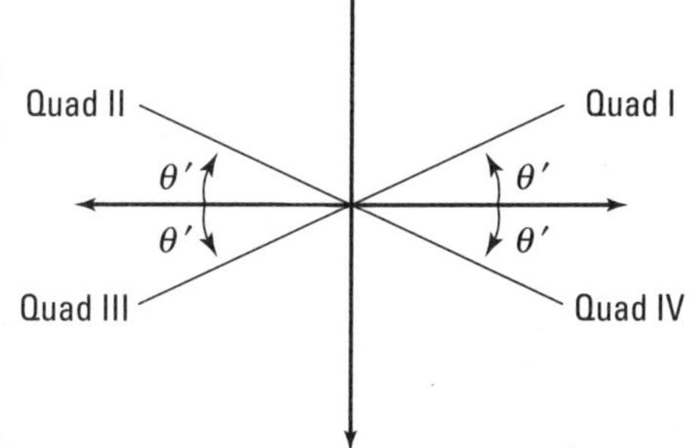

Figure 6-14: Finding the solution angle given the reference angle.

When you see a trig equation that asks you to solve for an unknown variable, you move backward from what you're given to arrive at a solution that makes sense. This solution should be in the form of an angle measurement, and the location of the angle should be in the correct quadrant. Knowledge of the unit circle comes in handy here, because you'll be thinking of angles that fulfill the requirements of the given equation.

Suppose you're asked to solve $2\cos x = 1$. To solve, you need to think about what angles on the unit circle whose cosine values when multiplied by 2 are 1, and follow these steps:

1. **Isolate the trig function on one side.**

   You solve for $\cos x$ by dividing both sides by 2: $\cos x = \frac{1}{2}$.

2. **Determine which quadrants your solutions lie in.**

   Keeping in mind that cosine is an x value (see the earlier section "Finding values for the six trig functions"), you draw four triangles — one in each quadrant — with the x-axis legs labeled ½. Figure 6-15 shows these four triangles.

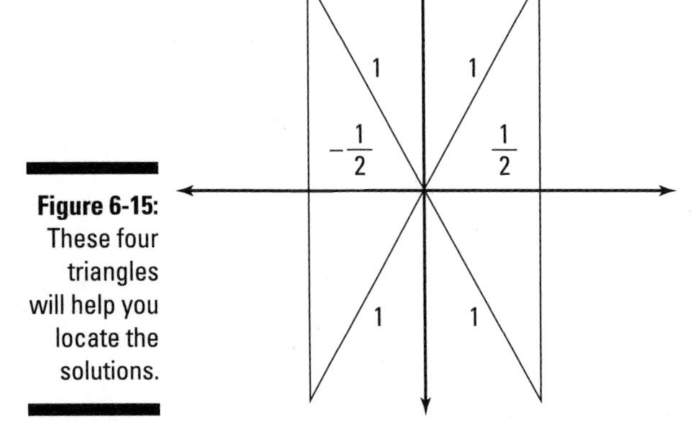

Figure 6-15: These four triangles will help you locate the solutions.

   The two triangles on the left have a value of –½ for the horizontal leg, not ½. Therefore, you can eliminate them; your solutions are in quadrants I and IV.

3. **Fill in the missing leg values for each triangle.**

   You've already marked the x-axis legs. Based on knowledge of the unit circle and special triangles, you know that the side parallel to the y-axis has to be $\frac{\sqrt{3}}{2}$, and the hypotenuse is 1. Figure 6-16 shows the two labeled triangles.

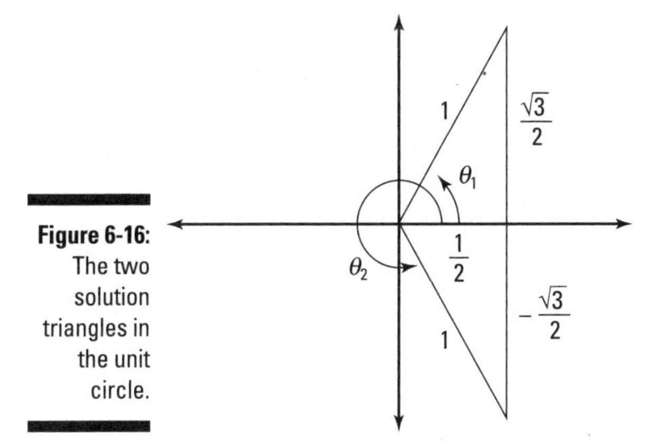

Figure 6-16: The two solution triangles in the unit circle.

4. **Determine the reference angle.**

   In the special right triangles, a side length of ½ is the short leg of a 30°-60°-90° right triangle. This means that the cosine (or the part along the *x*-axis) is the short leg, and the vertical leg is the long leg. So, the vertex of the angle at the center of the unit circle has a measure of 60°, making the reference angle $\pi/3$.

5. **Express the solutions in standard form.**

   The reference angle $\theta' = \pi/3$. The first quadrant solution is the same as the reference angle: $\theta' = \pi/3$. The fourth quadrant solution is $\theta = 2\pi - \theta' = 2\pi - \pi/3 = 5\pi/3$.

Combining reference angles with other solving techniques

You can incorporate reference angles (see the previous section) into some other pre-calc techniques to solve trig equations. One such technique is factoring. You've been factoring since algebra (and in Chapter 4), so this shouldn't be anything new. When confronted with an equation that's equal to 0 and a trig function that's being squared, or two different trig functions that are being multiplied together, you should try to use factoring to get your solution first. After factoring, you can use the zero product property (see Chapter 1) to set each factor equal to 0 and then solve them separately.

Try solving an example that involves factoring a trinomial — $2\sin^2 x + \sin x - 1 = 0$ — using the following steps:

1. **Let a variable equal the trig ratio and rewrite the equation to simplify.**

   Let $u = \sin x$ and rewrite the equation as $2u^2 + u - 1 = 0$.

2. **Check to make sure that the equation factors.**

   Remember to always check for greatest common factor first. Refer to the factoring information in Chapter 4.

3. **Factor the quadratic.**

   The equation $2u^2 + u - 1 = 0$ factors to $(u + 1)(2u - 1) = 0$.

4. **Switch the variables back to trig functions.**

   Rewriting your factored trig equation gives you $(\sin x + 1)(2\sin x - 1) = 0$.

5. **Use the zero product property to solve.**

   If $\sin x = -1$, $x = 3\pi/2$; if $\sin x = 1/2$, $x = \pi/6$ and $x = 5\pi/6$.

In pre-calc, you may be required to take the square root of both sides to solve a trig function. For example, if you're given an equation such as $4\sin^2\theta - 3 = 0$, follow these steps:

1. **Isolate the trig expression.**

   For $4\sin^2\theta - 3 = 0$, add 3 to each side and divide by 4 on both sides to get $\sin^2\theta = 3/4$.

2. **Take the square root of both sides.**

   Don't forget to take the positive and negative square roots, which gives you $\sin\theta = \pm\frac{\sqrt{3}}{2}$.

3. **Solve to find the reference angle.**

   The sine of θ is both positive and negative for this example, which means that the solutions, or angles, are in all four quadrants. The positive solutions are in quadrants I and II, and the negative solutions are in quadrants III and IV. Use the reference angle in quadrant I to guide you to all four solutions.

   If $\sin\theta = \frac{\sqrt{3}}{2}$, the y value in the first quadrant is the long leg of the 30°-60°-90° triangle. Therefore, the reference angle is $\pi/3$.

4. **Find the solutions.**

   Use the reference angle to find the four solutions:

   - $\theta = \pi/3$
   - $\theta = 2\pi/3$
   - $\theta = 4\pi/3$
   - $\theta = 5\pi/3$

   Note that two of these solutions come from the positive sign value and two come from the negative.

Not Just a Job for Noah: Making and Measuring Arcs

Knowing how to calculate the circumference of a circle and, in turn, the length of an *arc* — a portion of the circumference — is important in pre-calc because you can use that information to analyze the motion of an object moving in a circle.

An arc can come from a *central angle,* which is one whose vertex is located at the center of the circle. You can measure an arc in two different ways:

- **As an angle:** The measure of an arc as an angle is the same as the central angle that intercepts it.
- **As a length:** The length of an arc is directly proportional to the circumference of the circle and is dependent on both the central angle and the radius of the circle.

If you think back to geometry, you may remember that the formula for the circumference of a circle is $C = 2\pi r$, with r standing for the radius. Also recall that a circle has 360°. So, if you need to find the length of an arc, you need to figure out what part of the whole circumference (or what fraction) you're looking at.

You use the following formula to calculate the arc length; θ represents the measure of the angle in degrees, and s represents arc length, as shown in Figure 6-17:

$$s = \frac{\theta}{360} \cdot 2\pi r$$

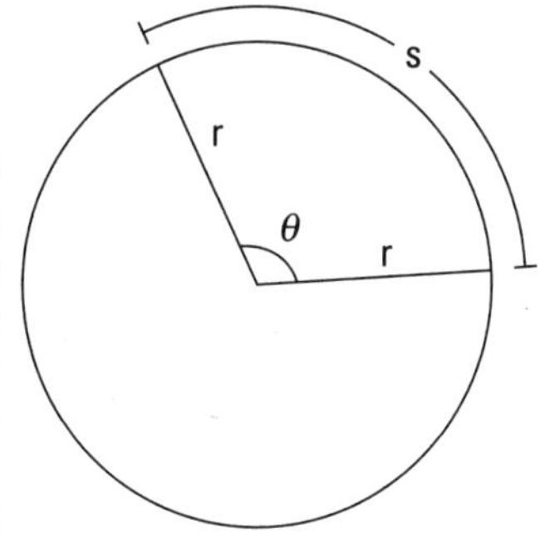

Figure 6-17: The variables involved in calculating arc length.

If the given angle is in radians, the 2π cancels and its arc length is

$$s = \frac{\theta}{\cancel{2\pi}} \cdot \cancel{2\pi} r \rightarrow s = \theta r$$

Time for an example. To find the length of an arc with an angle measurement of 40° if the circle has a radius of 10, use the following steps:

1. **Assign variable names to the values in the problem.**

   The angle measurement here is 40°, which is θ. The radius is 10, which is *r*.

2. **Plug the known values into the formula.**

   This gives you $s = \frac{40}{360} \cdot 2\pi(10)$.

3. **Simplify to solve the formula.**

   You first get $\frac{1}{9} \cdot 20\pi$, which multiplies to $\frac{20\pi}{9}$.

Figure 6-18 shows what this arc looks like.

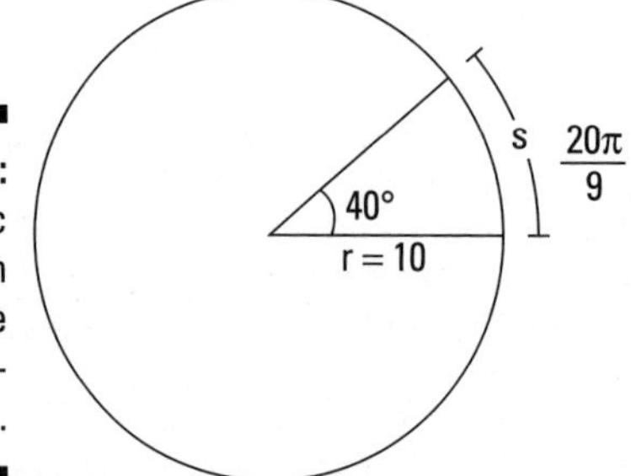

Figure 6-18: The arc length for an angle measurement of 40°.

Now try a different problem. Find the measure of the central angle of a circle in radians with an arc length of 28π and a radius of 16. This time, you must solve for θ (the formula is $s = r\theta$ when dealing with radians):

1. **Plug in what you know to the radian formula.**

   This gives you $28\pi = 16\theta$.

2. **Divide both sides by 16.**

   Your formula will look like this: $\frac{28\pi}{16} = \frac{16\theta}{16}$.

3. **Reduce the fraction.**

   You're left with $\frac{7\pi}{4} = \theta$. So, the solution is $\theta = \frac{7\pi}{4}$.

Chapter 7

Graphing and Transforming Trig Functions

In This Chapter

- Plotting and transforming the sine and cosine parent graphs
- Picturing and changing tangent and cotangent
- Charting and altering secant and cosecant

"Graph the trig function . . .". The beginning of this command sends shivers down the spines of many pre-calc students. But we're here to say that you have nothing to fear, because graphing functions can be easy. Graphing functions is simply a matter of inserting the x value (the *domain*) in place of the function's variable and solving the function to get the y value (the *range*). You continue with that calculation until you have enough points to plot. When do you know you have enough? When you get a clear line, ray, curve, or what have you on the graph.

You've dealt with functions before in math, but up until now, the input of a function was typically x. In trig functions, however, the input of the function is typically θ, which is basically just another variable to use. This chapter shows you how to graph trig functions by using various values for θ. We start with the *parent graphs* — the foundation upon which everything else for graphing is built. From there, you can stretch a trig function graph, move it around on the coordinate plane, or flip and shrink it, which we also cover in this chapter.

In Chapter 6, you use two ways of measuring angles: degrees and radians. Lucky for you, you now get to focus solely on radians when graphing trig functions. Mathematicians have always graphed in radians when working with trig functions, and we want to continue that tradition — until someone comes up with a better way, of course.

Drafting the Sine and Cosine Parent Graphs

The trig functions, especially sine and cosine, displayed their usefulness in the last chapter. After putting them under the microscope, you're now ready to begin graphing them. Just like with the parent functions in Chapter 3, after you discover the basic shape of the sine and cosine graphs, you can begin to graph more complicated versions, using the same transformations you discover in Chapter 3:

- Vertical and horizontal transformations
- Vertical and horizontal translations
- Vertical and horizontal reflections

Knowing how to graph trig functions allows you to measure the movement of objects that move back and forth or up and down in a regular interval, such as pendulums. Sine and cosine as functions are perfect ways of expressing this type of movement, because their graphs are repetitive and they oscillate (like a wave). The following sections illustrate these facts.

The sine graph

Sine graphs move in waves. The waves crest and fall over and over again forever, because you can keep plugging in values for θ for the rest of your life, if you really wanted to. This section shows you how to construct the parent graph for the sine function, $f(\theta) = \sin\theta$ (for more on parent graphs, refer to Chapter 3).

Because all the values of the sine function come from the unit circle, you should be pretty comfy and cozy with the unit circle before proceeding with this work. If you're not, we advise that you return to Chapter 6 for a brush-up.

You can graph any trig function in four (or five) steps. Here are the steps to constructing the graph of the parent function $f(\theta) = \sin\theta$:

1. **Find the values for domain and range.**

   No matter what you put into the sine function, you get an answer as output, because θ can rotate about the unit circle in either direction an infinite amount of times. This means that the domain of sine is all real numbers, or $(-\infty, \infty)$.

On the unit circle, the y values are your sine values — what you get after plugging the value of θ into the sine function. Because the radius of the unit circle is 1, the y values can't be more than 1 or less than negative 1 — your range for the sine function. So, in the x direction, the wave (or *sinusoid,* in math language) goes on forever, and in the y direction, the sinusoid oscillates only between –1 and 1, including these values. In interval notation, you write this as [–1, 1].

2. **Calculate the graph's x-intercepts.**

   When you graphed lines in algebra, the x-intercepts occurred when $y = 0$. In this case, sine is the y value. Find out where the graph crosses the x-axis by finding unit circle values where sine is 0. The graph crosses the x-axis three times: once at 0, once at π, and once at 2π. You now know that three of the coordinate points are (0, 0), (π, 0), and (2π, 0).

3. **Calculate the graph's maximum and minimum points.**

   To complete this step, use your knowledge of the range from Step 1. You know that the highest value of y is 1. Where does this happen? At $\pi/2$. You now have another coordinate point at ($\pi/2$, 1). You also can see that the lowest value of y is –1, when x is $3\pi/2$. Hence, you have another coordinate point: ($3\pi/2$, –1).

4. **Sketch the graph of the function.**

   Using the five key points as a guide, connect the points with a smooth, round curve. Figure 7-1 shows the parent graph of sine.

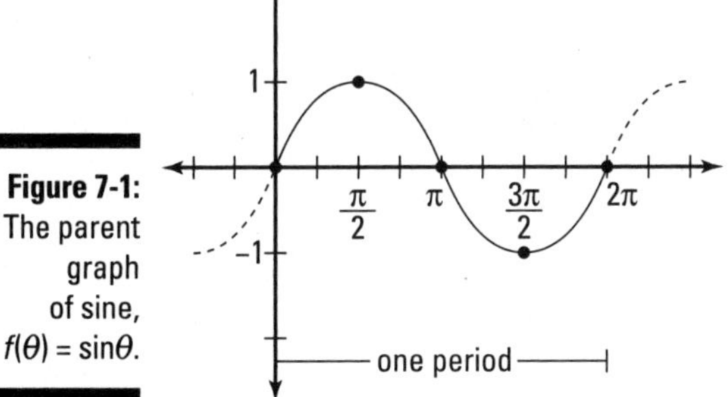

Figure 7-1: The parent graph of sine, $f(\theta) = \sin\theta$.

The parent graph of the sine function has a couple important characteristics worth noting:

- **It repeats itself every 2π radians.** This occurs because 2π radians is one trip around the unit circle — called the *period* of the sine graph — and after that you start to go around again. Usually, you're asked to draw the graph to show one period of the function, because in this period you capture all possible values for sine before it starts repeating over and over again. The graph of sine is called *periodic* because of this repeating pattern.
- **It's symmetrical about the origin (thus, in math speak, it's an *odd function*).** The sine function has 180° point symmetry about the origin. If you look at it upside down, the graph looks exactly the same. The official math definition of an *odd function,* though, is $f(-x) = -f(x)$ for every value of x in the domain. In other words, if you put in an opposite input, you'll get an opposite output. For example, $\sin(\pi/6)$ is $1/2$, but if you look at $\sin(-\pi/6)$, you get $-1/2$.

The cosine graph

The parent graph of cosine looks very similar to the sine function parent graph, but it has its own sparkling personality (like fraternal twins, we suppose). Cosine graphs follow the same basic pattern and have the same basic shape as sine graphs; the difference lies in the location of the maximums and minimums. These occur at different domain, or x values, ¼ of a period away from each other. Thus, the two graphs are shifts of ¼ of the period from each other.

Just as with the sine graph, you'll use the five key points of graphing trig functions to get the parent graph of the cosine function. If necessary, you can refer to the unit circle for the cosine values to start with (see the Cheat Sheet at the front of this book). As you work more with these functions, your dependence on the unit circle should decrease, until eventually you won't need it at all. Here are the steps:

1. **Find the values for domain and range.**

   As with sine graphs (see the previous section), the domain of cosine is all real numbers, and its range is $-1 \le y \le 1$, or $[-1, 1]$.
2. **Calculate the graph's x-intercepts.**

Referring to the unit circle, find where the graph crosses the x-axis by finding unit circle values of 0. It crosses the x-axis twice — once at $\pi/2$ and at $3\pi/2$. This gives you two coordinate points: $(\pi/2, 0)$ and $(3\pi/2, 0)$.

3. **Calculate the graph's maximum and minimum points.**

   Using your knowledge of the range for cosine from Step 1, you know the highest value that y can be is 1, which happens twice for cosine — once at 0 and once at 2π (see Figure 7-2), giving you two maximums: $(0, 1)$ and $(2\pi, 1)$. The minimum value that y can be is -1, which occurs at π. You now have another coordinate pair at $(\pi, -1)$.

4. **Sketch the graph of the function.**

   Figure 7-2 shows the full parent graph of cosine with the five key points plotted.

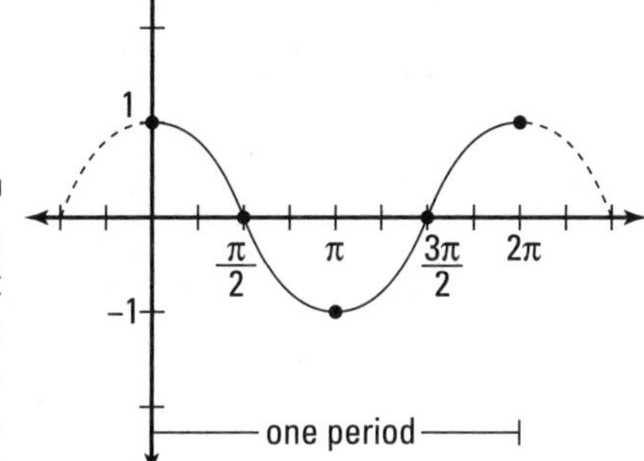

Figure 7-2: The parent graph of cosine, $f(\theta) = \cos\theta$.

The cosine parent graph has a couple characteristics worth noting:

- **It repeats every 2π radians.** This means it's a periodic function, so its waves will rise and fall in the graph (see the previous section for the full explanation).
- **It's symmetrical about the y-axis (in mathematical dialect, it's an *even function*).** Unlike the sine function, which has 180° symmetry, cosine has y-axis symmetry. In other words, you can fold the graph in half at the y-axis and it matches exactly. The formal definition of an even function is $f(x) = f(-x)$ — if you plug in the opposite input, you'll get the same output. For example, $\cos(\pi/6) = \frac{\sqrt{3}}{2}$, and $\cos(-\pi/6) = \frac{\sqrt{3}}{2}$. Even though the input sign changed, the output sign stayed the same. This always remains true for any θ value and its opposite for cosine.

Graphing Tangent and Cotangent

The graphs for the tangent and cotangent functions are quite different from the sine and cosine graphs. The graphs of sine and cosine are very similar to one another in shape and size. However, when you divide one function by the other, the graph you create looks nothing like either of the graphs it came from. (Tangent is defined as $\frac{\sin \theta}{\cos \theta}$, and cotangent is $\frac{\cos \theta}{\sin \theta}$.)

The graphs of tangent and cotangent can be tough for some students to grasp, but you can master them with practice. The hardest part of graphing tangent and cotangent comes from the fact that they both have asymptotes in their graphs (see Chapter 3), because they're rational functions.

The tangent graph has an asymptote wherever the cosine is 0, and the cotangent graph has an asymptote wherever the sine is 0. Keeping these asymptotes separate from one another will help you to draw your graphs.

The tangent and cotangent functions have parent graphs just like any other function. Using the graphs of these functions, you can make the same types of transformations that apply to the parent graphs of any function. The following sections plot the parent graphs of tangent and cotangent.

Tangent

The easiest way to remember how to graph the tangent function is to remember that $\tan\theta = \frac{\sin \theta}{\cos \theta}$ (see Chapter 6). Because $\cos\theta = 0$ for various values of θ, this does some interesting things to tangent's graph. When the denominator of a fraction is 0, the fraction is *undefined.* So, the graph of tangent will jump over an *asymptote,* which is where the function is undefined, at each of these places.

Table 7-1 presents θ, $\sin\theta$, $\cos\theta$, and $\tan\theta$. It shows the roots (or zeros), asymptotes (where the function is undefined), and the behavior of the graph in between certain key points on the unit circle.

Table 7-1 Finding Out where $\tan\theta$ Is Undefined

θ	0	$0<\theta<\pi/2$	$\pi/2$	$\pi/2<\theta<\pi$	π	$\pi<\theta<3\pi/2$	$3\pi/2$	$3\pi/2<\theta<2\pi$	2π
$\sin\theta$	0	positive	1	positive	0	negative	–1	negative	0
$\cos\theta$	1	positive	0	negative	–1	negative	0	positive	1
$\tan\theta$	0	positive	undefined	negative	0	positive	undefined	negative	0

To plot the parent graph of a tangent function, you start out by finding the vertical asymptotes. This gives you some structure from which you can fill in the missing points:

1. **Find the vertical asymptotes so you can find the domain.**

   In order to find the domain of the tangent function, you have to locate the vertical asymptotes. The first asymptote occurs when $\theta = \pi/2$, and they repeat every π radians (see the unit circle in the Cheat Sheet of this book). (***Note:*** The period of the tangent graph is π radians, which is different from that of sine and cosine.) Tangent, in other words, will have asymptotes when $\theta = \pi/2$ and $3\pi/2$.

   The easiest way to write this is $\theta \neq \pi/2 + \pi n$, where n is an integer. You write "$+ \pi n$" because the period of tangent is π radians, so if there's an asymptote at $\pi/2$ and you add or subtract π, you automatically find the next asymptote.

2. **Determine values for the range.**

   Recall that the tangent function can be defined as $\frac{\sin \theta}{\cos \theta}$. Both of these can be decimals, and when you divide decimals by decimals, the end value increases.

   There are no restrictions on the range of tangent; you aren't stuck between 1 and –1, like with sine and cosine. In fact, the ratios are any and all numbers. The range is $(-\infty, \infty)$.

3. **Calculate the graph's *x*-intercepts.**

   Tangent's parent graph has roots (it crosses the *x*-axis) at 0, π, and 2π. You can find these values by setting $\frac{\sin \theta}{\cos \theta}$ equal to 0 and then solving. The *x*-intercepts for the parent graph of tangent are located wherever the sine value is 0.

4. **Figure out what's happening to the graph between the interecepts and the asymptotes.**

   The first quadrant of tangent is positive and points upward toward the asymptote at $\pi/2$, because all sine and cosine values are positive in the first quadrant. Quadrant II is negative because sine is positive and cosine is negative. Quadrant III is positive because both sine and cosine are negative, and quadrant IV is negative because sine is negative while cosine is positive.

   Note: There are no maximum or minimum points on a tangent graph.

Figure 7-3 shows what the parent graph of tangent looks like when you put it all together.

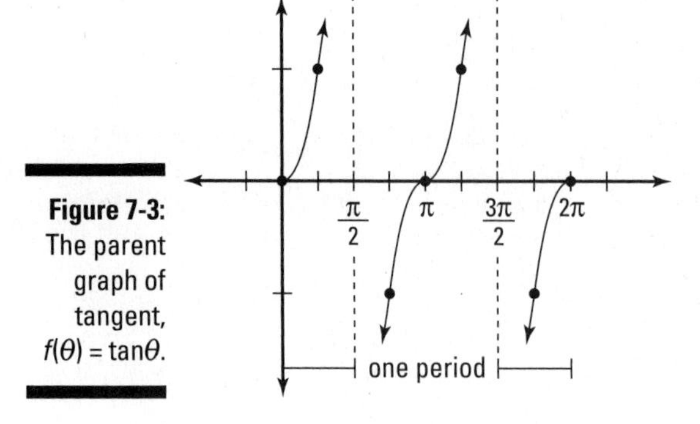

Figure 7-3: The parent graph of tangent, $f(\theta) = \tan\theta$.

Cotangent

The parent graphs of sine and cosine are very similar, because the values are exactly the same; they just occur for different values of θ. Similarly, the parent graphs of tangent and cotangent are comparable, because they both have asymptotes and x-intercepts. The only differences you can see are the values of θ where the asymptotes and x-intercepts occur. You can find the parent graph of the cotangent function, $\frac{\cos\theta}{\sin\theta}$, by using the same techniques you use to find the tangent parent graph (see the previous section).

Table 7-2 shows θ, $\cos\theta$, $\sin\theta$, and $\cot\theta$ so that you can see both the x-intercepts and the asymptotes in comparison. This will help you find the general shape of your graph so that you have a nice place to start.

Table 7-2 Spotting where $\cot\theta$ Is Undefined

θ	0	$0 < \theta < \pi/2$	$\pi/2$	$\pi/2 < \theta < \pi$	π
$\cos\theta$	1	positive	0	negative	–1
$\sin\theta$	0	positive	1	positive	0
$\cot\theta$	undefined	positive	0	negative	undefined
θ		$\pi < \theta < 3\pi/2$	$3\pi/2$	$3\pi/2 < \theta < 2\pi$	2π
$\cos\theta$		negative	0	positive	1
$\sin\theta$		negative	–1	negative	0
$\cot\theta$		positive	0	negative	undefined

To sketch the full parent graph of cotangent, follow these steps:

1. **Find the vertical asymptotes so you can find the domain.**

   Because cotangent is the quotient of cosine divided by sine, and sinθ is sometimes 0, the graph of the cotangent function may have asymptotes, just like with tangent. However, these asymptotes occur whenever the sinθ = 0. The asymptotes of cotθ are at 0, π, and 2π.

   The cotangent parent graph repeats every π units. Its domain is based on its vertical asymptotes: The first one comes at 0 and then repeats every π radians. The domain, in other words, is $\theta \neq 0 + \pi n$, where n is an integer.

2. **Find the values for the range.**

   Similar to the tangent function, you can define cotangent as $\frac{\cos\theta}{\sin\theta}$. Both of these can be decimals, and when you divide decimals by decimals, the end value increases. There are also no restrictions on the range of cotangent; the ratios are any and all numbers — $(-\infty, \infty)$.

3. **Determine the *x*-intercepts.**

   The roots (or zeros) of cotangent occur wherever the cosine value is 0. This occurs at $\pi/2$ and $3\pi/2$.

4. **Evaluate what happens to the graph between the *x*-intercepts and the asymptotes.**

   The positive and negative values in the four quadrants stay the same as in tangent, but the asymptotes change the graph. You can see the full parent graph for cotangent in Figure 7-4.

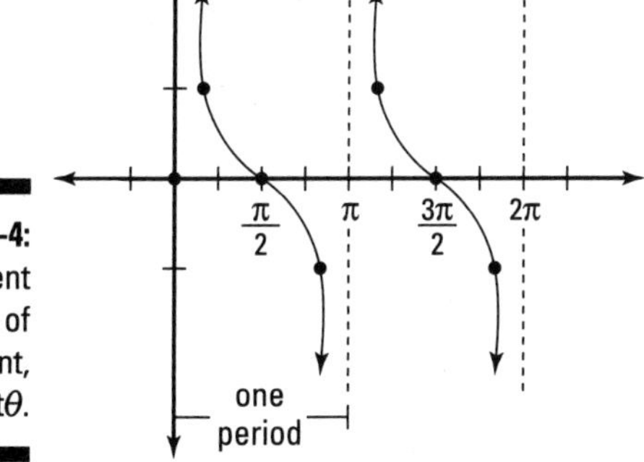

Figure 7-4: The parent graph of cotangent, $g(\theta) = \cot\theta$.

Putting Secant and Cosecant in Pictures

As with tangent and cotangent, the graphs of secant and cosecant have asymptotes. Do you know why? Because $\sec\theta = \frac{1}{\cos\theta}$, and $\csc\theta = \frac{1}{\sin\theta}$. Both sine and cosine have values of 0, which causes the denominators to be 0 and the functions to have asymptotes. These are important considerations when plotting the parent graphs, which we do in the sections that follow.

Secant

Secant is defined as $\frac{1}{\cos\theta}$, and you can graph it by using steps similar to those from the tanget and cotangent sections.

There are two places where the cosine graph crosses the x-axis on the interval $[0, 2\pi]$, so there will be two asymptotes in the secant graph, which divide the period interval into three smaller sections. There aren't any x-intercepts in the parent secant graph (it's hard to find them on any transformed graph, so usually you won't be asked to).

Follow these steps to picture the parent graph of secant:

1. **Find the asymptotes of the secant graph.**

   Because secant is the reciprocal of cosine (see Chapter 6), any place on the cosine graph where the value is 0 will create an asymptote on the secant graph. (And any point with 0 in the denominator is undefined.) Finding these points first will help to define the rest of the graph.

   The parent graph of cosine has values of 0 at $\frac{\pi}{2}$ and $\frac{3\pi}{2}$. So, the graph of secant has asymptotes at those same values. Figure 7-5 shows only the asymptotes.

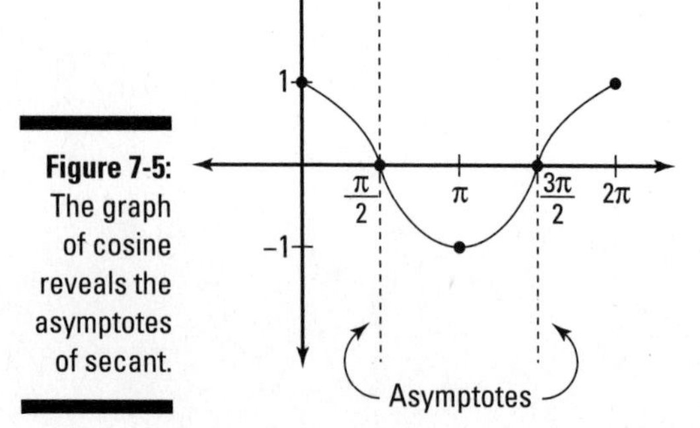

Figure 7-5: The graph of cosine reveals the asymptotes of secant.

2. **Calculate what happens to the graph at the first interval between the asymptotes.**

   The period of the parent cosine graph starts at 0 and ends at 2π. You need to figure out what the graph does in between

   - Zero and the first asymptote at $\pi/2$
   - The two asymptotes in the middle
   - The second asymptote and the end of the graph at 2π

   Start on the interval $(0, \pi/2)$. The graph of cosine goes from 1, into fractions, and all the way down to 0. Secant takes the reciprocal of all these values and ends on this first interval at the asymptote. The graph gets bigger and bigger rather than smaller, because as the fractions in the cosine function get smaller, their reciprocals in the secant function get bigger.

3. **Repeat for the second interval $(\pi/2, 3\pi/2)$.**

   If you refer to the cosine graph, you see that halfway between $\pi/2$ and $3\pi/2$, from the low point, the line has nowhere to go except closer to 0 in both directions. So, secant's graph (the reciprocal) gets bigger in a negative direction.

4. **Repeat for the last interval $(3\pi/2, 2\pi)$.**

   This interval is a mirror image of what happens in the first interval.

5. **Find the domain and range of the graph.**

   Its asymptotes are at $\pi/2$ and repeat every π, so the domain of secant is $\theta \neq \pi/2 + \pi n$, where n is an integer. The graph exists only for numbers ≥ 1 or ≤ -1. Its range, therefore, is $(-\infty, -1] \cup [1, \infty)$.

You can see the parent graph of secant in Figure 7-6.

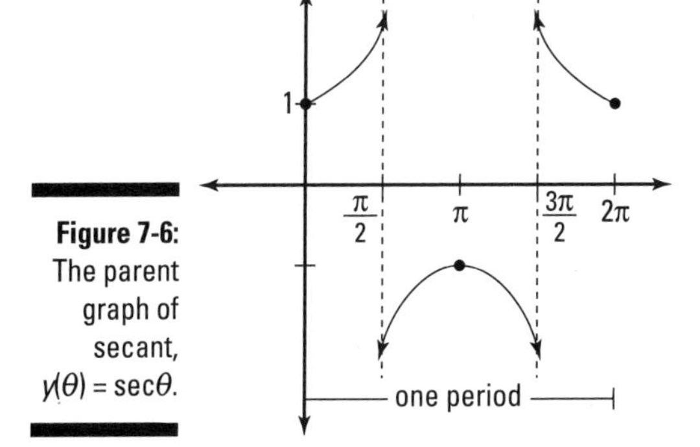

Figure 7-6: The parent graph of secant, $y(\theta) = \sec\theta$.

Cosecant

Cosecant is almost exactly the same as secant because it's the reciprocal of sine (as opposed to cosine). Anywhere sine has a value of 0, you'll see an asymptote in the cosecant graph. Because the sine graph crosses the x-axis three times on the interval $[0, 2\pi]$, there will be three asymptotes and two sub-intervals to graph.

The reciprocal of 0 is undefined, and the reciproal of an undefined value is 0. Because the graph of sine is never undefined, the reciprocal of sine can never be 0. For this reason, there are no x-intercepts for the parent graph of the cosecant function, so don't bother looking for them.

The following list explains how to graph cosecant:

1. **Find the asymptotes of the graph.**

   Because cosecant is the reciprocal of sine, any place on sine's graph where the value is 0 creates an asymptote on cosecant's graph. The parent graph of sine has values of 0 at 0, π, and 2π. So, cosecant has three asymptotes. Figure 7-7 shows these asymptotes.

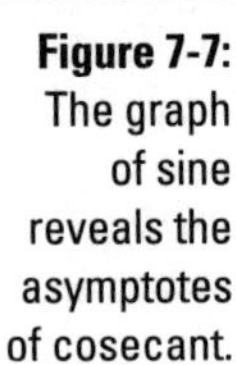

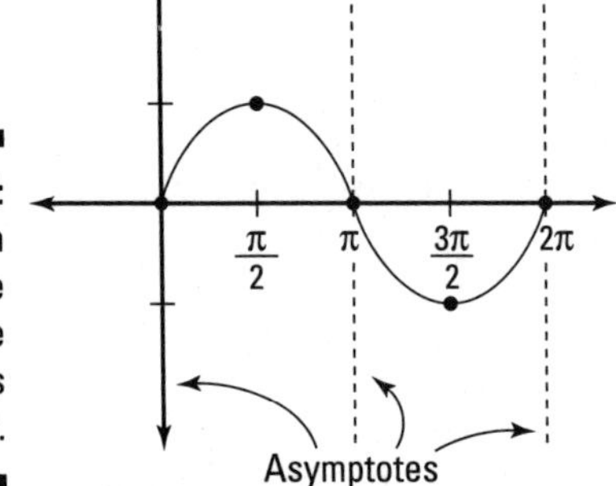

Figure 7-7: The graph of sine reveals the asymptotes of cosecant.

2. **Calculate what happens to the graph at the first interval between 0 and π.**

   The period of the parent sine graph starts at 0 and ends at 2π. You can figure out what the graph does in between the first asymptote at 0 and the second asymptote at π.

   The graph of sine goes from 0 to 1 and then back down again. Cosecant takes the reciprocal of these values, which causes the graph to get bigger.

3. **Repeat for the second interval (π, 2π).**

   If you refer to the sine graph, you see that it goes from 0 down to –1 and then back up again. So, cosecant's graph gets bigger in the negative direction.

4. **Find the domain and range of the graph.**

   Cosecant's asymptotes start at 0 and repeat every π. Its domain is $\theta \neq 0 + \pi n$, where n is an integer. The graph also exists for numbers ≥ 1 or ≤ -1. Its range, therefore, is $(-\infty, -1] \cup [1, \infty)$.

You can see the full graph in Figure 7-8.

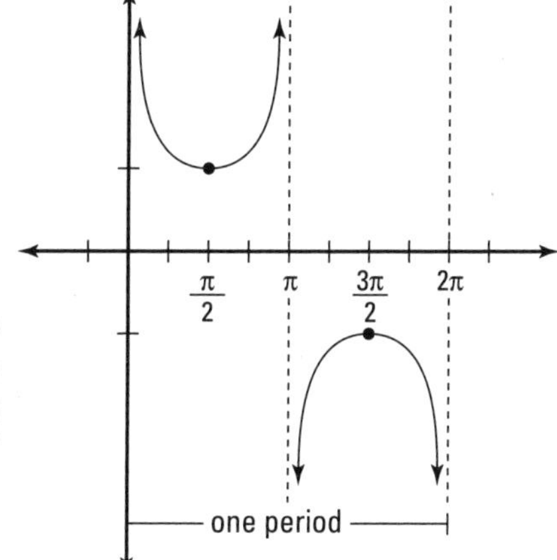

Figure 7-8: The parent graph of cosecant, $y(\theta) = \csc\theta$.

Transforming Trig Graphs

The basic parent graphs open the door to many advanced and complicated graphs, which ultimately have more real-world applications. Usually, functions that model real-world situations are stretched or shrunk, or even shifted to an entirely different location on the coordinate plane. The good news is that the *transformation* of trig functions follows the same set of general guidelines as the transformations you see in Chapter 3.

The rules for graphing complicated trig functions are actually pretty simple. When asked to graph a more complicated trig function, you can take the parent graph (which you know from the previous sections) and alter it in some way to find the more complex graph. Basically, you can change each parent graph of a trig function in four ways:

- You can change the parent graph of any trig graph by using a vertical transformation. ***Note:*** When dealing with the graph for sine and cosine functions, a vertical transformation will change the graph's height, also known as the *amplitude*.
- You can alter any parent graph with a horizontal transformation (now known as the *period*) and make it move faster or slower, which affects its horizontal length.
- You can shift any parent graph up, down, left, or right.
- You can reflect any parent graph across the x- or y-axis.

The following sections cover how to transform the parent trig graphs. However, before you move on to transforming these graphs, make sure you're comfortable with the parent graphs from the previous sections. Otherwise, it's very easy to get confused on nitpicky things.

Screwing with sine and cosine graphs

The sine and cosine graphs look similar to a spring. If you pull the ends of this spring, all the points will be farther apart from one another; in other words, the spring will be stretched. If you push the ends of the spring together, all the points will be closer together; in other words, the spring will be shrunk. So, the graphs of sine and cosine look and act a lot like a spring, but these springs can be changed both horizontally *and* vertically; aside from pulling the ends or pushing them together, you can make the spring taller or shorter. Now that's some spring!

We'll show you how to alter the parent graphs for sine and cosine using both vertical and horizontal stretches and shrinks. You'll also see how to move the graph around the coordinate plane using translations (which can be both vertical and horizontal).

Changing the amplitude

Multiplying a trig function by a constant changes the graph of the parent function; specifically, you change the amplitude of the graph. When measuring the height of a graph, you measure the distance between the maximum crest and the minimum wave. Smack dab in the middle of that measurement is a horizontal line called the *sinusoidal axis. Amplitude* is the measure of the distance from the sinusoidal axis to the maximum or the minimum. Figure 7-9 illustrates this point further.

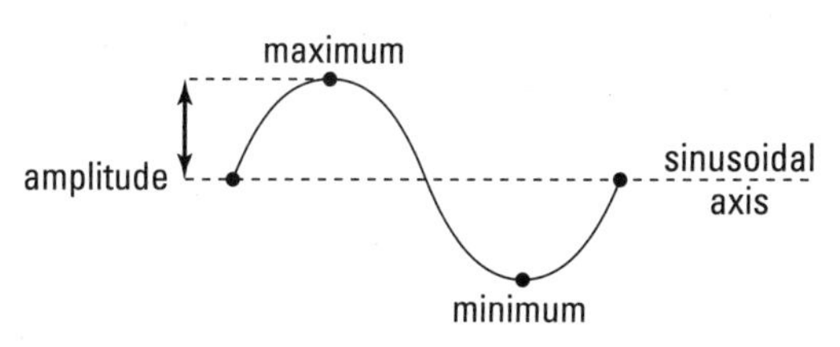

Figure 7-9: The sinusoidal axis and amplitude of a trig function graph.

By multiplying a trig function by certain values, you can make the graph taller or shorter:

- **Positive values of amplitudes greater than 1 make the height of the graph taller.** This makes sense, because when you amplify a sound, you make it louder. Basically, $2\sin\theta$ makes the graph taller than $\sin\theta$; $5\sin\theta$ makes it even taller, and so on. For example, if $g(\theta) = 2\sin\theta$, you multiply the height of the original sine graph by 2 at every point. Every place on the graph, therefore, will be twice as tall as the original.
- **Fraction values between 0 and 1 make the graph shorter.** You can say that $\frac{1}{2}\sin\theta$ is shorter than $\sin\theta$, and $\frac{1}{5}\sin\theta$ is even shorter. For example, if $h(\theta) = \frac{1}{5}\sin\theta$, you multiply the parent graph's height by $\frac{1}{5}$ at each point, making it that much shorter.

The change of amplitude affects the range of the function as well, because the maximum and minimum values of the graph change. Before you multiply a sine or cosine function by 2, for instance, its graph oscillated between –1 and 1; now it moves between –2 and 2.

Sometimes you multiply a trigonometric function by a negative number. This doesn't make the amplitude negative, however! Amplitude is a measure of distance, and distance can't be negative. You can't walk –5 feet, for instance, no matter how hard you try. Even if you walk backward, you still walk 5 feet. Similarly, if $k(\theta) = -5\sin\theta$, its amplitude is still 5. The negative sign just flips the graph upside down.

Table 7-3 shows a comparison of an original input (θ) and the value of $\sin\theta$ with $g(\theta)2\sin\theta$, $h(\theta) = \frac{1}{2}\sin\theta$, and $k(\theta) = -5\sin\theta$.

Table 7-3 Comparing How Different Amplitudes Affect sinθ

θ	$f(\theta) = \sin\theta$	$g(\theta) = 2\sin\theta$	$h(\theta) = \frac{1}{2}\sin\theta$	$k(\theta) = -5\sin\theta$
0	0	$2 \cdot 0 = 0$	$\frac{1}{2} \cdot 0 = 0$	$-5 \cdot 0 = 0$
$\pi/2$	1	$2 \cdot 1 = 2$	$\frac{1}{2} \cdot 1 = \frac{1}{2}$	$-5 \cdot 1 = -5$
π	0	$2 \cdot 0 = 0$	$\frac{1}{2} \cdot 0 = 0$	$-5 \cdot 0 = 0$
$3\pi/2$	-1	$2 \cdot -1 = -2$	$\frac{1}{2} \cdot -1 = -\frac{1}{2}$	$-5 \cdot -1 = 5$
2π	0	$2 \cdot 0 = 0$	$\frac{1}{2} \cdot 0 = 0$	$-5 \cdot 0 = 0$

Don't worry, you won't have to recreate Table 7-3 for any pre-calc reasons. We just want you to see the comparison between the parent function and the more complicated functions. Keep in mind that this table displays only values of the sine function and transformations of it; you can easily do the same thing for cosine.

Figure 7-10 illustrates what the graphs of sine look like after the transformations. Figure 7-10a is the graph of $f(\theta)$; Figure 7-10b is $g(\theta)$; Figure 7-11a is $h(\theta)$; and Figure 7-11b is $k(\theta)$.

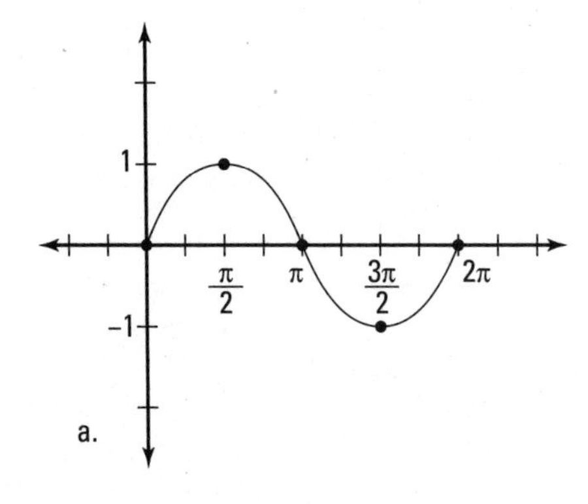

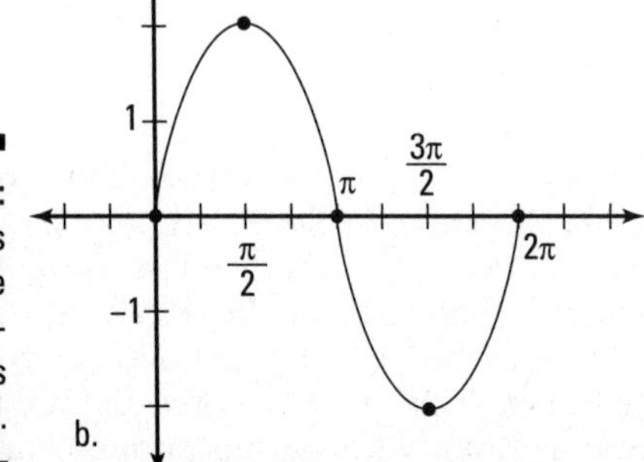

Figure 7-10: The graphs of example transformations of sine.

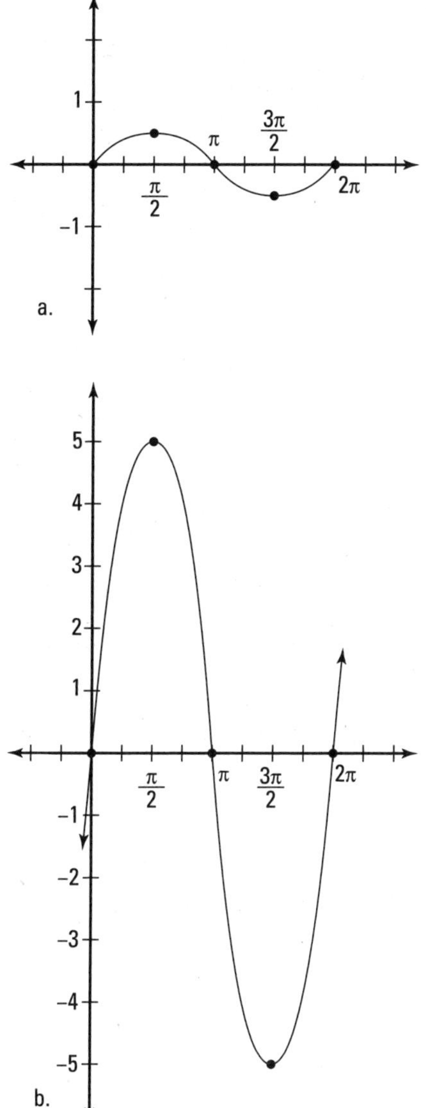

Figure 7-11: More examples of transformations of sine.

Altering the period

The *period* of the parent graphs of sine and cosine is 2π, which is once around the unit circle (see the earlier section "The sine graph"). Sometimes in trig, the variable (θ), not the function, gets multiplied by a constant. This action affects the period of the trig function graph. For example, $f(x) = \sin 2x$ makes the graph repeat itself twice in the same amount of time; in other words, the graph moves twice as fast. Think of it like pressing fast forward on a DVD. Figure 7-12 shows function graphs with various period changes.

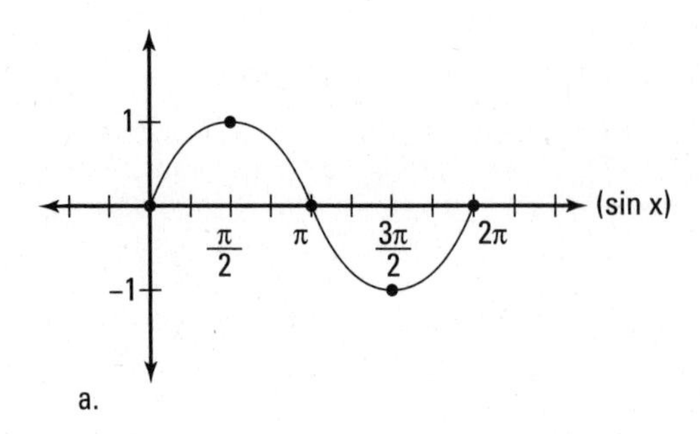

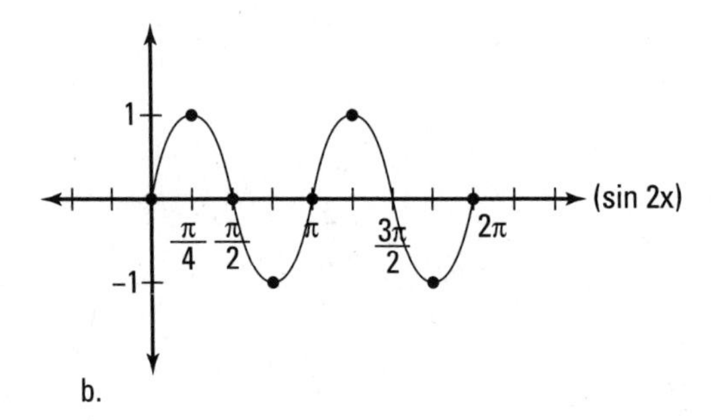

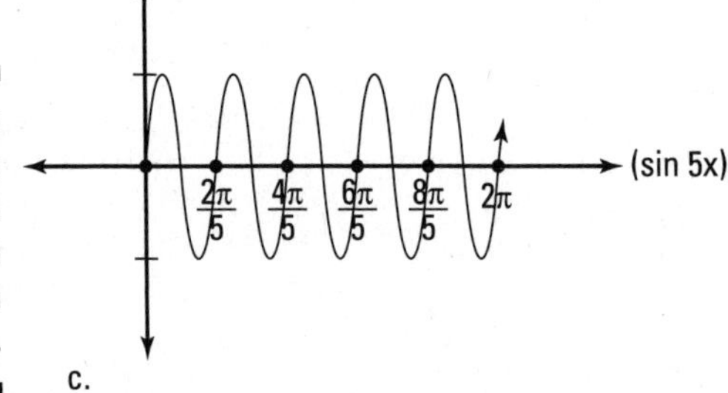

Figure 7-12: Creating period changes on function graphs.

To find the period of $f(x) = \sin 2x$, set $2 \cdot \text{period} = 2\pi$ (the period of the original sine function) and solve for the period. In this case, the period $= \pi$, so the graph will finish its trip at π. Each point along the *x*-axis also will move at twice the speed.

You can make the graph of a trig function move faster or slower with different constants:

- **Positive values of period greater than 1 make the graph repeat itself more and more frequently.** This is what you see in the example of *f(x)*.
- **Fraction values between 0 and 1 make the graph repeat itself less frequently.** For example, if $h(x) = \cos(\frac{1}{4}x)$, you can find its period by setting $\frac{1}{4} \cdot \text{period} = 2\pi$. Solving for period gets you 8π. Before, the graph finished at 2π; now it waits to finish at 8π, which slows it down by $\frac{1}{4}$.

You can have a negative constant multiplying the period. A negative constant affects how fast the graph moves, but in the opposite direction of the positive constant. For example, say $p(x) = \sin(3x)$ and $q(x) = \sin(-3x)$. The period of $p(x)$ is $\frac{2}{3}\pi$, while the period of $q(x)$ is $\frac{-2}{3}\pi$. The graph of $p(x)$ moves to the right of the *y*-axis, while the graph of $q(x)$ moves to the left. Figure 7-13 illustrates this point clearly.

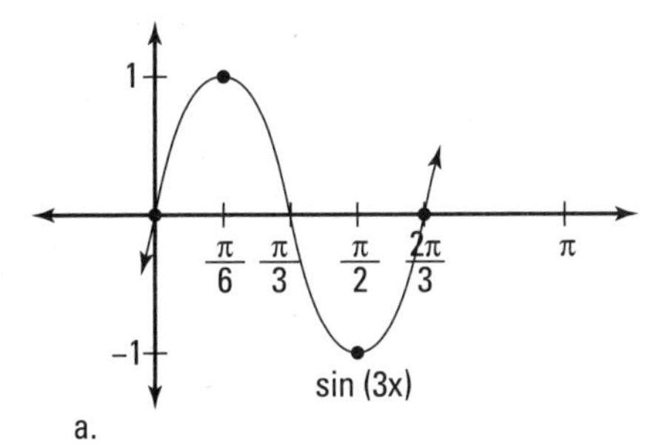

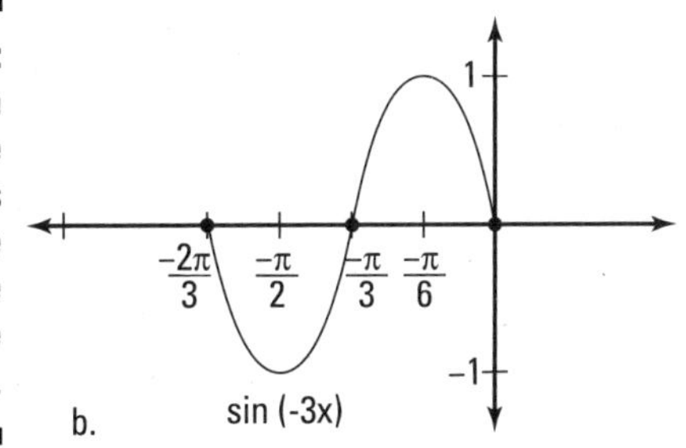

Figure 7-13: Graphs with negative periods move to the opposite side of the *y*-axis.

Don't confuse amplitude and period when graphing trig functions. For example, $f(x) = 2\sin x$ and $g(x) = \sin 2x$ affect the graph differently — $f(x) = 2\sin x$ makes it taller, and $g(x) = \sin 2x$ makes it move faster.

Shifting the waves on the coordinate plane

The movement of a parent graph around the coordinate plane is another type of transformation known as a *translation* or a *shift.* For this type of transformation, every point on the parent graph is moved somewhere else on the coordinate plane. A translation doesn't affect the overall shape of the graph; it only changes its location on the plane. In this section, we show you how to take the parent graphs of sine and cosine and shift them both horizontally and vertically.

Did you pick up the rules for shifting a function horizontally and vertically from Chapter 3? If not, go back and check them out, because they're important for sine and cosine graphs as well.

Most math books write the horizontal and vertical shifts as $\sin(x - h) + v$, or $\cos(x - h) + v$. The variable h represents the horizontal shift of the graph, and v represents the vertical shift of the graph. The sign makes a difference in the direction of the movement. For example,

- $f(x) = \sin(x - 3)$ moves the parent graph of sine to the right by 3.
- $g(x) = \cos(x + 2)$ moves the parent graph of cosine to the left by 2.
- $k(x) = \sin x + 4$ moves the parent graph of sine up 4.
- $p(x) = \cos x - 4$ moves the parent graph of cosine down 4.

For example, if you need to graph $y = \sin(\theta - \pi/4) + 3$, follow these steps:

1. **Identify the parent graph.**

   You're looking at sine, so draw its parent graph (see the earlier section "The sine graph"). The starting value for the parent graph of $\sin\theta$ is at $x = 0$.

2. **Shift the graph horizontally.**

   To find the new starting place, set what's inside the parentheses equal to the starting value of the parent graph: $\theta - \pi/4 = 0$, so $\theta = \pi/4$ is where this graph will start its period. You move every point on the parent graph to the right by $\pi/4$. Figure 7-14 shows what you have so far.

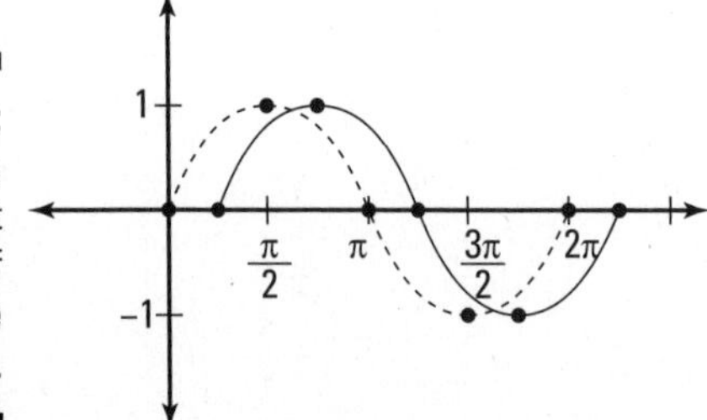

Figure 7-14: Shifting the parent graph of sine to the right by $\pi/4$.

3. **Move the graph vertically.**

   The sinusoidal axis of the graph moves up three positions in this function, so shift all the points of the parent graph this direction now. You can see this shift in Figure 7-15.

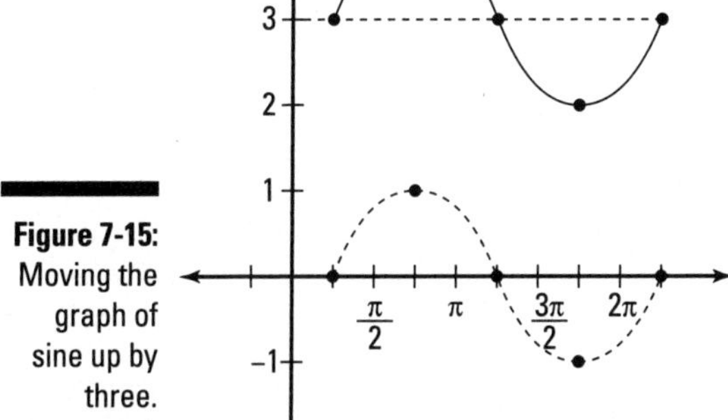

Figure 7-15: Moving the graph of sine up by three.

4. **State the domain and range of the transformed graph, if asked.**

   The domain and range of a function may be affected by a transformation. When this happens, you may be asked to state the new domain and range. Usually, you can visualize the range of the function easily by looking at the graph. Two factors that change the range are a vertical transformation (stretch or shrink) and a vertical translation.

   Keep in mind that the range of the parent sine graph is [–1, 1]. Shifting the parent graph up three units makes the range of $y = \sin(\theta - \frac{\pi}{4}) + 3$ shift up three units also. Therefore, the new range is [2, 4]. The domain of this function isn't affected; it's still $(-\infty, \infty)$.

Combining transformations in one fell swoop

When you're asked to graph a trig function with multiple transformations, we suggest that you do them in this order:

1. Change the amplitude.
2. Change the period.
3. Shift the graph horizontally.
4. Shift the graph vertically.

The equations that combine all the transformations into one are as follows:

$$f(x) = a \cdot \sin[p(x - h)] + v$$

$$f(x) = a \cdot \cos[p(x - h)] + v$$

The absolute value of the variable a is the amplitude. You take 2π and divide by p to find the period. The variable h is the horizontal shift, and v is the vertical shift.

The most important thing to know is that sometimes a problem will be written so that it looks like the period and the horizontal shift are both inside the trig function. For example, $f(x) = \sin(2x - \pi)$ makes it look like the period is twice as fast and the horizontal shift is π, doesn't it? But that isn't correct. All period shifts *must* be factored out of the expression to actually be period shifts, which in turn reveals the true horizontal shifts. You need to rewrite $f(x)$ as $\sin 2(x - \pi/2)$. This function tells you that the period is twice as fast, but that the horizontal shift is actually $\pi/2$ to the right.

Because this is so important, we want to present another example just so you grasp what we mean. With the following steps, graph $y = -3\cos(1/2\, x + \pi/4) - 2$:

1. **Write the equation in its proper form by factoring out the period constant.**

   This gives you $y = -3\cos[1/2\,(x + \pi/2)] - 2$.

2. **Graph the parent graph.**

   Graph the original cosine function as you know it (see the earlier section "The cosine graph").

3. **Change the amplitude.**

   This graph has an amplitude of 3, but the negative sign turns it upside down, which affects the graph's range. The range is now $[-3, 3]$. You can see the amplitude change in Figure 7-16.

4. **Alter the period.**

   The constant $1/2$ affects the period. Solving the equation $1/2 \cdot \text{period} = 2\pi$ gives you the period $= 4\pi$. The graph moves half as fast and finishes at 4π, which you can see in Figure 7-17.

5. **Shift the graph horizontally.**

   When you factored out the period constant in Step 1, you discovered that the horizontal shift is to the left $\pi/2$. This is shown in Figure 7-18.

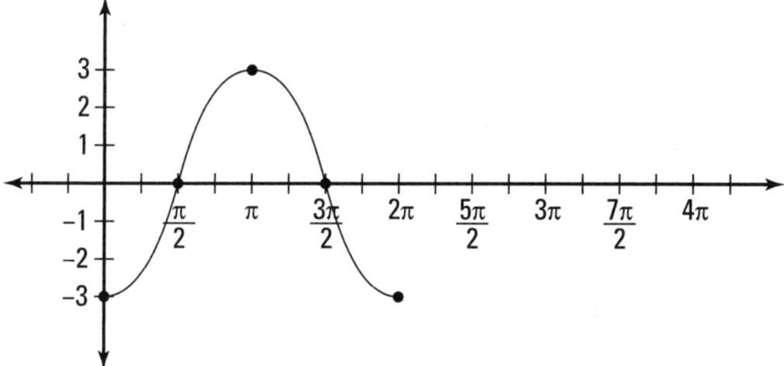

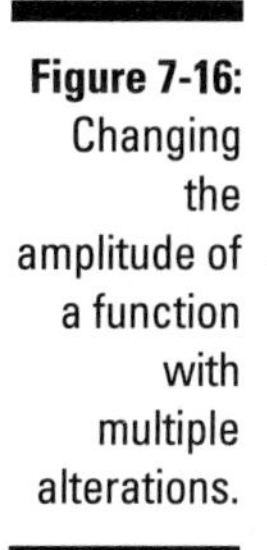

Figure 7-16: Changing the amplitude of a function with multiple alterations.

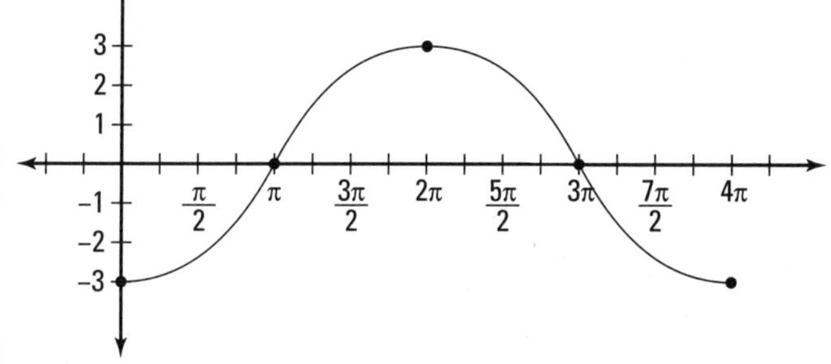

Figure 7-17: Changing the period to 4π.

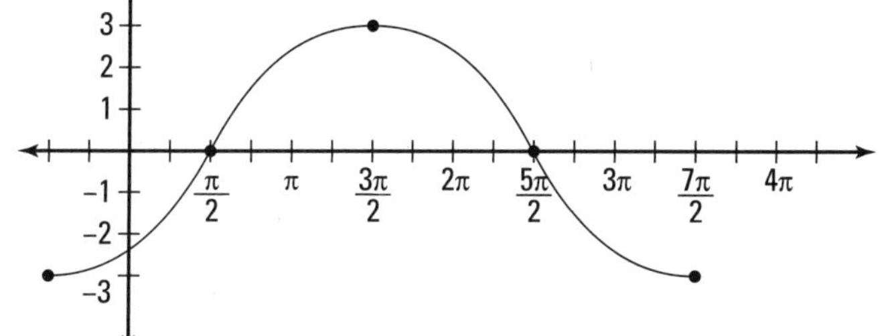

Figure 7-18: A horizontal shift to the left.

6. **Shift the graph vertically.**

   Due to the – 2 you see in Step 1, this graph moves down two positions, which you can see in Figure 7-19.

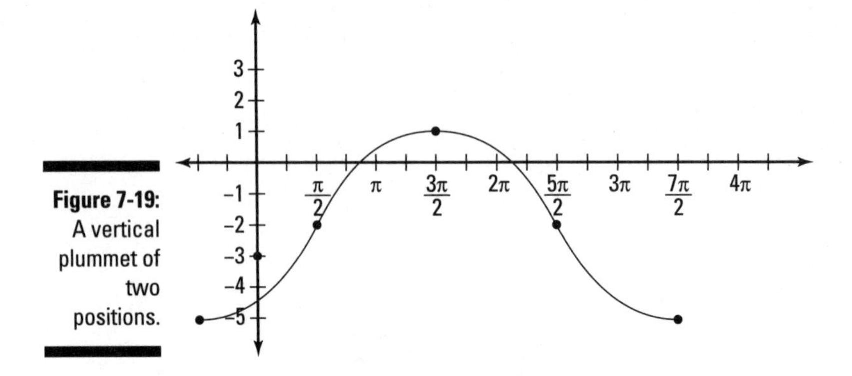

Figure 7-19: A vertical plummet of two positions.

7. **State the new domain and range.**

   The functions of sine and cosine are defined for all values of θ. The domain for the cosine function is all real numbers, or $(-\infty, \infty)$. The range of the graph in Figure 7-18 has been stretched because of the amplitude change, and shifted down.

   To find the range of a function that has been shifted vertically, you add or subtract the vertical shift (–2) from the altered range based on the amplitude. For this problem, the range of the transformed cosine function is $[-3 - 2, 3 - 2]$, or $[-5, 1]$.

Tweaking tangent and cotangent graphs

The transformations for sine and cosine work for tangent and cotangent, too (see the earlier transformations section in this chapter for more). Specifically, you can transform the graph vertically, change the period, shift the graph horizontally, or shift it vertically. As always, though, you should take each transformation one step at a time.

For example, to graph $f(\theta) = \frac{1}{2}\tan\theta - 1$, follow these steps:

1. **Sketch the parent graph for tangent (see the "Graphing Tangent and Cotangent" section).**

2. **Shrink or stretch the parent graph.**

   The vertical shrink is ½ for every point on this function, so each point on the tangent parent graph will be half as tall.

It's harder to see vertical changes for tangent and cotangent graphs, but they're there. Concentrate on the fact that the parent graph has points ($\frac{\pi}{4}$, 1) and ($\frac{-\pi}{4}$, –1), which in the transformed function become ($\frac{\pi}{4}$, $\frac{1}{2}$) and ($\frac{-\pi}{4}$, $\frac{-1}{2}$). As you can see in Figure 7-20, the graph really is half as tall!

3. **Change the period.**

   The constant $\frac{1}{2}$ doesn't affect the period. Why? Because it sits in front of the tangent function, which only affects vertical, not horizontal movement.

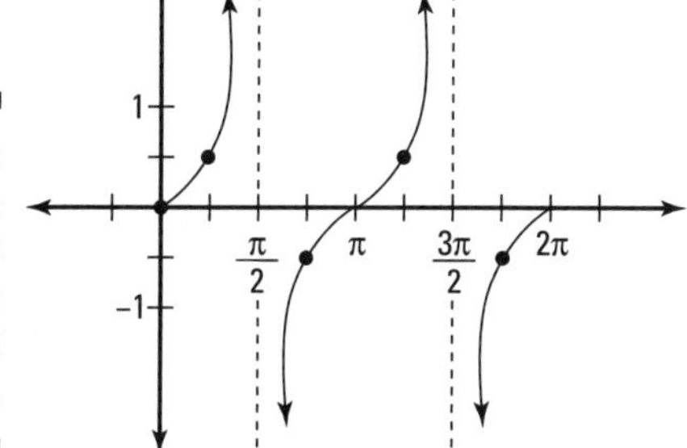

Figure 7-20: The amplitude is half as tall in the changed graph.

4. **Shift the graph horizontally and vertically.**

   This graph doesn't shift horizontally because there is no constant being added inside the grouping symbols (parentheses) of the function. So, you don't need to do anything here. The – 1 at the end of the function is a vertical shift that moves the graph down one position. Figure 7-21 shows the shift.

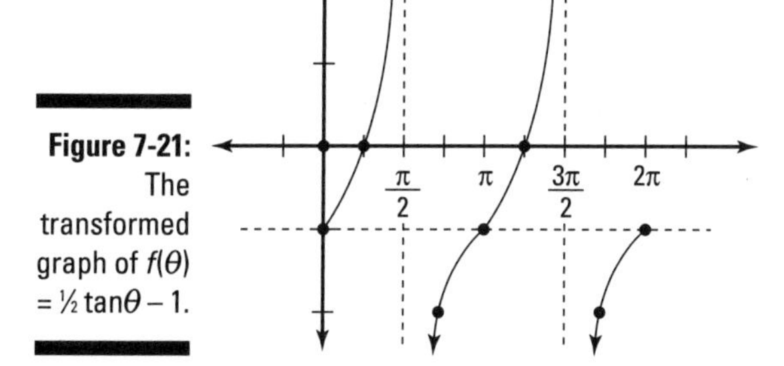

Figure 7-21: The transformed graph of $f(\theta) = \frac{1}{2}\tan\theta - 1$.

5. **State the transformed function's domain and range, if asked.**

   Because the range of the tangent function is all real numbers, transforming its graph doesn't affect the range, only the domain. The domain of the tangent function isn't all real numbers because of the asymptotes. The domain of the example function hasn't been affected by the transformations, however: $\theta \neq \frac{\pi}{2} + \pi n$, where n is an integer.

Now that you've graphed the basics, you can graph a function that has a period change, as in the function $y = \cot(2\pi x + \frac{\pi}{2})$. You see a lot of π in that one. Relax! You know this graph has a period change because you see a number inside the parentheses that's multiplied by the variable. This constant changes the period of the function, which in turn changes the distance between the asymptotes. In order for the graph to show this change correctly, you must factor this constant out of the parentheses. Take the transformation one step at a time:

1. **Sketch the parent graph for cotangent.**

   See the information in the "Cotangent" section to determine how to get the graph of cotangent.

2. **Shrink or stretch the parent graph.**

   No constant is multiplying the outside of the function; therefore, you can apply no amplitude change.

3. **Find the period change.**

   You factor out the 2π, which affects the period. The function will now read $y = \cot[2\pi(x + \frac{1}{4})]$.

   The period of the parent funtion for cotangent is π. Therefore, instead of dividing 2π by the period constant to find the period change (like you did for the sine and cosine graphs), you must divide π by the period constant. This will give you the period for the transformed cotangent function.

   Set $2\pi \cdot \text{period} = \pi$ and solve for the period. This gives you a period of $\frac{1}{2}$ for the transformed function. The graph of this function starts to repeat at $\frac{1}{2}$, which is different from $\frac{\pi}{2}$, so be careful when you're labeling your graph.

   Up until now, every trig function you've graphed has been a fraction of π (such as $\frac{\pi}{2}$), but this period isn't a fraction of π; it's just a rational number. When this happens, you must graph it as such. Figure 7-22 shows this step.

4. **Determine the horizontal and vertical shifts.**

   Because you've already factored the period constant, you can see that the horizontal shift is to the left $\frac{1}{4}$. Figure 7-23 shows this transformation on the graph.

   No constant is being added to or subtracted from this function on the outside, so the graph doesn't experience a vertical shift.

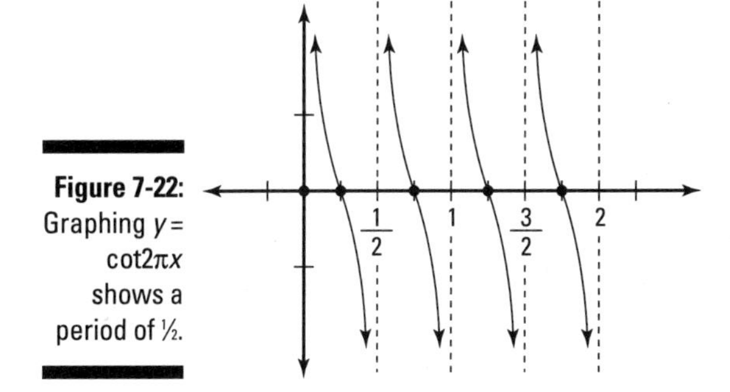

Figure 7-22: Graphing $y = \cot 2\pi x$ shows a period of ½.

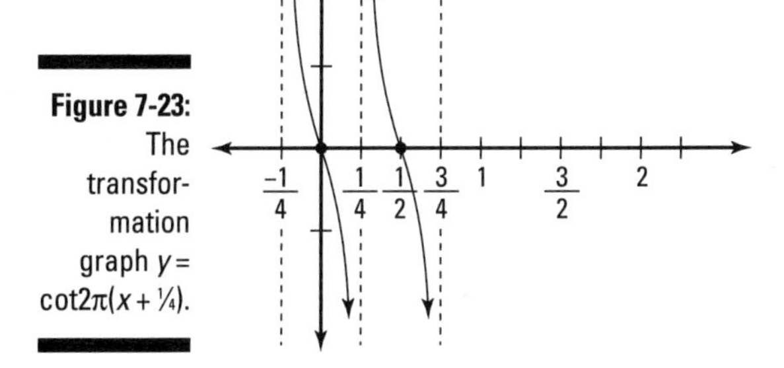

Figure 7-23: The transformation graph $y = \cot 2\pi(x + \frac{1}{4})$.

5. **State the transformed function's domain and range, if asked.**

   The horizontal shift affects the domain of this graph. To find the first asymptote, set $2\pi x + \frac{\pi}{2} = 0$ (setting the period shift equal to the original first asymptote). You find that $x = -\frac{1}{4}$ is your new asymptote. The graph will repeat every ½ radians because of its period. So, the domain is $\theta \neq -\frac{1}{4} + \frac{1}{2}n$, where n is an integer. The graph's range isn't affected: $(-\infty, \infty)$.

Transforming the graphs of secant and cosecant

To graph transformed secant and cosecant graphs, your best bet is to graph their reciprocal functions and transform them first. The reciprocal functions, sine and cosine, are easier to graph because they don't have as many complex parts (no asymptotes, basically). If you can graph the reciprocals first, you can deal with the more complicated pieces of the secant/cosecant graphs last.

For example, take a look at the graph $f(\theta) = \frac{1}{4}\sec\theta - 1$.

1. **Graph the transformed reciprocal function.**

   Look at the reciprocal function for secant, which is cosine. Pretend just for a bit that you're graphing $f(\theta) = \frac{1}{4}\cos\theta - 1$ (see the earlier section "Transforming Sine and Cosine Graphs"). Follow all the rules for the cosine graph to end up with a graph that looks like the one in Figure 7-24.

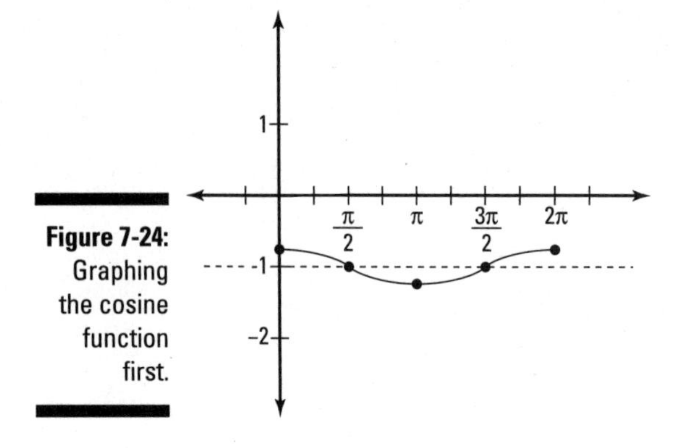

Figure 7-24: Graphing the cosine function first.

2. **Sketch the asymptotes of the transformed reciprocal function.**

   Wherever the transformed graph of $\cos\theta$ crosses its sinusoidal axis, you have an asymptote in $\sec\theta$. You see that $\cos\theta = 0$ when $\theta = \frac{\pi}{2}$ and $\frac{3\pi}{2}$.

3. **Find out what the graph looks like between each asymptote.**

   Now that you've identified the asymptotes, you simply figure out what happens on the intervals between them, like you do in Steps 2 through 4 of the section on the parent graph for secant. The finished graph ends up looking like the one in Figure 7-25.

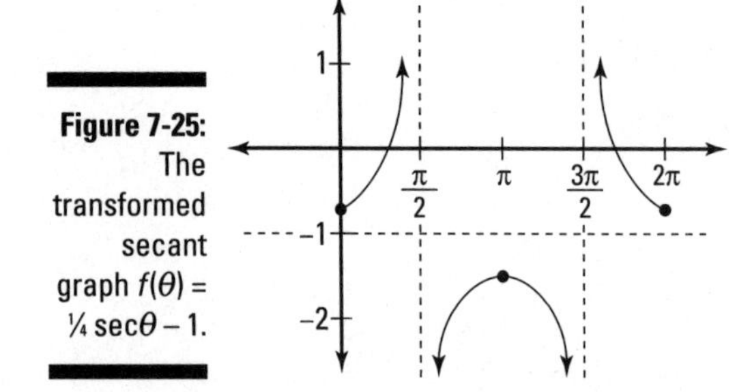

Figure 7-25: The transformed secant graph $f(\theta) = \frac{1}{4}\sec\theta - 1$.

4. **State the domain and range of the transformed function.**

   Because the new transformed function may have different asymptotes than the parent function for secant, and it may be shifted up or down, you may be required to state the new domain and range.

   This example, $f(\theta) = \frac{1}{4}\sec\theta - 1$, has asymptotes at $\pi/2$, $3\pi/2$, and so on, repeating every π radians. Therefore, the domain is restricted to not include these values, which is written $\theta \neq \pi/2 \pm \pi \cdot n$ where n is an integer. In addition, the range of this function changes because the transformed function is shorter than the parent function and has been shifted down two. The range has two separate intervals, $(-\infty, -5/4]$ and $[-3/4, \infty)$.

You can graph a transformation of the cosecant graph by using the same steps you use when graphing the secant function, only this time you use the sine function to guide you.

The shape of the transformed cosecant graph should be very similar to the secant graph, except the asymptotes will be in different places. For this reason, be sure you're graphing with the help of the sine graph to transform the cosecant graph, and the cosine function to guide you for the secant graph.

For the last example of this chapter, graph the transformed cosecant graph $g(\theta) = \csc(2x - \pi) + 1$:

1. **Graph the transformed reciprocal function.**

   Look first at the function $g(\theta) = \sin(2x - \pi) + 1$. The rules to transforming a sine function tell you to first factor out the 2 and get $g(\theta) = \sin 2(x - \pi/2) + 1$. There is a horizontal shrink of 2, a horizontal shift of $\pi/2$ to the right, and a vertical shift of up one. Figure 7-26 shows the transformed sine graph.

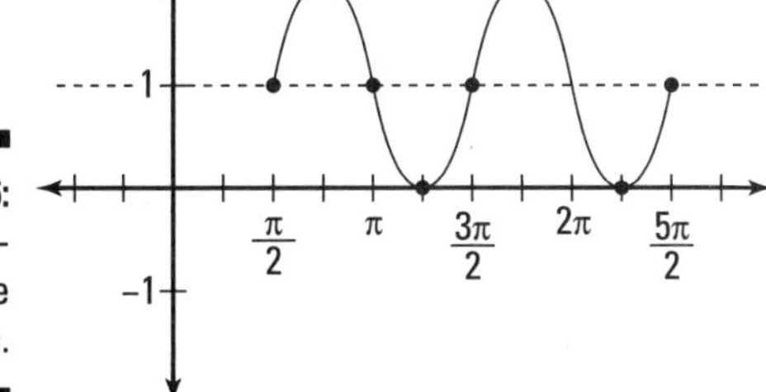

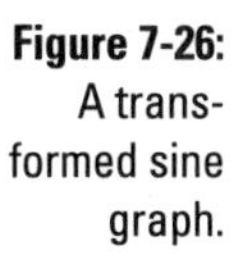

Figure 7-26: A transformed sine graph.

2. **Sketch the asymptotes of the reciprocal function.**

   The sinusoidal axis that runs through the middle of the sine function is the line $y = 1$. Therefore, an asymptote of the cosecant graph will exist everywhere the transformed sine function crosses this line. The asymptotes of the cosecant graph are at $\frac{\pi}{2}$ and π and repeat every $\frac{\pi}{2}$ radians.

3. **Figure out what happens to the graph between each asymptote.**

   You can use the transformed graph of the sine function to determine where the cosecant graph is positive and negative. Because the graph of the transformed sine function is positive in between $\frac{\pi}{2}$ and π, the cosecant graph is positive as well and extends up when getting closer to the asymptotes. Similarly, because the graph of the transformed sine function is negative inbetween π and $\frac{3\pi}{2}$, the cosecant is also negative in this interval. The graph alternates between positive and negative in equal intervals for as long as you want to sketch the graph.

   Figure 7-27 shows the transformed cosecant graph.

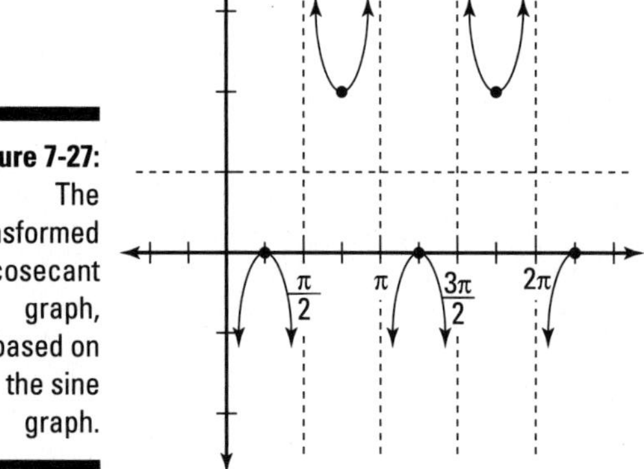

Figure 7-27: The transformed cosecant graph, based on the sine graph.

4. **State the new domain and range.**

   Just as with the transformed graph of the secant function (see the previous list), you may be asked to state the new domain and range for the cosecant function. The domain of the transformed cosecant function is all values of θ except for the values that are asymptotes. From the graph, you can see that the domain is all values of θ, where $\theta \neq \frac{\pi}{2} \pm \frac{\pi}{2} \cdot n$, where n is an integer. The range of the transformed cosecant function is also split up into two intervals: $(-\infty, 0] \cup [2, \infty)$.

Chapter 8

Using Trig Identities: The Basics

In This Chapter

- Reviewing the basics of solving trig equations
- Simplifying and proving expressions with fundamental trig identities
- Handling more complicated proofs

In the next two chapters, you work on simplifying expressions, using basic trig identities to prove more complicated identities, and solving equations that involve trig functions. Because we're fans of building momentum, we're going to start off slow with the basics. This chapter covers basic *identities,* which are statements that are always true and that you use throughout an entire equation to help you simplify the problem, one expression at a time, prior to solving it. And you're in luck, because trig has many of these identities.

The hard thing about simplifying trig expressions by using trig identities, though, is knowing when to stop. Enter *proofs,* which give you an end goal so that you know when you've hit a stopping point. You use proofs when you need to show that two expressions are equal, even though they look completely different. (If you thought you were done with proofs when you moved past geometry, think again!)

If you listen to one piece of advice, heed this: Know these basic identities forward, backward, and upside down, because you'll use them consistently in pre-calculus, and they'll make your life easier when things get complicated. When you commit the basic identities to memory but you can't remember the more complex identities we present in Chapter 9 (which is highly likely), you can simply use a combination of the basic identities to derive a "new" identity that suits your situation. This chapter shows you the way.

This chapter focuses on two main ideas that are central to your studies in pre-calc: You'll simplify expressions and prove complicated identities. These concepts share one common theme: They both involve the trig functions you're introduced to back in Chapter 6.

Keeping the End in Mind: A Quick Primer on Identities

The road to get to the end of each type of problem in this chapter is very similar. However, the end result of the two main ideas — simplifying expressions and proving complicated identities — will be different, and that's what we discuss in this section:

- **Simplifying:** Simplifying algebraic expressions with real numbers and variables is nothing new to you (we hope); you base your simplification process off of properties of algebra and real numbers that you *know* are always true. In this chapter, we give you some new rules to play by so that you can work with trig problems of all different sorts. Think of basic pre-calc identities as tools in your toolbox to help you build your trig house. One tool on its own doesn't do you any good. However, when you put the identities together, all the things you can do with them may surprise you! (For instance, you can make a trig expression with many different functions simplify into one function, or perhaps simplify to be real numbers [usually 0 or 1].)
- **Proving:** Trig proofs have two sides of an equation, and your job is to make one side look like the other. Usually, you make the more-complicated side look like the less-complicated side. Sometimes, though, if you can't get one side to look like the other, you can "cheat" and work on the other side for a while (or both at the same time). In the end, though, you need to show that one side transforms into the other.

Here's what you can't do in a trig proof: add, subtract, multiply, or divide from both sides. Because you're trying to *prove* that the two quantities are equal, you can't *assume* that they're equal to begin with; performing operations on both sides assumes the sides are equal.

Lining Up the Means to the End: Basic Trig Identities

If you're reading this book straight through, you'll likely recognize some of the identities in this chapter, because we brush over them in earlier chapters.

We use trig functions in Chapter 6 when we examine the ratios between the sides of right triangles. By definition, the reciprocal trig functions are identities, for instance, because they're true for all angle values. We reserved the full discussion of identities for this chapter because the identities aren't necessary to do the mathematical calculations in the earlier chapters, but now it's time to expand your horizons and work with some more complicated (yet still basic) identities.

In the following sections, we introduce you to the most basic (and most useful) identities. With this information, you can manipulate complicated trig expressions into expressions that are much simpler and more user friendly. This simplification process is one that takes a lot of practice. However, after you master simplifying trig expressions, proving complex identities and solving complicated equations will be a breeze.

If each step you take to simplify, prove, or solve a problem with trig is based on an identity (and executed correctly), you pretty much guarantee that you'll get the right answer; the particular route you take to get there doesn't matter. However, fundamental math skills still apply; you can't just throw out math rules willy-nilly. Some important, fundamental rules that people often forget when working with identities include the following:

- Dividing a fraction by another fraction is the same as multiplying by its reciprocal.
- To add or subtract two fractions, you must find the common denominator.
- Always factor out the greatest common factor and factor trinomials (see Chapter 4).

Reciprocal identities

When you're asked to simplify an expression involving cosecant, secant, or cotangent, you change the expression to functions that involve sine, cosine, or tangent, respectively. You do this so that you can cancel functions and simplify the problem. When you change functions in this manner, you're using the *reciprocal identities.* (Technically, the identities are trig functions that just happen to be considered identities as well because they help you simplify expressions.) The following list presents these reciprocal identities:

$$\csc\theta = \frac{1}{\sin\theta}$$

$$\sec\theta = \frac{1}{\cos\theta}$$

$$\cot\theta = \frac{1}{\tan\theta} = \frac{\cos\theta}{\sin\theta}$$

Every trig ratio can be written as a combination of sines and/or cosines, so changing all the functions in an equation to sines and cosines is the simplifying strategy that works most often. Always try to do this first, and then look to see if things cancel and simplify. Also, it's usually easier to deal with sines and cosines if you're looking for a common denominator for fractions. From there, you can use what you know about fractions to simplify as much as you can.

Simplifying an expression with reciprocal identities

Look for opportunities to use reciprocal identities whenever the problem you're given contains secant, cosecant, or cotangent. All these functions can be written in terms of sine and cosine, and sines and cosines are always the best place to start. For example, to simplify $\frac{\cos\theta \cdot \csc\theta}{\cot\theta}$, you can use reciprocal identities. Just follow these steps:

1. **Change all the functions into versions of the sine and cosine functions.**

   Because this problem involves a cosecant and a cotangent, you use the reciprocal identities to change $\csc\theta = \frac{1}{\sin\theta}$ and $\cot\theta = \frac{1}{\tan\theta} = \frac{\cos\theta}{\sin\theta}$.

   This gives you $\frac{\cos\theta\left(\frac{1}{\sin\theta}\right)}{\left(\frac{\cos\theta}{\sin\theta}\right)}$.

2. **Break up the complex fraction by rewriting the division bar that's present in the original problem as ÷.**

   This gives you $\cos\theta \cdot \left(\frac{1}{\sin\theta}\right) \div \left(\frac{\cos\theta}{\sin\theta}\right)$.

3. **Invert the last fraction and multiply.**

   This now sets up as $\cos\theta \cdot \left(\frac{1}{\sin\theta}\right) \cdot \left(\frac{\sin\theta}{\cos\theta}\right)$.

4. **Cancel the functions to simplify.**

   That leaves you with $\cancel{\cos\theta} \cdot \frac{1}{\cancel{\sin\theta}} \cdot \frac{\cancel{\sin\theta}}{\cancel{\cos\theta}} = 1$. The sines and cosines cancel and you end up getting 1 as your answer.

Working backward: Using reciprocal identities to prove equalities

Math teachers frequently ask you to prove complicated identities, because the process of proving those identities helps you to wrap your brain around the conceptual side of math. Oftentimes, you'll be asked to prove identities that involve the secant, cosecant, or cotangent functions. Whenever you see these functions in a proof, the reciprocal identities usually are the best places to start. Without the reciprocal identities, you could be going in circles all day, without ever actually getting anywhere.

For example, to prove $\tan\theta \cdot \csc\theta = \sec\theta$, you can work with the left side of the equality only. Follow these simple steps:

1. **Convert all functions to sines and cosines.**

   The left side of the equation now looks like $\frac{\sin\theta}{\cos\theta} \cdot \frac{1}{\sin\theta}$.

2. **Cancel all possible terms.**

   Cancelling gives you $\frac{\cancel{\sin\theta}}{\cos\theta} \cdot \frac{1}{\cancel{\sin\theta}}$, which simplifies to $\frac{1}{\cos\theta} = \sec\theta$.

3. **You can't leave the reciprocal function in the equality, so convert back again.**

   $\sec\theta = \frac{1}{\cos\theta}$, so $\sec\theta = \sec\theta$. Bingo!

Pythagorean identities

The *Pythagorean identities* are among the most useful identities because they simplify complicated expressions so nicely. When you see a trig function that's squared ($\sin^2$, $\cos^2$, and so on), keep these identities in mind. They're built from previous knowledge of right triangles and the alternate trig function values (which we explain in Chapter 6). Recall that the x-leg is $\cos\theta$, the y-leg is $\sin\theta$ and the hypotenuse is 1. Since you know that $\text{leg}^2 + \text{leg}^2 = \text{hypotenuse}^2$, thanks to the Pythagorean Theorem, then you will also know that $\sin^2\theta + \cos^2\theta = 1$. In this section, we show you where these important identities come from and then how to use them.

The three Pythagorean identities are

$$\sin^2 x + \cos^2 x = 1$$

$$\tan^2 x + 1 = \sec^2 x$$

$$1 + \cot^2 x = \csc^2 x$$

To limit the amount of memorizing you have to do, you can use the first Pythagorean identity to derive the other two:

- If you divide every term of $\sin^2\theta + \cos^2\theta = 1$ by $\sin^2\theta$, you get $\frac{\sin^2\theta}{\sin^2\theta} + \frac{\cos^2\theta}{\sin^2\theta} = \frac{1}{\sin^2\theta}$, which simplifies to the next Pythagorean identity:

  $1 + \cot^2\theta = \csc^2\theta$, because

  1. $\frac{\sin^2\theta}{\sin^2\theta} = 1$
  2. $\frac{\cos^2\theta}{\sin^2\theta} = \cot^2\theta$, due to the reciprocal identity
  3. $\frac{1}{\sin^2\theta} = \csc^2\theta$, for the same reason

- When you divide every term of $\sin^2\theta + \cos^2\theta = 1$ by $\cos^2\theta$, you get $\frac{\sin^2\theta}{\cos^2\theta} + \frac{\cos^2\theta}{\cos^2\theta} = \frac{1}{\cos^2\theta}$, which simplifies to the following Pythagorean identity:

  $\tan^2\theta + 1 = \sec^2\theta$, because

  1. $\frac{\sin^2\theta}{\cos^2\theta} = \tan^2\theta$
  2. $\frac{\cos^2\theta}{\cos^2\theta} = 1$
  3. $\frac{1}{\cos^2\theta} = \sec^2\theta$

Putting the Pythagorean identities into action

You normally use Pythagorean identities if you know one function and are looking for another. For example, if you know the sine ratio, you can use the first Pythagorean identity from the previous section to find the cosine ratio. In fact, you can find whatever you're asked to find if all you have is the value of one trig function and the understanding of what quadrant the angle θ is in.

For example, if you know that $\sin\theta = 24/25$ and $\pi/2 < \theta < \pi$, you can find $\cos\theta$ by following these steps:

1. **Plug what you know into the appropriate Pythagorean identity.**

   Because you're using sine and cosine, you use the first identity: $\sin^2\theta + \cos^2\theta = 1$. Plug in the values you know to get $(24/25)^2 + \cos^2\theta = 1$.

2. **Isolate the trig function with the variable on one side.**

   First square the sine value to get $\frac{576}{625}$, giving you $\frac{576}{625} + \cos^2\theta = 1$. Subtract $\frac{576}{625}$ from both sides (Hint: You need to find a common denominator): $\cos^2\theta = \frac{49}{625}$.

3. **Square root both sides to solve.**

   You now have $\cos\theta = \pm\frac{7}{25}$. But you can have only one solution because of the constraint $\frac{\pi}{2} < \theta < \pi$ you're given in the problem.

4. **Draw a picture of the unit circle so you can visualize the angle.**

   Because $\frac{\pi}{2} < \theta < \pi$, the angle lies in quadrant II, so the cosine of θ must be negative. You have your answer: $\cos\theta = -\frac{7}{25}$.

Using the Pythagorean identities to prove an equality

The Pythagorean identities pop up frequently in trig proofs. Pay attention and look for trig functions being squared. Try changing them to a Pythagorean identity and see if anything interesting happens. This section shows you how one proof could involve a Pythagorean identity.

After you change sines and cosines, the proof simplifies and makes your job that much easier. For example, follow these steps to prove $\frac{\sin x}{\csc x} + \frac{\cos x}{\sec x} = 1$:

1. **Convert all the functions in the equality to sines and cosines.**

   This gives you $\dfrac{\sin x}{\frac{1}{\sin x}} + \dfrac{\cos x}{\frac{1}{\cos x}} = 1$.

2. **Use the properties of fractions to simplify.**

   Dividing by a fraction is the same as multiplying by its reciprocal, so

   $\sin x \cdot \frac{\sin x}{1} + \cos x \cdot \frac{\cos x}{1} = 1$.

   $\sin^2 x + \cos^2 x = 1$.

3. **Identify the Pythagorean identity on the left side of the equality.**

   Because $\sin^2 x + \cos^2 x = 1$, you can say that $1 = 1$.

Even-odd identities

Because sine, cosine, and tangent are functions (trig functions), they can be defined as even or odd functions as well (see Chapter 3). Sine and tangent are both odd functions, and cosine is an even function. In other words,

$\sin(-x) = -\sin x$

$\cos(-x) = \cos x$

$\tan(-x) = -\tan x$

These identities will all make appearances in problems that ask you to simplify an expression, prove an identity, or solve an equation (see Chapter 6). The big red flag this time? The fact that the variable inside the trig function is negative. When tan(–*x*), for example, appears somewhere in an expression, it should usually be changed to –tan*x*.

Simplifying expressions with even-odd identities

Mostly, you use even-odd identities for graphing purposes, but you may see them in simplifying problems as well. (You can find the graphs of the trigonometric equations in Chapter 7 if you need a refresher.) You use an even-odd identity to simplify any expression where –*x* (or whatever variable you see) is inside the trig function.

In the following list, we show you how to simplify $(1 + \sin(-x))(1 - \sin(-x))$:

1. **Get rid of all the –*x* values inside the trig functions.**

   You see two sin(–*x*) functions, so you replace them both with –sin*x* to get $(1 + (-\sin x))(1 - (-\sin x))$.

2. **Simplify the new expression.**

   First adjust the two negative signs within the parentheses to get $(1 - \sin x)(1 + \sin x)$, and then FOIL these two binomials to get $1 - \sin^2 x$.

3. **Look for any combination of terms that could give you a Pythagorean identity.**

   Whenever you see a function squared, you should think of the Pythagorean identities. Looking back at the section "Pythagorean identities," you see that $1 - \sin^2 x$ is the same as $\cos^2 x$. Now the expression is fully simplified as $\cos^2 x$.

Proving an equality with even-odd identities

When asked to prove an identity, if you see a negative variable inside a trig function, you automatically use an even-odd identity. First, replace all trig functions with (–θ) inside the parentheses. Then simplify the trig expression to make one side look like the other side. Here's just one example of how this works. Prove $\frac{\cos(-\theta) - \sin(-\theta)}{\sin\theta} - \frac{\cos(-\theta) + \sin(-\theta)}{\cos\theta} = \sec\theta\csc\theta$ with the following steps:

1. **Replace all negative angles and their trig functions with the even-odd identity that matches.**

   $$\frac{\cos\theta + \sin\theta}{\sin\theta} - \frac{\cos\theta - \sin\theta}{\cos\theta} = \sec\theta\csc\theta$$

2. **Simplify the new expression.**

   Because the right side doesn't have any fractions in it, eliminating the fractions from the left side is an excellent place to start. In order to subtract fractions, you first must find a common denominator. However, before doing that, look at the first term of the fraction. This fraction can be split up into the sum of two fractions, as can the second fraction. By doing this first, certain terms will simplify and make your job much easier when it actually comes time to work with the fractions.

   Therefore, you get $\frac{\cos\theta}{\sin\theta} + \frac{\sin\theta}{\sin\theta} - \left(\frac{\cos\theta}{\cos\theta} - \frac{\sin\theta}{\cos\theta}\right) = \sec\theta\csc\theta$, which quickly simplifies to $\frac{\cos\theta}{\sin\theta} + \cancel{1} - \cancel{1} + \frac{\sin\theta}{\cos\theta} = \sec\theta\csc\theta$, which simplifies even further to $\frac{\cos\theta}{\sin\theta} + \frac{\sin\theta}{\cos\theta} = \sec\theta\csc\theta$.

   Now you must find a common denominator. For this example, the common denominator is $\sin\theta \cdot \cos\theta$. Multiplying the first term by $\frac{\cos\theta}{\cos\theta}$ and the second term by $\frac{\sin\theta}{\sin\theta}$ gives you $\frac{\cos^2\theta}{\sin\theta \cdot \cos\theta} + \frac{\sin^2\theta}{\sin\theta \cdot \cos\theta} = \sec\theta\csc\theta$. You can rewrite this as $\frac{\cos^2\theta + \sin^2\theta}{\sin\theta \cdot \cos\theta} = \sec\theta\csc\theta$. Here is a Pythagorean identity in its finest form! $\text{Sin}^2\theta + \cos^2\theta$ is the most frequently used of the Pythagorean identities. This equation then simplifies to $\frac{1}{\sin\theta \cdot \cos\theta} = \sec\theta\csc\theta$.

   Using the reciprocal identities, you get $\frac{1}{\sin\theta} \cdot \frac{1}{\cos\theta} = \sec\theta\csc\theta$. So, $\sec\theta \cdot \csc\theta = \sec\theta \cdot \csc\theta$.

Co-function identities

If you take the graph of sine and shift it to the left or right, it looks exactly like the cosine graph (see Chapter 7). The same is true for tangent and cotangent, as well as secant and cosecant. That's the basic premise of *co-function identities* — they say that the sine and cosine functions have the same values, but those values are shifted slightly on the coordinate plane when you look at one function compared to the other. You have experience with all six of the trig functions, as well as their relationships to one another. The only difference is that in this section, we introduce them formally as *identities*.

The following list of co-function identities illustrates this point:

$\sin x = \cos(\pi/2 - x)$

$\cos x = \sin(\pi/2 - x)$

$\tan x = \cot(\pi/2 - x)$

$\cot x = \tan(\pi/2 - x)$

$\sec x = \csc(\pi/2 - x)$

$\csc x = \sec(\pi/2 - x)$

Putting co-function identities to the test

The co-function identities are great to use whenever you see $\pi/2$ inside the grouping parentheses. You may see functions in the expressions such as $\sin(\pi/2 - x)$. If the quantity inside the trig function looks like $(\pi/2 - x)$ or $(90° - \theta)$, you'll know to use the co-function identities.

For example, to simplify $\frac{\cos x}{\cos\left(\frac{\pi}{2} - x\right)}$, follow these steps:

1. **Look for co-function identities and substitute.**

   First realize that $\cos(\pi/2 - x)$ is the same as $\sin x$ because of the co-function identity. That means you can substitute $\sin x$ in for $\cos(\pi/2 - x)$ to get $\frac{\cos x}{\sin x}$.

2. **Look for other substitutions you can make.**

   $\frac{\cos x}{\sin x}$ is the same as $\cot x$ because of the reciprocal identity for cotangent.

Proving an equality by employing the co-function identities

Co-function identities also pop up in trig proofs. The expression $\pi/2 - x$ making an appearance in parentheses inside any trig function lets you know to use a co-function identity for the proof. To prove $\frac{\csc\left(\frac{\pi}{2} - \theta\right)}{\tan(-\theta)} = -\csc\theta$, follow these steps:

1. **Replace any trig functions with $\pi/2$ in them with the appropriate co-function identity.**

   Replacing $\csc(\pi/2 - \theta)$ with $\sec(\theta)$, you get $\frac{\sec\theta}{\tan(-\theta)} = -\csc\theta$.

2. **Simplify the new expression.**

   You have many trig identities at your disposal, and you may use any of them at any given time. Now is the perfect time to use an even-odd identity for tangent:

   $\frac{\sec\theta}{-\tan\theta} = -\csc\theta$

Then use the reciprocal identity for secant and the sines and cosines definition for tangent to get $\frac{\frac{1}{\cos\theta}}{\frac{-\sin\theta}{\cos\theta}} = -\csc\theta$.

Finally, rewrite this complex fraction as the division of two simpler fractions: $\frac{1}{\cancel{\cos\theta}} \cdot \frac{\cancel{\cos\theta}}{-\sin\theta} = -\csc\theta$. Cancel anything that's both in the numerator and the denominator and then simplify. This gives you $\frac{1}{-\sin\theta} = -\csc\theta$.

Rewrite the last line of the proof as $-\csc\theta = -\csc\theta$. (This should always be the last line of your proof for technical reasons.)

Periodicity identities

Periodicity identities illustrate how if you shift the graph of a trig function by one period to the left or right, you end up with the same function. (We cover periods and periodicity identities when we show you how to graph trig functions in Chapter 7.) The periods of sine, cosine, secant, and cosecant repeat every 2π; tangent and cotangent, on the other hand, repeat every π.

The following identities show how the different trig functions repeat:

$\sin(x + 2\pi) = \sin x$

$\cos(x + 2\pi) = \cos x$

$\tan(x + \pi) = \tan x$

$\cot(x + \pi) = \cot x$

$\sec(x + 2\pi) = \sec x$

$\csc(x + 2\pi) = \csc x$

Seeing how the periodicity identities work to simplify equations

Similar to the co-function identities, you use the periodicity identities when you see $(x + 2\pi)$ or $(x - 2\pi)$ inside a trig function. Because adding (or subtracting) 2π radians from an angle gives you a new angle in the same position, you can use that idea to form an identity. For tangent and cotangent only, adding or subtracting π radians from the angle gives you the same result, because the period of the tangent and cotangent functions is π.

For example, to simplify $\sin(2\pi + \theta) + \cos(2\pi + \theta) \cdot \cot(\pi + \theta)$, follow these steps:

1. **Replace all trig functions with 2π inside the parentheses with the appropriate periodicity identity.**

   For this example, $\sin\theta + \cos\theta \cdot \cot\theta$.

2. **Simplify the new expression.**

   Start with $\sin\theta + \cos\theta \cdot \frac{\cos\theta}{\sin\theta}$.

   To find a common denominator to add the fractions, multiply the first term by $\frac{\sin\theta}{\sin\theta}$. The new fraction is $\frac{\sin^2\theta}{\sin\theta} + \frac{\cos^2\theta}{\sin\theta}$. Add them together to get $\frac{\sin^2\theta + \cos^2\theta}{\sin\theta}$. You can see a Pythagorean identity in the numerator, so replace $\sin^2\theta + \cos^2\theta$ with 1. Therefore, the fraction becomes $\frac{1}{\sin\theta} = \csc\theta$.

Proving an equality with the periodicity identities

Using the periodicity identities also comes in handy when you need to prove an equality that includes the expression $(\theta + 2\pi)$, or the addition (or subtraction) of the period. For example, to prove $[\sec(2\pi + x) - \tan(\pi + x)][\csc(2\pi + x) + 1] = \cot x$, follow these steps:

1. **Replace all trig functions with the appropriate periodicity identity.**

   You're left with $(\sec x - \tan x)(\csc x + 1)$.

2. **Simplify the new expression.**

   For this example, the best place to start is to FOIL (see Chapter 4):

   $(\sec x \cdot \csc x + \sec x \cdot 1 - \tan x \cdot \csc x - \tan x \cdot 1)$

   Now convert all terms to sines and cosines to get $\frac{1}{\cos x} \cdot \frac{1}{\sin x} + \frac{1}{\cos x} - \frac{\sin}{\cos x} \cdot \frac{1}{\sin x} - \frac{\sin x}{\cos x}$. Then find a common denominator and add the fractions:

   $\frac{1}{\cos x \sin x} + \frac{\sin x}{\cos x \sin x} - \frac{\sin x}{\cos x \sin x} - \frac{\sin^2 x}{\cos x \sin x} = \frac{1 - \sin^2 x}{\cos x \sin x}$.

3. **Apply any other applicable identities.**

   You have a Pythagorean identity in the form of $1 - \sin^2 x$, so replace it with $\cos^2 x$. Cancel one of the cosines in the numerator (because it's squared) with the cosine in the denominator to get $\frac{\cos^2 x}{\cos x \sin x} = \frac{\cos x}{\sin x}$. Finally, this simplifies to $\cot x = \cot x$.

Tackling Difficult Trig Proofs: Some Techniques to Know

Historically speaking, most of our students have struggled with (and hated) proofs, so we've dedicated this section entirely to them (the proofs *and* the students). So far in this chapter, we've shown you proofs that require only a few basic steps to complete. Now, we show you how to tackle the more complicated proofs. The techniques here are based on ideas you've dealt with before in your math journey. Okay, so a few trig functions are thrown into it, but why should that scare you?

One tip will always help you when you're faced with complicated trig proofs that require multiple identities: *Always* check your work and review all the identities you know to make sure that you haven't forgotten something you can simplify.

The goal is to make one side of the given equation look like the other through a series of steps, all of which are based on identities, properties, and definitions. No cheating by making up something that doesn't exist to get done! All decisions you make must be based on the rules. Here's an overview of the techniques we show you in this section:

- **Fractions in proofs:** These types of proofs come equipped with every rule you've ever learned regarding fractions, plus you have to deal with identities on top of that.
- **Factoring:** Degrees higher than 1 on a trig function often are great indicators that you need to do some factoring.
- **Square roots:** When roots show up in a proof, sooner or later you'll probably have to square both sides to get things moving.
- **Working on both sides at once:** Sometimes, you may get stuck while working on one side of a proof. At that point, we recommend transitioning to the other side.

Dealing with dreaded denominators

Most students hate fractions. We don't know why, they just do! But in dealing with trig proofs, fractions will pop up. So allow us to throw you into the deep end of the pool. Even if you don't mind fractions, we suggest you still read this section because it shows you specifically how to work with fractions in trig proofs. There are three main types of proofs you'll work with that have fractions:

- Proofs where you end up creating fractions
- Proofs that start off with fractions
- Proofs that require multiplying by a conjugate to deal with a fraction

We break down each one of these in this section, with an example proof so you can see what to do.

Creating fractions when working with reciprocal identities

We often like to mention how converting all the functions to sines and cosines makes a trig proof easier. When terms are multiplying, this usually allows you to cancel and simplify to your heart's content so that one side of the equation ends up looking just like the other, which is the goal. But when the terms are adding or subtracting, you may create fractions where there were none before. This is especially true when dealing with secant and cosecant, because converting them to, respectively, $\frac{1}{\cos\theta}$ and $\frac{1}{\sin\theta}$ creates fractions. The same is true for tangent when you change it to $\frac{\sin\theta}{\cos\theta}$ and cotangent becomes $\frac{\cos\theta}{\sin\theta}$.

Here's an example that illustrates our point. Prove that $\sec^2 t + \csc^2 t = \sec^2 t \cdot \csc^2 t$ by following these steps:

1. **Convert all the trig functions to sines and cosines.**

   On the left side, you now have $\frac{1}{\cos^2 t} + \frac{1}{\sin^2 t} = \sec^2 t \cdot \csc^2 t$.
2. **Find the LCD of the two fractions.**

   This multiplication gives you $\frac{\sin^2 t}{\sin^2 t \cdot \cos^2 t} + \frac{\cos^2 t}{\sin^2 t \cdot \cos^2 t} = \sec^2 t \cdot \csc^2 t$.
3. **Add the two fractions.**

   The addition simplifies the expression to $\frac{\sin^2 t + \cos^2 t}{\sin^2 t \cdot \cos^2 t} = \sec^2 t \cdot \csc^2 t$.
4. **Simplify the expression with a Pythagorean identity in the numerator.**

   Now you're down to $\frac{1}{\sin^2 t \cdot \cos^2 t} = \sec^2 t \cdot \csc^2 t$.
5. **Use reciprocal identities to invert the fraction.**

   Both sides now have multiplication: $\csc^2 t \cdot \sec^2 t = \sec^2 t \cdot \csc^2 t$.

Some pre-calculus teachers will let you stop there; others, however, make you rewrite the equation so that the left and right sides match exactly. Every teacher has his or her own way of proving trig identities. Make sure that you adhere to his/her expectations; otherwise, it may cause you to lose points on a test.

6. **Use the properties of equality to rewrite.**

   The commutative property of multiplication (see Chapter 1) says that $a \cdot b = b \cdot a$, so $\sec^2 t \cdot \csc^2 t = \sec^2 t \cdot \csc^2 t$.

Starting off with fractions

When the expression you're given begins with fractions, most of the time it's your job to add (or subtract) them to get things to simplify. Here's one example of a proof where doing just that gets the ball rolling. To simplify $\frac{\cos t}{1+\sin t}+\frac{\sin t}{\cos t}$, you have to find the LCD to add the two fractions. With that as the beginning step, follow along:

1. **Find the LCD of the two fractions you must add.**

   The least common denominator is $(1 + \sin t) \cdot \cos t$, so multiply the first term by $\frac{\cos t}{\cos t}$ and multiply the second term by $\frac{1+\sin t}{1+\sin t}$:

   $\frac{\cos t}{1+\sin t}\left(\frac{\cos t}{\cos t}\right)+\frac{\sin t}{\cos t}\left(\frac{1+\sin t}{1+\sin t}\right)$.

2. **Multiply or distribute in the numerators of the fractions.**

   This gives you $\frac{\cos^2 t}{(1+\sin t)(\cos t)}+\frac{\sin t+\sin^2 t}{(1+\sin t)(\cos t)}$.

3. **Add the two fractions.**

   You now have one fraction: $\frac{\cos^2 t+\sin t+\sin^2 t}{(1+\sin t)(\cos t)}$.

4. **Look for any trig identities and substitute.**

   You can rewrite the numerator as $\frac{\cos^2 t+\sin^2+\sin t}{(1+\sin t)(\cos t)}$, which is equal to $\frac{1+\sin t}{(1+\sin t)(\cos t)}$ because $\cos^2 t + \sin^2 t = 1$ (a Pythagorean identity).

5. **Cancel or reduce the fraction.**

   After the top and the bottom are completely factored (see Chapter 4), you can cancel terms: $\frac{\cancel{(1+\sin t)}}{\cancel{(1+\sin t)}(\cos t)}=\frac{1}{\cos t}$.

6. **Change any reciprocal trig functions.**

   In this case, $\frac{1}{\cos t} = \sec t$.

Multiplying by a conjugate

When one side of a proof is a fraction with a binomial in its denominator, always consider multiplying by the conjugate before you do anything else. Most of the time, this technique allows you to simplify things. When it doesn't, you're left to your own devices (and the other techniques presented in this section) to make the proof work.

For example, to rewrite $\frac{\sin\theta}{\sec\theta - 1}$ without a fraction, follow these steps:

1. **Multiply by the conjugate of the denominator.**

   The conjugate of $a + b$ is $a - b$, and vice versa. So, you have to multiply by $\sec\theta + 1$ on the top and bottom of the fraction. This gives you $\frac{\sin\theta(\sec\theta + 1)}{(\sec\theta - 1)(\sec\theta + 1)}$.

2. **FOIL the conjugates.**

   Now you have the fraction $\frac{\sin\theta(\sec\theta + 1)}{\sec^2\theta - 1}$. If you've been following along all chapter, the bottom should look awfully familiar. One of those Pythagorean identities? Yep!

3. **Change any identities to their simpler form.**

   Using the identity on the bottom, you get $\frac{\sin\theta(\sec\theta + 1)}{\tan^2\theta}$.

4. **Change every trig function to sines and cosines.**

   Here it gets more complex: $\frac{\sin\theta\left(\frac{1}{\cos\theta} + 1\right)}{\left(\frac{\sin^2\theta}{\cos^2\theta}\right)}$.

5. **Change the big division bar to a division sign, and then invert the fraction so you can multiply instead.**

   Check this step out: $\sin\theta\left(\frac{1}{\cos\theta} + 1\right) \div \left(\frac{\sin^2\theta}{\cos^2\theta}\right) = \sin\theta\left(\frac{1}{\cos\theta} + 1\right) \cdot \frac{\cos^2\theta}{\sin^2\theta}$.

6. **Cancel what you can from the expression.**

   The sine on the top cancels one of the sines on the bottom, leaving you with the following:

   $$\cancel{\sin\theta}\left(\frac{1}{\cos\theta} + 1\right) \cdot \frac{\cos^2\theta}{\sin^{\cancel{2}}\theta} = \left(\frac{1}{\cos\theta} + 1\right) \cdot \frac{\cos^2\theta}{\sin\theta}$$

7. **Distribute and watch what happens!**

   Through cancellations, you go from $\frac{1}{\cancel{\cos\theta}} \cdot \frac{\cos^{\cancel{2}}\theta}{\sin\theta} + \frac{\cos^2\theta}{\sin\theta}$ to $\frac{\cos\theta}{\sin\theta} + \frac{\cos\theta}{\sin\theta} \cdot \cos\theta$, which finally simplifies to $\cot\theta + \cot\theta \cdot \cos\theta$. And if you're asked to take it even a step further, you can factor to get $\cot\theta(1 + \cos\theta)$.

Going solo on each side

Sometimes doing work on both sides of a proof, one at a time, leads to a quicker solution. This holds true because in order to prove a very complicated identity, you may need to complicate the expression even further before it can begin to simplify. However, you should take this action only in dire circumstances after every other technique has failed (but don't tell your teacher we told you so!).

The main idea is that you work on the left side first, stop when you just can't go any further, and then switch to working on the right side. By doing this, your goal is to make the two sides of the proof meet in the middle somewhere. For example, to prove $\frac{1+\cot\theta}{\cot\theta} = \tan\theta + \csc^2\theta - \cot^2\theta$, follow these steps (see the work at the end of the numbered steps):

1. **Break up the fraction by writing each term in the numerator over the term in the denominator, separately.**

   The rules of fractions state that when only one term sits in the denominator, you can do this because each part on top is being divided by the bottom.

   You now have $\frac{1}{\cot\theta} + \frac{\cot\theta}{\cot\theta} = \tan\theta + \csc^2\theta - \cot^2\theta$.

2. **Use reciprocal rules to simplify.**

   The first fraction on the left side is the reciprocal of $\tan\theta$, and $\cot\theta$ divided by itself is 1.

   You now have $\tan\theta + 1 = \tan\theta + \csc^2\theta - \cot^2\theta$.

   You've come to the end of the road on the left side. The expression is now so simplified that it would be hard to expand it again to look like the right side, so you should turn to the right side and simplify it.

3. **Look for any applicable trig identities on the right side.**

   You use a Pythagorean identity to identify that $\csc^2\theta = 1 + \cot^2\theta$.

   You now have $\tan\theta + 1 = \tan\theta + 1 + \cot^2\theta - \cot^2\theta$.

4. **Cancel where possible.**

   Aha! The right side has $\cot^2\theta - \cot^2\theta$, which is 0! Cancel them to leave only $\tan\theta + 1 = \tan\theta + 1$.

$$
\begin{array}{ccc}
\dfrac{1+\cot\theta}{\cot\theta} & = & \tan\theta + \csc^2\theta - \cot^2\theta \\
\downarrow & & \uparrow \\
\dfrac{1}{\cot\theta} + \dfrac{\cot\theta}{\cot\theta} & = & \tan\theta + \csc^2\theta - \cot^2\theta \\
\downarrow & & \uparrow \\
\tan\theta + 1 & = & \tan\theta + \csc^2\theta - \cot^2\theta \\
\downarrow & & \uparrow \\
\tan\theta + 1 & = & \tan\theta + (1+\cot^2\theta) - \cot^2\theta \\
\downarrow & & \uparrow \\
\tan\theta + 1 & \underrightarrow{=} & \tan\theta + 1
\end{array}
$$

5. **Rewrite the proof starting on one side and ending up like the other side.**

$$
\begin{array}{lcl}
\dfrac{1+\cot\theta}{\cot\theta} & = & \tan\theta + \csc^2\theta - \cot^2\theta \\
\dfrac{1}{\cot\theta} + \dfrac{\cot\theta}{\cot\theta} & & \\
\tan\theta + 1 & & \\
\tan\theta + 1 + \cot^2\theta - \cot^2\theta & & \\
\tan\theta + \csc^2\theta - \cot^2\theta \checkmark & &
\end{array}
$$

Some pre-calc teachers will never accept working on both sides of an equation as a valid proof. If you're unfortunate enough to encounter a teacher like this, we suggest that you still work on both sides of the equation, but for your eyes only. Be sure to rewrite your work for your teacher by simply going down one side and up the other (like we did in Step 5).

Chapter 9

Pre-Calc, Here I Come! Advanced Identities Lead the Way

In This Chapter

- Applying the sum and difference formulas of trig functions
- Utilizing double-angle formulas
- Cutting angles in two with half-angle formulas
- Changing from products to sums and back
- Tossing aside exponents with power-reducing formulas

Prior to the invention of calculators (not as long ago as you may imagine), people had only one way to calculate the exact trig values for angles not shown on the unit circle: using advanced identities. Even now, most calculators give you only an approximation of the trig value, not the exact one. Exact values are important to trig calculations and to their applications (and to teachers, of course). Engineers designing a bridge, for example, don't want an *almost* correct value — and neither should you, for that matter.

This chapter is the meat and potatoes of pre-calc identities: It contains the bulk of the formulas that you need to know for calculus. It builds on the basic identities we discuss in Chapter 8. Advanced identities provide you with opportunities to calculate values that you couldn't calculate before — like finding the exact value of the sine of 15°, or figuring out the sine or cosine of the sum of angles without actually knowing the value of the angles. This information is truly helpful when you get to calculus, which takes these calculations to another level (a level at which you integrate and differentiate by using these identities).

Finding Trig Functions of Sums and Differences

Long ago, some fantastic mathematicians found identities that hold true when adding and subtracting angle measures from special triangles (30°-60°-90° right triangles and 45°-45°-90° right triangles; see Chapter 6). The focus is to find a way to rewrite an angle as a sum or difference. Those mathematicians were curious; they could find the trig values for the special triangles but wanted to know how to deal with other angles that aren't part of the special triangles on the unit circle. They could solve problems with multiples of 30° and 45°, but there were so many more angles that could be formed that they knew nothing about!

Constructing these angles was simple; however, evaluating trig functions for them proved to be a bit more difficult. So, they put their collective minds together and discovered the identities we discuss in this section. (Their only problem: They still couldn't find plenty of other trig values by using the sum $(a + b)$ and difference $(a - b)$ formulas.)

This section takes the information that we cover in earlier chapters, such as calculating trig values of special angles, and takes it to the next level. We introduce you to advanced identities that allow you to find trig values of angles that are multiples of 15°.

Note: You'll never be asked to find the sine of 87°, for example, without a calculator in this section, because it can't be written as the sum or difference of special angles. The 30°-60°-90° and the 45°-45°-90° triangles can always be boiled down to the same special ratios of their sides, but other triangles cannot. So, if you can break down the given angle into the sum or difference of two known angles, you have it made in the shade because you can use the sum or difference formula to find the trig value you're looking for. (If you can't express the angle as the sum or difference of special angles, you have to find some other way to solve the problem.)

For the most part, when you're presented with advanced identity problems in pre-calculus, you'll be asked to work with angles in radians. Of course, sometimes you'll have to work with degrees as well. We start with calculations in degrees because they're easier to manipulate. We then switch to radians and show you how to make the formulas work with them, too.

Searching out the sine of $(a \pm b)$

Using the special right triangles (see Chapter 6), which have points on the unit circle that are easy to identify, you can find the sine of 30° and 45° angles (among others). However, no point on the unit circle allows you to find trig values at angle measures that aren't special (such as the sine of 15°) directly. The sine does exist for such an angle (meaning it's a real number value) at a point on the circle; it just isn't one of the nicely labeled points. Don't despair, because this is where advanced identities help you out.

If you look really closely, you'll notice that 45° – 30° = 15°, and 45° + 30° = 75°. For the angles you can rewrite as the sum or difference of special angles, here are the sum and difference formulas for sine:

$$\sin(a + b) = \sin a \cdot \cos b + \cos a \cdot \sin b$$

$$\sin(a - b) = \sin a \cdot \cos b - \cos a \cdot \sin b$$

You can't rewrite the $\sin(a + b)$ as $\sin a + \sin b$. You can't distribute the sine into the values inside the parentheses, because sine isn't a multiplication operation; therefore, the distributive property doesn't apply (like it does to real numbers). Sine is a function, not a number or variable.

You have more than one way to combine unit circle angles to get a requested angle. You can write sin75° as sin(135° – 60°) or sin(225° – 150°). After you find a way to rewrite an angle as a sum or difference, roll with it. Use the one that works for you!

Calculating in degrees

Measuring angles in degrees for the sum and difference formulas is easier than measuring in radians, because adding and subtracting degrees is much easier than adding and subtracting radians. Adding and subtracting angles in radians requires finding a common denominator. Moreover, evaluating trig functions requires you to work backward from a common denominator to split the angle into two fractions with different denominators. (If the angle in the problem given to you is in radians, we show you the way in the next section.)

For example, follow these steps to find the sine of 135°:

1. **Rewrite the angle, using the special angles from right triangles (see Chapter 6).**

   One way to rewrite 135° is 90° + 45°.

2. **Choose the appropriate sum or difference formula.**

   We add in the example from Step 1, so it makes sense to use the sum formula and not the difference formula: $\sin(a + b) = \sin a \cdot \cos b + \cos a \cdot \sin b$.

3. **Plug the information you know into the formula.**

   You know that $\sin 135° = \sin(90° + 45°)$. This means $a = 90°$ and $b = 45°$. The formula gives you $\sin 90° \cos 45° + \cos 90° \sin 45°$.

4. **Use the unit circle (see Chapter 6) to look up the sine and cosine values you need.**

   You now have $1 \cdot \frac{\sqrt{2}}{2} + 0 \cdot \frac{\sqrt{2}}{2}$.

5. **Multiply and simplify to find the final answer.**

   You end up with $\frac{\sqrt{2}}{2} + 0 = \frac{\sqrt{2}}{2}$.

Calculating in radians

You can put the concept of sum and difference formulas to work using radians. This is different than solving equations because here you're asked to find the trig value of a specific angle that isn't readily marked on the unit circle (but still a multiple of 15° or $\pi/12$ radians). Prior to choosing the appropriate formula (Step 2 from the previous section), you simply break the angle into either the sum or the difference of two angles from the unit circle. Refer to the unit circle and notice the angles in radians in Figure 9-1. You see that all the denominators are different, which makes adding and subtracting them a nightmare. You must find a common denominator so that adding and subtracting is a dream. The common denominator is 12, as you can see in Figure 9-1.

Figure 9-1 comes in handy only for sum and difference formulas, because finding a common denominator is something you do only when you're adding or subtracting fractions.

For example, follow these steps to find the exact value of $\sin(\pi/12)$:

1. **Rewrite the angle in question, using the special angles in radians with common denominators.**

   From Figure 9-1, you want a way to add or subtract two angles so that, in the end, you get $\pi/12$. In this case, you can rewrite $\pi/12$ as $3\pi/12 - 2\pi/12$.

2. **Choose the appropriate sum/difference formula.**

   Because we rewrote the angle with subtraction, you need to use the difference formula.

3. **Plug the information you know into the chosen formula.**

   You know that $\sin \pi/12 = \sin(3\pi/12 - 2\pi/12)$. Substituting $a = 3\pi/12$ and $b = 2\pi/12$, you get $\sin\frac{3\pi}{12}\cos\frac{2\pi}{12} - \cos\frac{3\pi}{12}\sin\frac{2\pi}{12}$.

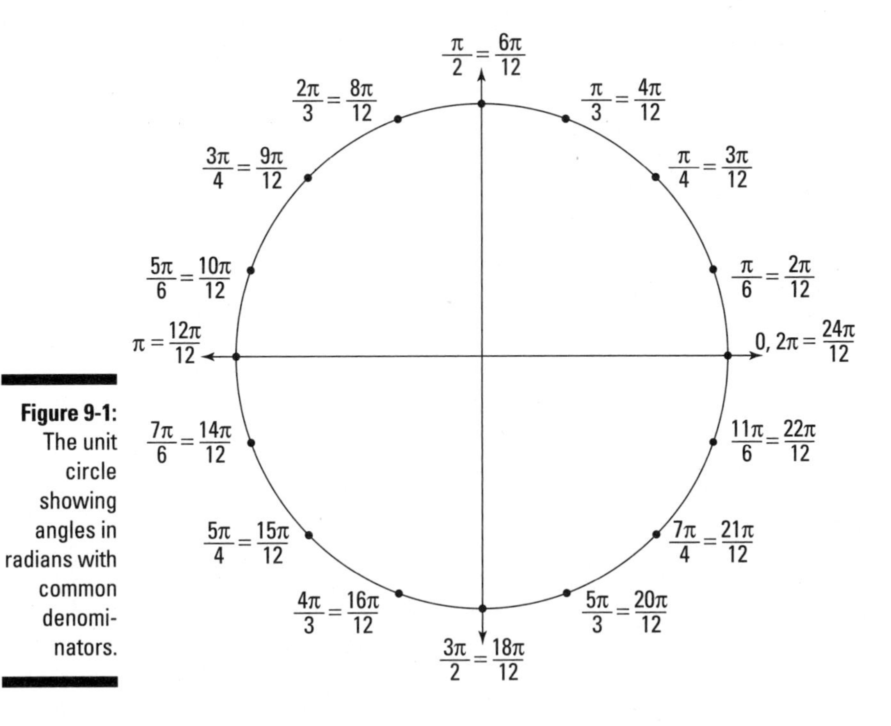

Figure 9-1: The unit circle showing angles in radians with common denominators.

4. **Reduce the fractions in the formula to ones you're more comfortable with.**

   In our example, you can reduce to $\sin\frac{\pi}{4}\cos\frac{\pi}{6} - \cos\frac{\pi}{4}\sin\frac{\pi}{6}$. Now you'll have an easier time referring to the unit circle to get your equation.

5. **Use the unit circle to look up the sine and cosine values that you need.**

   You now have $\frac{\sqrt{2}}{2}\cdot\frac{\sqrt{3}}{2} - \frac{\sqrt{2}}{2}\cdot\frac{1}{2}$.

6. **Multiply and simplify to get your final answer.**

   You end up with $\frac{\sqrt{6}}{4} - \frac{\sqrt{2}}{4} = \frac{\sqrt{6}-\sqrt{2}}{4}$.

Applying the sine sum and difference formulas to proofs

The goal when dealing with trig proofs in this chapter is the same as the goal when dealing with them in Chapter 8: You need to make one side of a given equation look like the other. You can work on both sides to get a little further if need be, but make sure you know how your teacher wants the proof to look. This section contains info on how to deal with sum and difference formulas in a proof.

When asked to prove $\sin(x + y) + \sin(x - y) = 2\sin x \cdot \cos y$, for example, follow these steps:

1. **Look for identities in the equation.**

   In our example, you can see the sum identity, $\sin(a + b) = \sin a \cdot \cos b + \cos a \cdot \sin b$, and the difference identity, $\sin(a - b) = \sin a \cdot \cos b - \cos a \cdot \sin b$ for sin.

2. **Substitute for the identities.**

   This gives you $\sin x \cdot \cos y + \cos x \cdot \sin y + \sin x \cdot \cos y - \cos x \cdot \sin y = 2\sin x \cdot \cos y$.

3. **Simplify to get the proof.**

   Two terms cancel to leave you with $\sin x \cdot \cos y + \sin x \cdot \cos y = 2\sin x \cdot \cos y$.

   Combine the like terms to get $2\sin x \cdot \cos y = 2\sin x \cdot \cos y$.

Calculating the cosine of $(a \pm b)$

After you familiarize yourself with the sum and difference formulas for sine, you can easily apply your newfound knowledge to calculate the sums and differences of cosines, because the formulas look very similar to each other. When working with sums and differences for sines and cosines, you're simply plugging given values in for variables. Just make sure you use the correct formula based on the information you're given in the question.

Here are the formulas for the sum and difference of cosines:

$$\cos(a + b) = \cos a \cdot \cos b - \sin a \cdot \sin b$$

$$\cos(a - b) = \cos a \cdot \cos b + \sin a \cdot \sin b$$

Applying the formulas to find the sum or difference of two angles

The sum and difference formulas for cosine (and sine) can do more than calculate a trig value for an angle not marked on the unit circle (at least angles that are multiples of 15°). They can also be used to find the sum or difference of two angles based on information given about the two angles. For such problems, you'll be given two angles (we'll call them A and B), the sine or cosine of A and B, and the quadrant(s) in which the two angles are located.

Use the following steps to find the exact value of $\cos(A + B)$, given that $\cos A = -3/5$ (A is in quadrant II of the coordinate plane) and $\sin B = -7/25$ (B is in QIII):

1. **Choose the appropriate formula and substitute the information you know to determine the missing information.**

   If $\cos(A + B) = \cos A \cdot \cos B - \sin A \cdot \sin B$, you know that $\cos(A + B) = -3/5 \cdot \cos B - \sin A \cdot -7/25$.

   To proceed any further, you need to find cosB and sinA.

2. **Draw pictures representing right triangles in the quadrant(s).**

   You need to draw one triangle for angle A in QII and one for angle B in QIII. Using the definition of sine as $\frac{opp}{hyp}$ and cosine as $\frac{adj}{hyp}$, Figure 9-2 shows these triangles. Notice that the value of a leg is missing in each triangle.

3. **To find the missing values, use the Pythagorean Theorem (once for each triangle; see Chapter 6).**

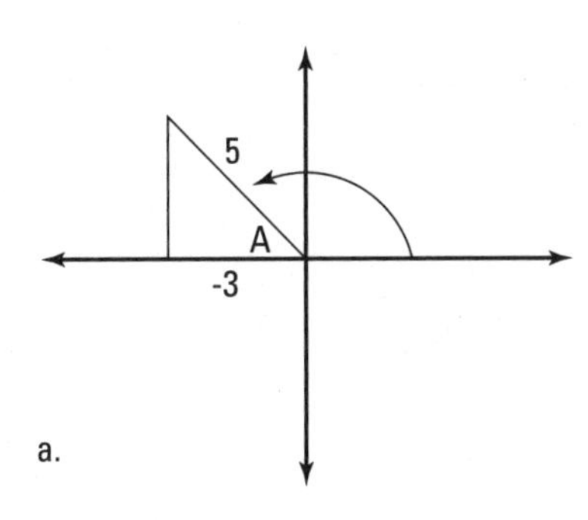

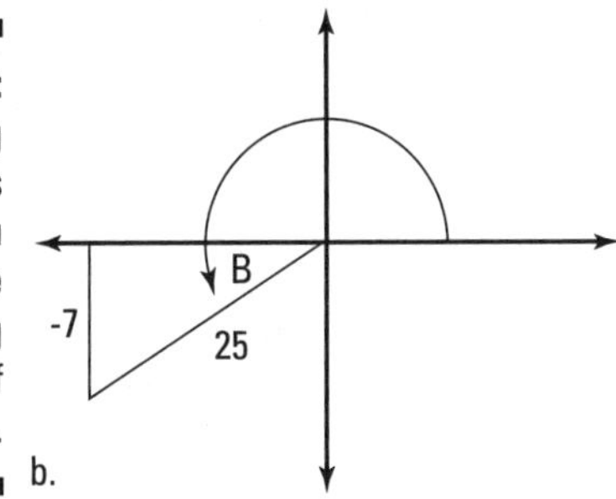

Figure 9-2: Drawing pictures helps you visualize the missing pieces of info.

   The missing leg in Figure 9-2a is 4, and the missing leg in Figure 9-2b is –24.

4. **Determine the missing trig ratios to use in the sum/difference formula.**

   You use the definition of cosine $\frac{adj}{hyp}$ to find that $\cos B = -24/25$ and the definition of sine $\frac{opp}{hyp}$ to find that $\sin A = 4/5$.

5. **Substitute the missing trig ratios into the sum/difference formula and simplify.**

   You now have $\cos(A + B) = {-3/5} \cdot {-24/25} - 4/5 \cdot {-7/25}$. Follow the order of operations to get $\cos(A + B) = 72/125 - (-28/125) = \frac{72 + 28}{125} = 100/125$.
   This simplifies to $\cos(A + B) = 4/5$.

Applying the cosine sum and difference formulas to proofs

You can prove the co-function identities from Chapter 8 by using the sum and difference formulas for cosine. For example, to prove $\cos(\pi/2 - x) = \sin x$, follow these steps:

1. **Outline the given information.**

   You start with $\cos(\pi/2 - x) = \sin x$.

2. **Look for sum and/or difference identities for cosine.**

   In this case, the left side of the equation is the difference formula for cosine. This means you can break up $\cos(\pi/2 - x)$ by using the difference formula for cosines to get $\cos\frac{\pi}{2}\cos x + \sin\frac{\pi}{2}\sin x = \sin x$.

3. **Refer to the unit circle and substitute all the information you know.**

   Using the unit circle, the previous equation simplifies to $0 \cdot \cos x + 1 \cdot \sin x$, which equals $\sin x$ on the left-hand side. Your equation now says $\sin x = \sin x$. Ta-dah!

Taming the tangent of $(a \pm b)$

As with sine and cosine (see the previous sections of this chapter), you can rely on formulas to find the tangent of a sum or a difference of angles. The main difference is that you can't read tangents directly from the coordinates of points on the unit circle, as you can with sine and cosine, because each point represents $(\cos\theta, \sin\theta)$; we explain more about this topic in Chapter 6. All hope isn't lost, however, because tangent is defined as $\frac{sin}{cos}$; because the sine of the angle is the y coordinate and the cosine is the x coordinate, you can express the tangent in terms of x and y on the unit circle as y/x.

Here are the formulas you need to find the tangent of a sum or difference of angles:

$$\tan(a + b) = \frac{\tan a + \tan b}{1 - \tan a \cdot \tan b}$$

$$\tan(a - b) = \frac{\tan a - \tan b}{1 + \tan a \cdot \tan b}$$

We suggest that you memorize these sweet little formulas, because then you won't have to use the sum and difference formulas for sine and cosine in the middle of a tangent problem, saving you time in the long run. If you choose not to memorize these two formulas, you can derive them by remembering that $\tan(a + b) = \frac{\sin(a+b)}{\cos(a+b)}$. It would be the same for the difference formula for tangent: $\tan(a-b) = \frac{\sin(a-b)}{\cos(a-b)}$.

Applying the formulas to solve a common problem

The sum and difference formulas for tangent work in similar ways to the sine and cosine formulas. You can use the formulas to solve a variety of problems. In this section, we show you how to find the tangent of an angle that isn't marked on the unit circle. You can do so as long as the angle can be written as the sum or difference of special angles.

For example, to find the exact value of $\tan 105°$, follow these steps (***Note:*** We don't mention the quadrant because the angle 105 degrees is in quadrant II. In the previous example, the angle wasn't given; a trig value was given, and each trig value has two angles on the unit circle that yield the value, so you need to know which quadrant the problem is talking about):

1. **Rewrite the given angle, using the information from special right-triangle angles (see Chapter 6).**

   Refer to the unit circle, noting that it's built from the special right triangles (see the Cheat Sheet in the front of this book), to find a combination of angles that add or subtract to get 105°. You can choose from 240° – 145°, 330° – 225°, and so on. In this example, we choose 60° + 45°. So, $\tan(105°) = \tan(60° + 45°)$.

2. **Choose the appropriate sum/difference formula.**

   Because we rewrote the angle with addition, you need to use the sum formula for tangent.

3. **Plug the information you know into the appropriate formula.**

   You now have $\frac{\tan 60° + \tan 45°}{1 - \tan 60° \cdot \tan 45°}$.

4. **Use the unit circle to look up the sine and cosine values that you need.**

   To find $\tan 60°$, you must locate 60° on the unit circle and use the sine and cosine values of its corresponding point to calculate the tangent:

   $$\tan 60° = \frac{\sin 60°}{\cos 60°} = \frac{\frac{\sqrt{3}}{2}}{\frac{1}{2}} = \frac{\sqrt{3}}{\cancel{2}} \cdot \frac{\cancel{2}}{1} = \sqrt{3}$$

Follow the same process for tan45°:

$$\tan 45^\circ = \frac{\sin 45^\circ}{\cos 45^\circ} = \frac{\frac{\sqrt{2}}{2}}{\frac{\sqrt{2}}{2}} = \frac{\cancel{\sqrt{2}}}{\cancel{2}} \cdot \frac{\cancel{2}}{\cancel{\sqrt{2}}} = 1$$

5. **Substitute the trig values from Step 4 into the formula.**

   This gives you $\frac{\sqrt{3}+1}{1-\left(\sqrt{3}\right)(1)}$, which simplifies to $\frac{\sqrt{3}+1}{1-\sqrt{3}}$.

6. **Rationalize the denominator.**

   You can't leave the square root on the bottom of the fraction. Because the denominator is a binomial (the sum or difference of two terms), you must multiply by its conjugate. The conjugate of $a + b$ is $a - b$, and vice versa. So, the conjugate of $1 - \sqrt{3}$ is $1 + \sqrt{3}$:

   $$\frac{\left(\sqrt{3}+1\right)}{\left(1-\sqrt{3}\right)} \cdot \frac{\left(1+\sqrt{3}\right)}{\left(1+\sqrt{3}\right)}.$$

   FOIL (see Chapter 4) both binomials to get

   $$\frac{\sqrt{3}+3+1+\sqrt{3}}{1+\sqrt{3}-\sqrt{3}-3}.$$

7. **Simplify the rationalized fraction to find the exact value of tangent.**

   Combine like terms to get $\frac{2\sqrt{3}+4}{-2}$. Make sure you fully simplify this fraction to get $-\sqrt{3}-2$.

Applying the sum and difference formulas to proofs

The sum and difference formulas for tangent are very useful if you want to prove a few of the basic identities from Chapter 8. For example, you can prove the co-function identities by using the difference formula and the periodicity identities by using the sum formula. If you see a sum or a difference inside a tangent function, you can try the appropriate formula to simplify things.

For instance, you can prove that the identity $\tan(\pi/4 + \theta) = \frac{1+\tan\theta}{1-\tan\theta}$ by following these steps:

1. **Look for identities for which you can substitute.**

   On the left side of the proof is the sum identity for tangent: $\tan(a + b) = \frac{\tan a + \tan b}{1 - \tan a \cdot \tan b}$. Working on the left side only gives you

   $$\frac{\tan\frac{\pi}{4} + \tan\theta}{1 - \tan\frac{\pi}{4}\tan\theta} = \frac{1+\tan\theta}{1-\tan\theta}.$$

2. **Use any applicable unit circle values to simplify the proof.**

   From the unit circle (see Chapter 6), you see that $\tan\frac{\pi}{4} = 1$, so you can plug in that value to get $\frac{1+\tan\theta}{1-(1)(\tan\theta)} = \frac{1+\tan\theta}{1-\tan\theta}$.

   From there, simple multiplication gives you $\frac{1+\tan\theta}{1-\tan\theta} = \frac{1+\tan\theta}{1-\tan\theta}$.

Doubling an Angle's Trig Value without Knowing the Angle

You use a *double-angle formula* to find the trig value of twice an angle. Sometimes you know the original angle, sometimes you don't. Working with double-angle formulas comes in handy when you need to solve trig equations or when you're given the sine, cosine, tangent, or other trig function of an angle and need to find the exact trig value of twice that angle without knowing the measure of the original angle. Isn't this your happy day?

Note: If you know the original angle in question, finding the sine, cosine, or tangent of twice that angle is easy; you can look it up on the unit circle (see the Cheat Sheet) or use your calculator to find the answer. However, if you don't have the measure of the original angle and you must find the exact value of twice that angle, the process isn't that simple. Read on!

Finding the sine of a doubled angle

To fully understand and be able to stow away the double-angle formula for sine, you should first understand where it comes from. (The double-angle formulas for sine, cosine, and tangent are extremely different from one another, although they can all be derived by using the sum formula.)

1. **To find $\sin 2x$, you must realize that it's the same as $\sin(x + x)$.**
2. **Use the sum formula for sine (see the section "Searching out the sine of $(a \pm b)$") to get $\sin x \cdot \cos x + \cos x \cdot \sin x$.**
3. **Simplify to get $\sin 2x = 2\sin x \cdot \cos x$.**

   This is called the double-angle formula for sine. If you're given an equation with more than one trig function and asked to solve for the angle, your best bet is to express the equation in terms of one trig function only. You often can achieve this by using the double-angle formula.

To solve $4\sin 2x \cdot \cos 2x = 1$, notice that it doesn't equal 0, so you can't factor it. Even if you subtract 1 from both sides to get 0, it can't be factored. So, that must mean there's no solution, right? Not quite. You have to check the identities first. The double-angle formula, for instance, says that $2\sin x \cdot \cos x = \sin 2x$. You can rewrite some things here:

1. **List the given information.**

   You have $4\sin 2x \cdot \cos 2x = 1$.

2. **Rewrite the equation to find a possible identity.**

   We go with $2 \cdot (2\sin 2x \cdot \cos 2x) = 1$.

3. **Apply the correct formula.**

   The double-angle formula for sine gives you $2 \cdot [\sin(2 \cdot 2x)] = 1$.

4. **Simplify the equation and isolate the trig function.**

   Break it down to $2 \cdot \sin 4x = 1$, which becomes $\sin 4x = \frac{1}{2}$.

5. **Find all the solutions for the trig equation.**

   This gives you $4x = \frac{\pi}{6} + 2\pi k$ and $4x = \frac{5\pi}{6} + 2\pi k$ where k is an integer. This tells you that each reference angle has four solutions, and you use the notation $+ 2\pi k$ to represent the circle. Then, you can divide everything (including the $2\pi k$) by 4, which gives you the solutions:

   - $x = \frac{\pi}{24} + \frac{\pi}{2} \cdot k$
   - $x = \frac{5\pi}{24} + \frac{\pi}{2} \cdot k$

This is the general solution, but there will be a time when you'll have to use this information to get to a solution on an interval.

Finding the solutions on an interval is a curveball thrown at you in pre-calc. For the previous problem, you can find a total of eight angles on the interval $[0, 2\pi)$. Because a coefficient was in front of the variable, you're left with, in this case, four times as many solutions, and you must state them all. You have to find the common denominator to add the fractions. In this case, $\frac{\pi}{2}$ becomes $\frac{12\pi}{24}$:

The first solution is $\frac{\pi}{24}$.

To find the second one: $\frac{\pi}{24} + \frac{\pi}{2}$ gives you $\frac{13\pi}{24}$.

To find the third one: $\frac{13\pi}{24} + \frac{\pi}{2}$ is $\frac{25\pi}{24}$.

To find the fourth: $\frac{25\pi}{24} + \frac{\pi}{2}$ is $\frac{37\pi}{24}$.

Doing this one more time gets you $\frac{49\pi}{24}$, which is really where you started (because you moved $\frac{48\pi}{24}$ from the original, which really is 2π — the period of the sine function). Meanwhile,

$\frac{5\pi}{24}$ is another solution.

$\frac{5\pi}{24} + \frac{\pi}{2}$ gives you $\frac{17\pi}{24}$.

$\frac{17\pi}{24} + \frac{\pi}{2}$ is $\frac{29\pi}{24}$.

$\frac{29\pi}{24} + \frac{\pi}{2}$ is $\frac{41\pi}{24}$.

You stop there because one more would get you back to the beginning again.

Calculating cosines for two

You can use three different formulas to find the value for $\cos 2x$ — the double-angle of cosine — so your job is to choose which one fits into the problem best. The double-angle formula for cosine comes from the sum formula, just like the double-angle formula for sine. If you can't remember the double-angle formula but you can remember the sum formula, just simplify $\cos(2x)$, which is the same as $\cos(x + x)$. Because using the sum formula for cosine yields $\cos 2x = \cos^2 x - \sin^2 x$, you have two additional ways to express this by using Pythagorean identities (see Chapter 8):

You can replace $\sin^2 x$ with $(1 - \cos^2 x)$ and simplify.

You can replace $\cos^2 x$ with $(1 - \sin^2 x)$ and simplify.

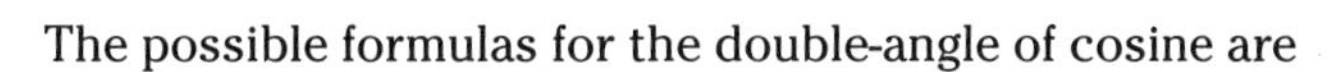

The possible formulas for the double-angle of cosine are

$\cos 2x = \cos^2 x - \sin^2 x$

$\cos 2x = 2\cos^2 x - 1$

$\cos 2x = 1 - 2\sin^2 x$

Looking at what you're given and what you're asked to find usually will lead you toward the right formula. And hey, if you don't pick the right one at first, you have two more to try!

Here's an example problem: If $\sec x = \frac{-15}{8}$. find the exact value of $\cos 2x$ if x is in quadrant II of the coordinate plane. Follow these steps to solve:

1. **Use the reciprocal identity (see Chapter 8) to change secant to cosine.**

   Because secant doesn't appear in any of the possible formula choices, you have to complete this step first. This means that $\cos x = \frac{-8}{15}$.

2. **Choose the appropriate double-angle formula.**

   Because you now know the cosine value, you should choose the second double-angle formula for this problem: $\cos 2x = 2\cos^2 x - 1$.

3. **Substitute the information you know into the formula.**

   You can plug cosine into the equation to get $\cos 2x = 2 \cdot (^{-8}/_{15})^2 - 1$.

4. **Simplify the formula to solve.**

   The exact value is $\cos 2x = 2 \cdot {}^{64}/_{225} - 1 = {}^{128}/_{225} - 1 = {}^{-97}/_{225}$.

Squaring your cares away

Yep, we've said it before and we'll say it again: When a square root appears inside a trig proof, you have to square both sides at some point to get where you need to go. For example, say you have to prove $2\sin^2 x - 1 = \sqrt{1 - \sin^2 2x}$. The square root on the right means that you should try squaring both sides:

1. **Square both sides.**

   You have $(2\sin^2 x - 1)^2 = \left(\sqrt{1 - \sin^2 2x}\right)^2$. This gives you $4\sin^4 x - 4\sin^2 x + 1 = 1 - \sin^2 2x$.

2. **Look for identities.**

   You can see a double angle on the right side: $\sin^2 2x = (\sin 2x)^2$. That gives $1 - (2\sin x\cos x)^2$, which is the same as $4\sin^4 x - 4\sin^2 x + 1 = 1 - 4\sin^2 x\cos^2 x$.

3. **Change all sines to cosines or vice versa.**

   Because you have more sines, change the $\cos^2 x$ by using the Pythagorean identity to get $4\sin^4 x - 4\sin^2 x + 1 = 1 - 4\sin^2 x(1 - \sin^2 x)$.

4. **Distribute the equation.**

   You end up with $4\sin^4 x - 4\sin x + 1 = 1 - 4\sin^2 x + 4\sin^4 x$.

   Using the commutative and associative properties of equality (from Chapter 1), you get $4\sin^4 x - 4\sin x + 1 = 4\sin^4 x - 4\sin x + 1$.

Having twice the fun with tangents

The double-angle formula for tangent isn't as exciting as the formulas for cosine (see the "Calculating cosines for two" section), because there aren't as many of them. That should make you happy, though, because you have less places to get confused. The double-angle formula for tangent is used less often than the double-angle for sine or cosine; however, you shouldn't overlook it just because it isn't as popular as its cooler counterparts! (Be advised, though, that instructors drill more heavily on sine and cosine.)

The double-angle formula for tangent is derived by simplifying $\tan(x + x)$ with the sum formula. However, the simplification process is much more complicated here because it involves fractions. So, we advise you to just memorize the formula.

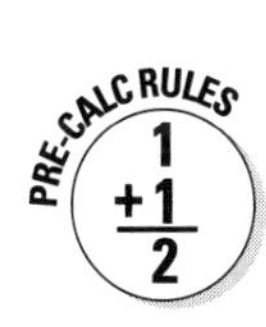

The double-angle identity for tangent is

$$\tan 2x = \frac{2\tan x}{1 - \tan^2 x}$$

When solving equations for tangent, remember that the period for the tangent function is π. This is important — especially when you must deal with more than one angle in an equation — because you usually need to find all the solutions on the interval $[0, 2\pi)$. When you're solving a double-angle equation, there will be twice as many solutions in that interval than for a single angle.

Follow these steps to find the solutions for $2\tan 2x + 2 = 0$ on the interval $[0, 2\pi)$:

1. **Isolate the trig function.**

   Subtract 2 from both sides to get $2\tan 2x = -2$. Divide both sides of the equation by 2 next: $\tan 2x = -1$.

2. **Solve for the double-angle by using inverse trig functions.**

   On the unit circle (see the Cheat Sheet), the tangent is negative in the second and fourth quadrants. Moreover, the tangent is –1 at $2x = \frac{3\pi}{4} + \pi \cdot k$ and $2x = \frac{7\pi}{4} + \pi \cdot k$, where k is an integer.

   Note: You have to add $\pi \cdot k$ to each solution to find *all* the solutions of the equation (see the earlier section "Finding the sine of a doubled angle").

3. **Isolate the variable.**

   Divide both sides of the equation by 2 to find x. (Remember that you have to divide both the angle and the period by 2.) This gives you $x = \frac{3\pi}{8} + \frac{\pi}{2} \cdot k$ and $x = \frac{7\pi}{8} + \frac{\pi}{2} \cdot k$.

4. **Find all the solutions on the required interval.**

   Adding $\frac{\pi}{2}$ to $\frac{\pi}{8}$ and $\frac{7\pi}{8}$ until you repeat yourself will give you all the solutions to the equation. Of course, first you must find a common denominator — in this case, 8:

   - $\frac{3\pi}{8} + \frac{\pi}{2} = \frac{3\pi}{8} + \frac{4\pi}{8} = \frac{7\pi}{8}$
   - $\frac{7\pi}{8} + \frac{\pi}{2} = \frac{7\pi}{8} + \frac{4\pi}{8} = \frac{11\pi}{8}$
   - $\frac{11\pi}{8} + \frac{\pi}{2} = \frac{11\pi}{8} + \frac{4\pi}{8} = \frac{15\pi}{8}$
   - $\frac{15\pi}{8} + \frac{\pi}{2} = \frac{9\pi}{8} + \frac{24\pi}{8} = \frac{19\pi}{8}$

   However, $\frac{19\pi}{8}$ is co-terminal with $\frac{3\pi}{8}$, so you're right back where you started. Now you've found all the solutions.

Taking Trig Functions of Common Angles Divided in Two

Some time ago, those pesky trigonometricians (we made up that word, by the way) found ways to calculate half of an angle with an identity. As you find out how to do with the sum and difference identities earlier in this chapter, you can use *half-angle identities* to evaluate a trig function of an angle that isn't on the unit circle by using one that is. For example, 15°, which isn't on the unit circle, is half of 30°, which is on the unit circle. Cutting special angles on the unit circle in half gives you a variety of new angles that can't be achieved by using the sum and difference formulas or the double-angle formulas. Although the half-angle formulas won't give you all the angles of the unit circle, they certainly get you closer than you were before.

The trick is knowing which type of identity serves your purpose best. Half-angle formulas are the better option when you need to find the trig values for any angle that can be expressed as half of another angle on the unit circle. For example, to evaluate a trig function of $\pi/8$, you can use the half-angle formula of $\pi/4$. Because there is no combination of sums or differences of special angles to get $\pi/8$, you know to use a half-angle formula.

You also can find the values of trig functions for angles like $\pi/16$ or $\pi/12$, each of which are exactly half of angles on the unit circle. Of course, these aren't the only types of angles the identities work for. You can continue to halve the trig-function value of half of any angle on the unit circle for the rest of your life (if you "have" nothing better to do). For example, 15° is half of 30°, and 7.5° is half of 15°.

The half-angle formulas for sine, cosine, and tangent are as follows:

$$\sin(\alpha/2) = \pm\sqrt{\frac{1-\cos\alpha}{2}}$$

$$\cos(\alpha/2) = \pm\sqrt{\frac{1+\cos\alpha}{2}}$$

$$\tan(\alpha/2) = \frac{1-\cos\alpha}{\sin\alpha} = \frac{\sin\alpha}{1+\cos\alpha}$$

In the half-angle formula for sine and cosine, notice that there is ± in front of each radical (square root). Whether your answer is positive or negative will depend on which quadrant the new angle (the half angle) is in. The half-angle formula for tangent does not have a ± sign in front, so the above does not apply to tangent.

For example, to find $\sin 165°$, follow these steps:

1. **Rewrite the trig function and the angle as half of a unit circle value.**

   First realize that 165° is half of 330°, so you can rewrite the sine function as $\sin(^{330}/_{2})$.

2. **Determine the sign of the trig function.**

   Because 165° is in quadrant II of the coordinate plane, its sine value should be positive.

3. **Substitute the angle value into the right identity.**

   The angle value 330° plugs in for α in the positive half-angle formula for sine. This gives you $\sin\left(\frac{330}{2}\right) = \sqrt{\frac{1-\cos 330}{2}}$.

4. **Replace $\cos\alpha$ with its actual value.**

   Use the unit circle to find the $\cos 330°$. Substituting that value into the equation gives you $\sqrt{\frac{1-\frac{\sqrt{3}}{2}}{2}}$.

5. **Simplify the half-angle formula to solve.**

   This is a three-step approach:

   a. Find the common denominator for the two fractions on top (including $^{2}/_{1}$) to get $\sqrt{\frac{\frac{2-\sqrt{3}}{2}}{2}}$.

   b. Use the rules for dividing fractions to get $\sqrt{\frac{2-\sqrt{3}}{4}}$.

   c. Finally, the square of the bottom simplifies to 2, and you end up with $\frac{\sqrt{2-\sqrt{3}}}{2}$.

A Glimpse of Calculus: Traveling from Products to Sums and Back

You've now reached the time-travel portion of the chapter, because all the information from here on comes into play mainly in calculus. In calc, you'll have to integrate functions, which is much easier to do when you're dealing with sums rather than products. The information in this section will help you prepare for the switch. Here, we show you how to express products as sums and how to transport from sums to products.

The information in this section is theoretical and applies specifically to calculus. We wish we had some really great real-world examples relating to this topic, but alas, it's just theory that you need to know to get ready for calc.

Expressing products as sums (or differences)

Integration of two things being multiplied together is extremely difficult, especially when you must deal with a mixture of trig functions. If you can break up a product into the sum of two different terms, each with its own trig function, doing the math will be much easier. But you don't have to worry about any of that right now. In pre-calculus, problems of this type usually say "express the product as a sum or difference." For the time being, you'll make the conversion from a product and that will be the end of the problem.

You have three product-to-sum formulas to digest: sine · cosine, cosine · cosine, and sine · sine. The following list breaks down these formulas:

- $\sin a \cos b = \frac{1}{2}\left[\sin(a+b) + \sin(a-b)\right]$

  If you're asked to find cosine · sine, you can rewrite it as sine · cosine because multiplication is commutative ($a \cdot b = b \cdot a$). That means that $\sin a \cdot \cos b = \cos b \cdot \sin a$. Notice that the order of the letters doesn't really matter because you just substitute the new ones into the formula (remember, though, that if you change the order on one side, you must change it on the other). For example, $\cos\frac{7\pi}{6} \cdot \sin\frac{\pi}{6} = \sin\frac{\pi}{6} \cdot \cos\frac{7\pi}{6}$.

  Suppose that you're asked to find $6\cos q \cdot \sin 2q$ as a sum. Rewrite this expression as $6\sin 2q \cdot \cos q$ and then plug what you know into the formula to get

  $$6\left(\frac{1}{2}\left[\sin(2q+q) + \sin(2q-q)\right]\right)$$

  $$= 6 \cdot \frac{1}{2}\left[\sin(3q) + \sin(q)\right]$$

  $$= 3\left[\sin(3q) + \sin(q)\right]$$

  That's all you can do.

- $\cos a \cdot \cos b = \frac{1}{2}\left[\cos(a+b) + \cos(a-b)\right]$

  For example, to express $\cos 6\theta \cdot \cos 3\theta$ as a sum, rewrite it as the following:

  $$\frac{1}{2}\left[\cos(6\theta + 3\theta) + \cos(6\theta - 3\theta)\right] = \frac{1}{2}\left[\cos 9\theta + \cos 3\theta\right]$$

  And that's it!

- $\sin a \cdot \sin b = \frac{1}{2}\left[\cos(a-b) - \cos(a+b)\right]$

  To express $\sin 5x \cdot \cos 4x$ as a sum, rewrite it as the following:

  $\frac{1}{2}\left[\cos(5x-4x) - \cos(5x+4x)\right] = \frac{1}{2}(\cos x - \cos 9x)$

Transporting from sums (or differences) to products

On the flip side of the previous section, you need to familiarize yourself with a set of formulas that change sums to products. These formulas won't be as common as some of the other formulas discussed in this chapter, but they may rear their not-so-pretty heads every once in a while. Sum to product formulas are useful to help you find the sum of two trig values that aren't on the unit circle. Of course, this works only if the sum or difference of the two angles ends up being an angle from the special triangles from Chapter 6.

Here are the sum/difference-to-product identities:

- $\sin x + \sin y = 2\sin\left(\frac{x+y}{2}\right)\cos\left(\frac{x-y}{2}\right)$
- $\sin x - \sin y = 2\cos\left(\frac{x+y}{2}\right)\sin\left(\frac{x-y}{2}\right)$
- $\cos x + \cos y = 2\cos\left(\frac{x+y}{2}\right)\cos\left(\frac{x-y}{2}\right)$
- $\cos x - \cos y = -2\sin\left(\frac{x+y}{2}\right)\sin\left(\frac{x-y}{2}\right)$

For example, say you're asked to find $\sin 105° + \sin 15°$ without a calculator. You're stuck, right? Well, not exactly. Because you're asked to find the sum of two trig functions whose angles aren't special angles, you can change this to a product by using the sum to product formulas. Follow these steps:

1. **Change the sum to a product.**

   Use $\sin x + \sin y = 2\sin\left(\frac{x+y}{2}\right)\cos\left(\frac{x-y}{2}\right)$ because you're asked to find the sum of two sine functions. This gives you $2\sin\frac{105+15}{2} \cdot \cos\frac{105-15}{2}$.

2. **Simplify the result.**

   Combining like terms and dividing gives you $2\sin 60 \cdot \cos 45$. Eureka! You've found it! Those are unit circle values, so continue to the next step.

3. Use the unit circle to simplify further.

$\sin 60^\circ = \frac{\sqrt{3}}{2}$, and $\cos 45^\circ = \frac{\sqrt{2}}{2}$. Substituting those values in, you get $2 \cdot \frac{\sqrt{3}}{2} \cdot \frac{\sqrt{2}}{2}$. Multiplying these values, you get $\frac{\sqrt{6}}{2}$.

Eliminating Exponents on Trig Functions with Power-Reducing Formulas

Power-reducing formulas allow you to get rid of exponents on trig functions so you can solve for an angle's measure. This will come in very handy when you get to calculus. (You're just going to have to trust us that you need to know this information!)

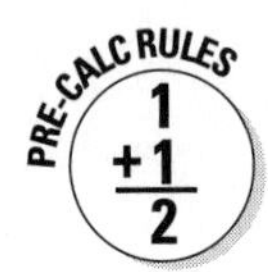

In the future, you'll definitely be asked to rewrite an expression using only the first power of a given trig function — either sine, cosine, or tangent — with the help of power-reducing formulas, because exponents can really complicate trig functions in calculus when you're attempting to integrate functions. In some cases, when the function is raised to the fourth power or higher, you may have to apply the power-reducing formulas more than once to eliminate all the exponents. You can use three power-reducing formulas to accomplish the elimination task:

$$\sin^2 u = \frac{1 - \cos 2u}{2}$$

$$\cos^2 u = \frac{1 + \cos 2u}{2}$$

$$\tan^2 u = \frac{1 - \cos 2u}{1 + \cos 2u}$$

For example, follow these steps to express $\sin^4 x$ without exponents:

1. **Apply the power-reducing formula to the trig function.**

   First, realize that $\sin^4 x = (\sin^2 x)^2$. Because the problem requires the reduction of $\sin^4 x$, you must apply the power-reducing formula twice. The first application gives you the following:

   $$(\sin^2 x)^2 = \left(\frac{1 - \cos 2x}{2}\right)^2 = \frac{(1 - \cos 2x)(1 - \cos 2x)}{4}$$

2. **FOIL the numerator.**

   You now have $(\sin^2 x)^2 = \frac{1 - 2\cos 2x + \cos^2 2x}{4} = \frac{1}{4}(1 - 2\cos 2x + \cos^2 2x)$.

3. **Apply the power-reducing formula again (if necessary).**

   Because the equation contains $\cos^2 2x$, you must apply the power-reducing formula for cosine.

   Because writing a power-reducing formula inside a power-reducing formula is very confusing, find out what $\cos^2 2x$ is by itself first and then plug it back in:

   $$\cos^2 2x = \frac{1+\cos 2(2x)}{2} = \frac{1+\cos 4x}{2}$$

   Plugging in $\frac{1+\cos 2(2x)}{2} = \frac{1+\cos 4x}{2}$ gives you

   $$\frac{1}{4}\left[1 - 2\cos 2x + \left(\frac{1+\cos 4x}{2}\right)\right].$$

4. **Simplify to get your result.**

   Factor out ½ from everything inside the brackets so that you don't have fractions both outside and inside the brackets. This gives you

   $\frac{1}{8}[2 - 4\cos 2x + 1 + \cos 4x]$. Combine like terms to get

   $\frac{1}{8}(3 - 4\cos 2x + \cos 4x)$.

Chapter 10

Solving Oblique Triangles with the Laws of Sines and Cosines

In This Chapter

- Mastering the Law of Sines
- Wielding the Law of Cosines
- Utilizing two methods to find the area of triangles

In order to *solve* a triangle, you need to find the measures of all three angles and the lengths of all three sides. Three of these pieces of information will be given to you as part of the math problem, so you need to find only the other three. Up until now throughout this book, we work in depth with right triangles. In Chapter 6, we help you find the lengths of missing sides by using the Pythagorean Theorem; find missing angles by using right-triangle trigonometry; and evaluate trig functions for specific angles. But what happens if you need to solve a triangle that *isn't* right?

You can connect any three points in a plane to form a triangle. Unfortunately, in the real world, these triangles won't always be right triangles. Finding missing angles and sides of *oblique triangles* can be more confusing, because there is no right angle. And without a right angle, there is no hypotenuse, which means the Pythagorean Theorem is useless. But don't worry; this chapter shows you the beaten path. The Law of Sines and the Law of Cosines are two methods that you can use to solve for missing parts of oblique triangles. The proofs of both laws are long and complicated, and you need not concern yourself with them. Instead, use these laws as formulas, where you can plug in information given to you and use algebra to solve for the missing pieces. Whether you use sines or cosines to solve the triangle, the types of information (sides or angles) given to you, and their location on the triangle are factors that help you decide which method is the best to use.

You may be wondering why we don't discuss the Law of Tangents in this chapter. The reason is simple: You can solve every oblique triangle with either the Law of Sines or the Law of Cosines, which are far less complicated than the Law of Tangents. Textbooks rarely refer to it, and teachers often steer clear.

The techniques we present here have tons of real-world applications, too. You can deal with everything from sailing a boat to putting out a forest fire by using triangles. For example, if two forest fire stations get a call for a fire, they can use the Law of Cosines to figure out which station is closest to the fire.

Before attempting to solve a triangle, *always* draw a picture that has the sides and angles clearly labeled. This helps you to visualize which pieces of information you still need. (Can't remember how to label a triangle? Think back to geometry.) Also, the information you're given to solve a triangle may not be the right combination required to use the Law of Sines formula (see the following section) — this happens when you're given all three sides of the triangle or two sides and the angle between them. You can use the Law of Sines in all other cases, so if you have the time, try to use the Law of Sines first. If the Law of Sines isn't an option, you'll know because solving for one of the variables will be impossible. When that happens, the Law of Cosines is there to save the day.

Whether you're using the Law of Sines or the Law of Cosines to solve for missing parts of a triangle, try not to do any of the calculations using your calculator until the very end. This will give you less rounding error in your final answers. For example, instead of evaluating the sines of all three angles and using the decimal approximations from the beginning, solve the equations and plug the final (extremely complicated) numeric expression into your calculator all at one time.

Solving a Triangle with the Law of Sines

You use the *Law of Sines* to find the missing parts of a triangle when you're given any three pieces of information involving at least one angle and at least one side directly opposite from it. There are three cases when this happens:

- **ASA** (Angle Side Angle): You're given two angles and the side in between them.
- **AAS** (Angle Angle Side): You're given two angles and a consecutive side.
- **SSA** (Side Side Angle): You're given two sides and a consecutive angle.

The formula for the Law of Sines is $\frac{a}{\sin A} = \frac{b}{\sin B} = \frac{c}{\sin C}$.

In order to solve for an unknown variable in the Law of Sines, you set two of the fractions equal to each other and use cross multiplication. When you're setting up for the cross multiplication, it doesn't matter which two parts you set equal to each other, but be careful not to have one equation with two unknown variables.

The bummer about problems that use the Law of Sines is that they take some time and careful work. Even though it's tempting to try to solve everything at once, take it one small step at a time. And don't overlook the obvious in order to blindly stick with the formula. If you're given two angles and one side, for instance, it's easy to find the third angle because all angles in a triangle must add up to 180°. Fill in the formula with what you know and get crackin'!

In the following sections, we show you how to solve a triangle in different situations, using the Law of Sines.

When you solve for an angle by using the Law of Sines, you have to assume that it's possible that a second set of solutions (or none at all) may exist. We cover that conundrum in this section as well. (In case you're really curious, that consideration applies only when you're working with a problem where you know two side measures and one angle measure of a triangle.)

When you know two angle measures

In this section, we take a look at the first two cases where you can use the Law of Sines to solve a triangle: Angle Side Angle (ASA) and Angle Angle Side (AAS). Whenever you're given two angles, find the third one immediately and work from there. In both of these cases, you'll find exactly one solution for the triangle under question.

A side sandwich: ASA

An *ASA* triangle means that you're given two angles and the side between them in a problem. For example, a problem could state that A = 32°, B = 47°, and $c = 21$, as in Figure 10-1. You also could be given ∠A, ∠C, and *b,* or ∠B, ∠C, and *a.* Figure 10-1 has all the given and unknown parts labeled for you.

Figure 10-1: A labeled ASA triangle.

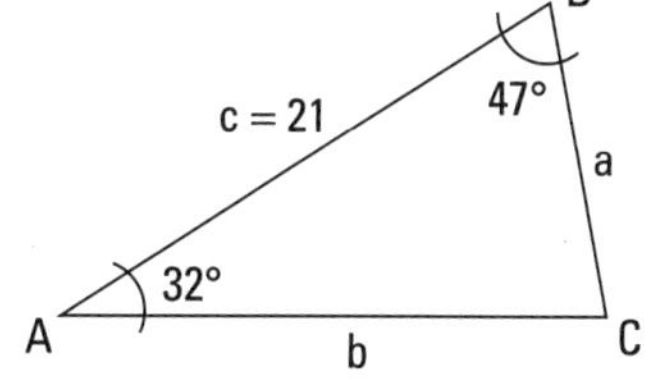

To find the missing information with the Law of Sines, follow these steps:

1. **Determine the measure of the third angle.**

   As a rule, ∠A + ∠B + ∠C = 180°. So, 32° + 47° + ∠C = 180°. 180° – 79° = ∠C = 101°.

2. **Set up the Law of Sines formula, filling in what you know.**

   You now have $\frac{a}{\sin 32°} = \frac{b}{\sin 47°} = \frac{21}{\sin 101°}$.

3. **Set two of the parts equal to each other and cross multiply.**

   We chose the first and third fractions, which looks like $\frac{a}{\sin 32°} = \frac{21}{\sin 101°}$. Cross multiplying, you have $a \cdot \sin 101° = 21 \cdot \sin 32°$.

4. **Find the decimal approximation of the missing side, using your calculator.**

   Because $\sin 101°$ is just a number, you can divide both sides of the equation by it to isolate the variable. So, $a = \frac{21 \cdot \sin 32°}{\sin 101°} \approx 11.34$.

5. **Repeat Steps 3 and 4 to solve for the other missing side.**

   Setting the second and third fractions equal to each other, you have $\frac{b}{\sin 47°} = \frac{21}{\sin 101°}$. This becomes $b \cdot \sin 101° = 21 \cdot \sin 47°$ when you cross multiply. Isolating the variable, you have $b = \frac{21 \cdot \sin 47°}{\sin 101°}$, or $b \approx 15.65$.

6. **State all the parts of the triangle as your final answer.**

   Some answers may be approximate, so make sure you maintain the proper signs:

   - A = 32° $a \approx 11.34$
   - B = 47° $b \approx 15.65$
   - C = 101° $c = 21$

Leaning toward the angle side: AAS

In many trig problems, you're given two angles and a side that isn't between them. This is called an *AAS* problem. For example, you could be given B = 68°, C = 29°, and $b = 15.2$, as shown by Figure 10-2. Notice that if you start at side b and move counterclockwise around the triangle, you come to ∠C and then ∠B. This is a good way to check if a triangle is an example of AAS.

After you find the third angle, an AAS problem just becomes a special case of ASA. Here are the steps to solve:

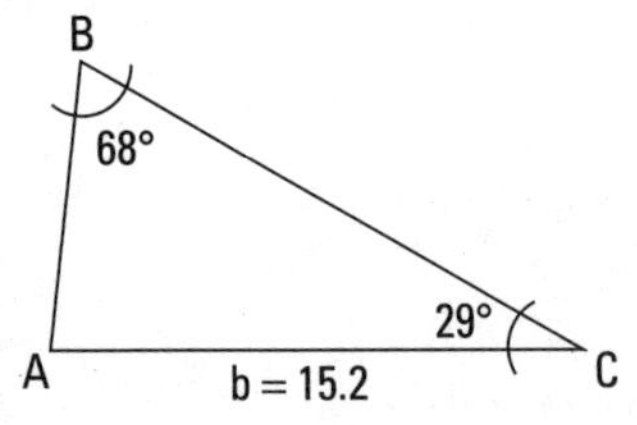

Figure 10-2: A labeled AAS triangle.

1. **Determine the measure of the third angle.**

   You can say that $68° + 29° + \angle A = 180°$. Then $\angle A = 83°$.

2. **Set up the Law of Sines formula, filling in what you know.**

   You now have $\frac{a}{\sin 83°} = \frac{15.2}{\sin 68°} = \frac{c}{\sin 29°}$.

3. **Set two of the parts equal to each other and then cross multiply.**

   We chose to use *a* and *b*, so $\frac{a}{\sin 83°} = \frac{15.2}{\sin 68°}$. To cross multiply, you have $a \cdot \sin 68° = 15.2 \cdot \sin 83°$.

4. **Solve for the missing side.**

   You divide by $\sin 68°$, so $a = \frac{15.2 \cdot \sin 83°}{\sin 68°}$, or $a \approx 16.27$.

5. **Repeat Steps 3 and 4 to solve for the other missing side.**

   Setting *b* and *c* equal to each other, you have $\frac{15.2}{\sin 68°} = \frac{c}{\sin 29°}$. When you cross multiply, you have $15.2 \cdot \sin 29° = c \cdot \sin 68°$. Divide by $\sin 68°$ to isolate the variable. You now have $c = \frac{15.2 \cdot \sin 29°}{\sin 68°}$, or $c \approx 7.95$.

6. **State all the parts of the triangle as your final answer.**

   Your final answer sets up as follows:

   - $A = 83°$ $a \approx 16.27$
   - $B = 68°$ $b = 15.2$
   - $C = 29°$ $c \approx 7.95$

When you know two consecutive side lengths (SSA)

In some trig problems you may be given two sides of a triangle and an angle that isn't between them. This is the classic case of *SSA*. In this scenario, you could have one solution, two solutions, or no solutions.

Wondering why? Recall from geometry that you can't prove that two triangles are congruent using SSA, because these conditions can build you two triangles that aren't the same. Figure 10-3 shows two triangles that fit SSA but aren't congruent.

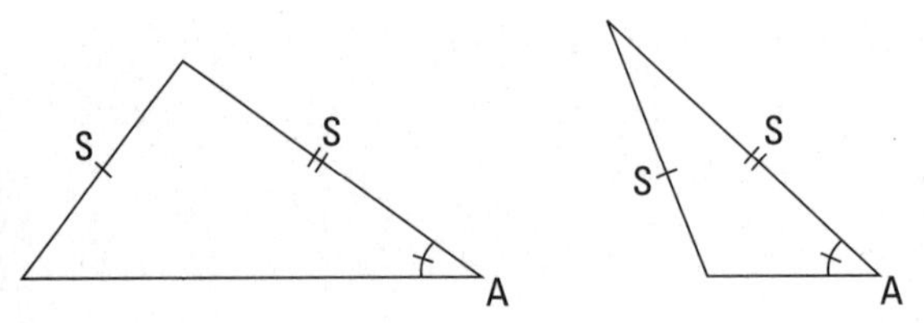

Figure 10-3: Non-congruent triangles that follow the SSA format.

If you begin with an angle and then continue around to draw the other two sides, you'll find that sometimes you can't make a triangle with those measurements. And sometimes, you can make two different triangles. Unfortunately, the latter means actually solving two different triangles.

Most SSA cases have only one solution. This is because if you use what you're given to sketch the triangle, most of the time you'll have only one way to draw it. When you're faced with an SSA problem, you may be tempted to figure out how many solutions you need to find before you start the solving process. Not so fast! In order to determine the number of possible solutions in an SSA problem, you should start solving first. You'll either arrive at one solution or find that there are no solutions (because you get an error message in your calculator). If you find one solution, you can look for the second set of solutions. If you get a negative angle in the second set, you'll know that the triangle only has one set of solutions.

In our opinion, the best approach is to always assume that you'll find two solutions, because remembering all the rules that determine the number of solutions probably will take up far too much time and energy (which is why we don't even go into them here; they're too complicated and too variable-heavy). If you treat every SSA problem as if it has two solutions until you gather enough information to prove otherwise, you'll be twice as likely to find all the appropriate solutions.

Preparing for the worst: Two solutions

It helps to gain some experience with solving a triangle that has more than one solution. The first set of solutions that you find in such a situation will always be an acute triangle. The second set of solutions will be an obtuse triangle. Remember to always look for two solutions for any problem.

For example, say you're given $a = 16$, $c = 20$, and $A = 48°$. Figure 10-4a shows you what the picture may look like. However, couldn't the triangle also look like Figure 10-4b? Both situations follow the constraints of the given information of the triangle. If you start by drawing your picture with the given angle, the side next to the angle has a length of 20, and the side across from the angle is 16 units long. There are two different ways this could happen. Angle C could be an acute angle or an obtuse angle; the given information isn't restrictive enough to tell you which one it is. Therefore, you have to find both sets of solutions.

Figure 10-4: Two possible representations of an SSA triangle.

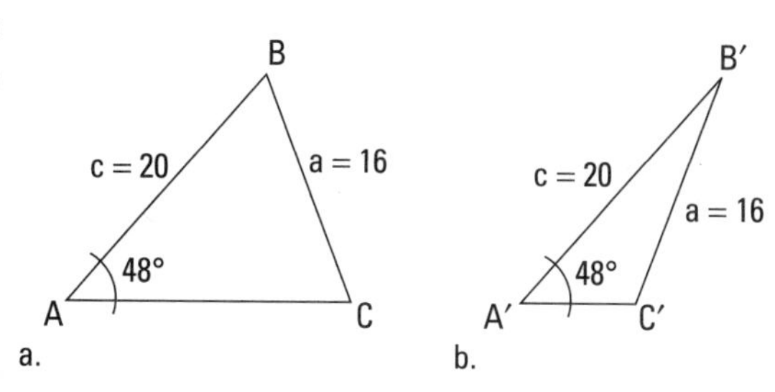

Solving this triangle by using steps similar to those described for both the ASA and AAS cases gives you the two possible solutions shown in Figure 10-4. Because you have two missing angles, you need to find one of them first, which is why the steps here are different than the other two cases:

1. **Fill in the Law of Sines formula with what you know.**

   This sets up as $\frac{16}{\sin 48°} = \frac{b}{\sin B} = \frac{20}{\sin C}$.

2. **Set two fractions equal to each other so that you have only one unknown.**

   If you decided to solve for C, you'd set the first and third fractions equal to each other so you'd have $\frac{16}{\sin 48°} = \frac{20}{\sin C}$.

3. **Cross multiply and isolate the sine function.**

   This gives you $20 \cdot \sin 48° = 16 \cdot \sin C$. To isolate the sine function, you divide by 16:

   $\sin C = \frac{20 \cdot \sin 48°}{16}$

4. **Take the inverse sine of both sides.**

   $\sin^{-1}(\sin C) = \sin^{-1}\left(\frac{20 \cdot \sin 48°}{16}\right)$. The right-hand side goes right into your handy calculator to give you $C \approx 68.27°$.

5. **Determine the third angle.**

   You know that $48° + 68.27 + \angle B = 180°$, so $B \approx 63.73°$.

6. **Plug the final angle back into the Law of Sines formula to find the third side.**

   This gives you $16 \cdot \sin 63.73° = b \cdot \sin 48°$.

   Finally, $b = \frac{16 \sin 63.73°}{\sin 48°} \approx 19.31$.

Of course, this isn't the only solution to the triangle. Refer to Step 4, where you solved for $\angle C$, and then look at Figure 10-5.

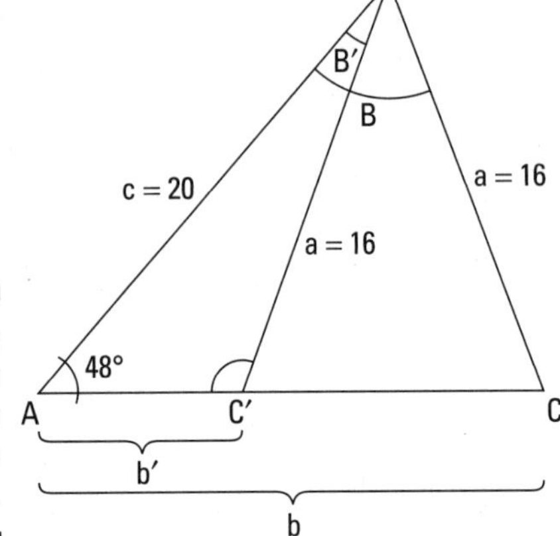

Figure 10-5: The two possible triangles overlapping.

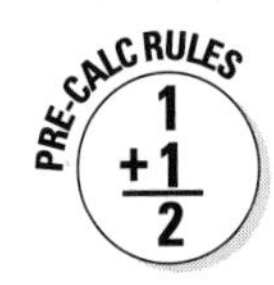

Triangle ABC is the solution that you solve for in the previous steps. Triangle AB'C' is the second set of solutions you must look for. A certain trig identity isn't used in solving or simplifying trig expressions because it isn't helpful for those, but it is useful for solving triangles. This identity says that $\sin(180° - \theta) = \sin\theta$.

In the case of the previous example, $\sin(68.27°) = \sin(180° - 68.27°) = \sin(111.73°)$. Notice that while $\sin 68.27° \approx 0.9319$, $\sin 111.73° \approx 0.9319$ as well. However, if you plug $\sin^{-1}(0.9319)$ into your calculator to solve for θ, 68.27° is the only solution you get. Subtracting this value from 180° gives you the other ambiguous solution for $\angle C$, which is usually denoted as C' so you don't confuse it with the first solution.

The following steps build on these actions so you can find all the solutions for this SSA problem:

1. **Use the trig identity $\sin(180° - \theta) = \sin\theta$ to find the second angle of the second triangle.**

   Because $C \approx 68.27°$, subtract this value from 180° to find that $C' \approx 111.73°$.

2. **Find the measure of the third angle.**

   If $A = 48°$ and $C' \approx 111.73°$, $B' \approx 20.27°$, because they all must add to 180°.

3. **Plug these angle values into the Law of Sines formula.**

   You now have $\frac{16}{\sin 48°} = \frac{b'}{\sin 20.27°} = \frac{20}{\sin 111.73°}$.

4. **Set two parts equal to each other in the formula.**

   You need to find b'. Set the first fraction equal to the second to get $\frac{16}{\sin 48^\circ} = \frac{b'}{\sin 20.27^\circ}$.

5. **Cross multiply to solve for the variable.**

   You set $b' \cdot \sin 48^\circ = 16 \cdot \sin 20.27^\circ$. Isolate b' to get $b' = \frac{16 \cdot \sin 20.27^\circ}{\sin 48^\circ}$, so $b' \approx 7.46$.

6. **List *all* the answers to the two triangles (see the previous numbered list).**

   Originally, you were given that $a = 16$, $c = 20$, and $A = 48^\circ$. The answers that you found are as follows:

   - **First triangle:** $B \approx 63.73^\circ$, $C \approx 68.27^\circ$, $b = 19.31$
   - **Second triangle:** $B' \approx 20.27^\circ$, $C' \approx 111.73^\circ$, $b' = 7.46$

Arriving at the ideal: One solution

If you don't get an error message in your calculator when attempting to solve a triangle, you know you can find at least one solution. But how do you know if you'll find only one? The answer is, you don't. Keep solving as if there are two solutions, and in the end, you will see there is only one.

For example, say you're asked to solve a triangle if $a = 19$, $b = 14$, and $A = 35^\circ$. Figure 10-6 shows what this triangle looks like.

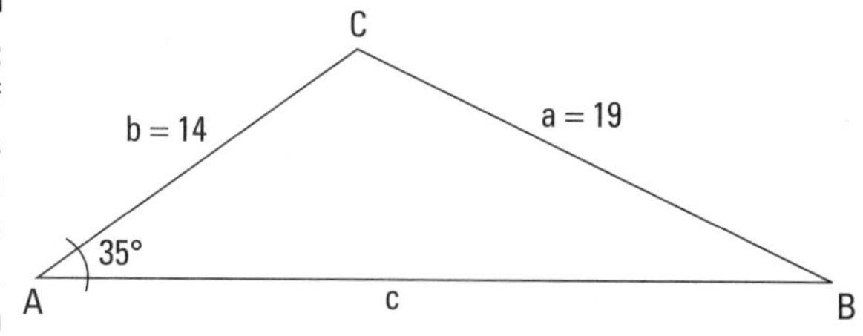

Figure 10-6: The setup of an SSA triangle with only one solution set.

Because you know only one of the angles of the triangle, you use the two given sides and the given angle to find one of the missing angles first. That will lead you to the third angle and then the third side. Follow these steps to solve this triangle:

1. **Fill in the Law of Sines formula with what you know.**

   You have $\frac{19}{\sin 35^\circ} = \frac{14}{\sin B} = \frac{c}{\sin C}$.

2. **Set two parts of the formula equal to each other.**

   Because you're given *a, b,* and A, solve for ∠B first. If you try to solve for side *c* or ∠C first, you'll have two unknown variables in your equation, which would create a dead end.

   Following this advice, you have $\frac{19}{\sin 35^\circ} = \frac{14}{\sin B}$.

3. **Cross multiply the equation.**

   You set $19 \cdot \sin B = 14 \cdot \sin 35^\circ$.

4. **Isolate the sine function.**

   This gives you $\sin B = \frac{14 \cdot \sin 35^\circ}{19}$.

5. **Take the inverse sine of both sides of the equation.**

   This sets up as $\cancel{\sin^{-1}(\sin} B) = \sin^{-1}\left(\frac{14 \cdot \sin 35^\circ}{19}\right)$, which simplifies to $B = \sin^{-1}\left(\frac{14 \cdot \sin 35^\circ}{19}\right) \approx 25^\circ$.

6. **Determine the measure of the third angle.**

   You know that $35^\circ + 25^\circ + \angle C = 180^\circ$, so $C \approx 120^\circ$.

7. **Set the two parts equal to each other so that you have only one unknown.**

   You now have $\frac{19}{\sin 35^\circ} = \frac{c}{\sin 120^\circ}$.

8. **Cross multiply and then isolate the variable to solve.**

   You start with $19 \cdot \sin 120^\circ = c \cdot \sin 35^\circ$, so $c = \frac{19 \cdot \sin 120^\circ}{\sin 35^\circ}$, or $c \approx 28.69$.

9. **Write out all six pieces of information devised from the formula.**

   Your answer sets up as follows:

   - $A = 35^\circ$ $a = 19$
   - $B = 25^\circ$ $b = 14$
   - $C = 120^\circ$ $c \approx 28.69$

10. **Look for a second set of solutions.**

    The first thing you did in this example was to find B. You see from Step 5 that B is approximately 25°. If the triangle has two solutions, the measure of B' is 180° – 25°, or 155°. Then, to find the measure of angle C', you start with $\angle A + \angle B' + \angle C' = 180^\circ$. This simplifies to $35^\circ + 155^\circ + \angle C' = 180^\circ$, or $\angle C' = -10^\circ$.

    Angles can't have negative measures, so this tells you that the triangle has only one solution. Don't you feel better knowing that you exhausted the possibility?

Kind of a pain: No solutions

If a problem gives you an angle and two consecutive sides of a triangle, you may find that the second side won't be long enough to reach the third side of the triangle. In this situation, no solution exists for the problem. However, you may not be able to tell this just by looking at the picture — you really need to solve the problem to know for sure. So, begin to solve the triangle just as you do in the previous sections.

For example, say $b = 19$, $A = 35°$, and $a = 10$. Figure 10-7 shows what the picture should look like.

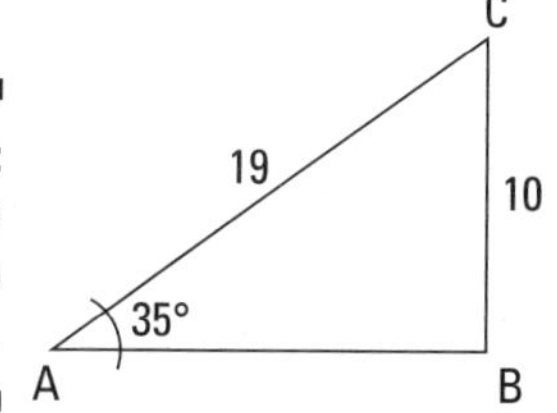

Figure 10-7: A triangle with no solution.

If you start solving this triangle by using the methods from previous setups, something very interesting will happen: Your calculator will give you an error message when you try to find the unknown angle. This is because the sine of an angle must be between –1 and 1. If you try to take the inverse sine of a number outside this interval, the value for the angle will be undefined (meaning it doesn't exist). The following steps illustrate this occurrence:

1. **Fill in the Law of Sines formula with what you know.**

   You have $\frac{10}{\sin 35°} = \frac{19}{\sin B} = \frac{c}{\sin C}$.

2. **Set two fractions equal and cross multiply.**

   You start with $\frac{10}{\sin 35°} = \frac{19}{\sin B}$ and end up with $19 \cdot \sin 35° = 10 \cdot \sin B$.

3. **Isolate the sine function.**

   This gives you $\sin B = \frac{19 \cdot \sin 35°}{10}$, or $\sin B \approx 1.09$.

4. **Take the inverse sine of both sides to find the missing angle.**

   Notice that you get an error message when you try to plug this into your calculator. This happens because $\sin B \approx 1.09$, but the sine of an angle can't be larger than 1 or less than –1. Therefore, the measurements given can't form a triangle, meaning that the problem has no solution.

Conquering a Triangle with the Law of Cosines

You use the *Law of Cosines* formulas to solve a triangle if you're given one of the following situations:

- Two sides and the included angle (SAS)
- All three sides of the triangle (SSS)

In order to solve for the angles of a triangle by using the Law of Cosines, you first need to find the lengths of all three sides. You have three formulas at your disposal to find missing sides, and three formulas to find missing angles. If a problem gives you all three sides to begin with, you're all set because you can manipulate the side formulas to come up with the angle formulas (we explain how in the first section that follows). If a problem gives you two sides and the angle between them, you first find the missing side and then find the missing angles.

To find a missing side of a triangle, use the following formulas, which comprise the Law of Cosines:

$$a^2 = b^2 + c^2 - 2bc \cdot \cos A$$

$$b^2 = a^2 + c^2 - 2ac \cdot \cos B$$

$$c^2 = a^2 + b^2 - 2ab \cdot \cos C$$

The side formulas are very similar to one another, with only the letters changed around. So, if you can remember just two of them, you can change the order to quickly find the other. The following sections put the Law of Cosines formulas into action to solve SSS and SAS triangles.

When you use the Law of Cosines to solve a triangle, you'll find only one set of solutions (one triangle), so don't waste any time looking for a second set. With this formula, you're solving SSS and SAS triangles, from triangle congruence postulates in geometry. You can use these congruence postulates because they lead to only one triangle every time. (For more on geometry rules, check out *Geometry For Dummies,* by Wendy Arnone, PhD [Wiley].)

SSS: Finding angles using only sides

Some textbooks provide three formulas students can use to solve for an angle by using the Law of Cosines. However, you don't have to memorize the three angle formulas to solve SSS problems. If you remember the formulas to find missing sides using the Law of Cosines (see the introduction to this section), you can use algebra to solve for an angle. Here's how, for example, to solve for angle A:

1. $a^2 = b^2 + c^2 - 2bc \cdot \cos A$ (beginning formula)
2. $a^2 - b^2 = c^2 - 2bc \cdot \cos A$ (subtract b^2 from both sides)
3. $a^2 - b^2 - c^2 = -2bc \cdot \cos A$ (subtract c^2 from both sides)
4. $\frac{a^2 - b^2 - c^2}{-2bc} = \cos A$ (divide both sides by $-2bc$)
5. $\frac{b^2 + c^2 - a^2}{2bc} = \cos A$ (distribute the negative and rearrange terms)
6. $\cos^{-1}\left(\frac{b^2 + c^2 - a^2}{2bc}\right) = A$ (take the inverse cosine of both sides)

The same process applies to finding angles B and C, so you end up with these formulas for finding angles:

$$A = \cos^{-1}\left(\frac{b^2 + c^2 - a^2}{2bc}\right)$$

$$B = \cos^{-1}\left(\frac{a^2 + c^2 - b^2}{2ac}\right)$$

$$C = \cos^{-1}\left(\frac{a^2 + b^2 - c^2}{2ab}\right)$$

Suppose you have three pieces of wood that measure all different lengths. One board is 12 feet long, another is 9 feet long, and the last is 4 feet long. If you want to build a sandbox using these pieces of wood, at what angles must you lay down all the pieces so that each side meets? If each piece of wood is one side of the triangular sandbox, you must use the Law of Cosines to solve for the three missing angles.

Let $a = 12$, $b = 4$, and $c = 9$; it doesn't matter which angle you find first. Follow these steps to solve:

1. **Decide which angle you want to solve for first and then plug the sides into the formula.**

   We solve for $\angle A$: $A = \cos^{-1}\left(\frac{4^2 + 9^2 - 12^2}{2 \cdot 4 \cdot 9}\right) = \cos^{-1}\left(\frac{-47}{72}\right) \approx 130.75°$.

2. **Solve for the other two angles.**

   $\angle B = \cos^{-1}\left(\frac{12^2 + 9^2 - 4^2}{2 \cdot 12 \cdot 9}\right) = \cos^{-1}\left(\frac{209}{216}\right) \approx 14.63°$.

   $\angle C = \cos^{-1}\left(\frac{12^2 + 4^2 - 9^2}{2 \cdot 12 \cdot 4}\right) = \cos^{-1}\left(\frac{79}{96}\right) \approx 34.62°$.

3. **Check your answers by adding the angles you found.**

   You find that 130.75° + 14.63° + 34.62° = 180°.

Picturing your solutions, the angle across from the 12-foot board (A) needs to be 130.75°; the angle across from the 4-foot board (B) needs to be 14.63°; and the angle across from the 9-foot board (C) needs to be 34.62°. See Figure 10-8.

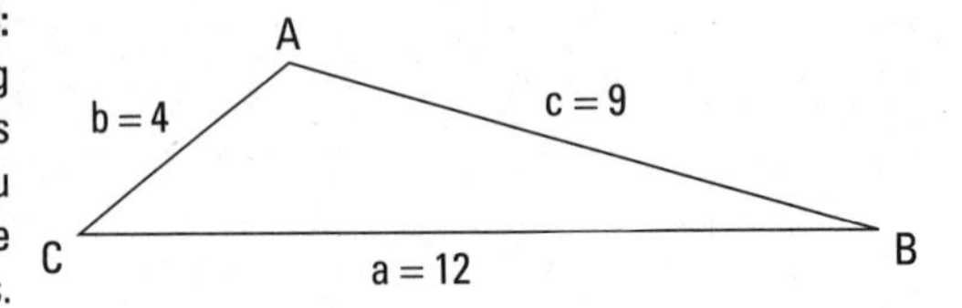

Figure 10-8: Determining angles when you know three side lengths.

SAS: Tagging the angle in the middle (and the two sides)

If a problem gives you the lengths of two sides of a triangle and the measure of the angle in between them, you use the Law of Cosines to find the other side (which you need to do first). When you have the third side, you can easily use all the side measures to calculate the remaining angle measures.

For example, if $a = 12$, $b = 23$, and C = 39°, you solve for side c first and then solve for ∠A and ∠B. Follow these simple steps:

1. **Sketch a picture of the triangle and clearly label all given sides and angles.**

   By drawing a picture, you can make sure that the Law of Cosines is the method you should use to solve the triangle. Figure 10-9 has all the parts labeled.

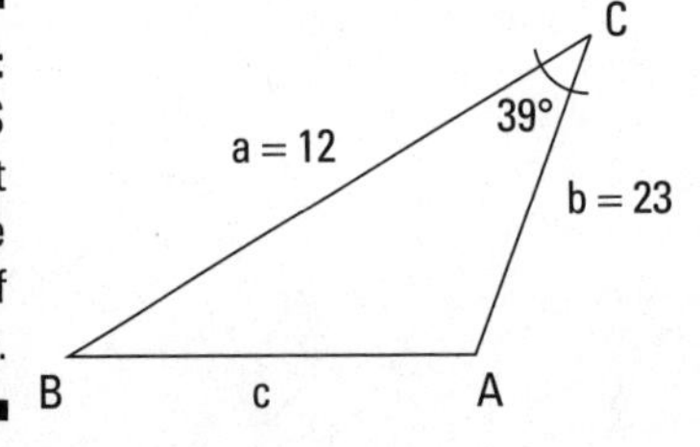

Figure 10-9: An SAS triangle that calls for the Law of Cosines.

2. **Decide which side formula you need to use first.**

   Because sides a and b are given, you use the following formula to find side c:

   $c^2 = a^2 + b^2 - 2ab \cdot \cos C$

3. **Plug the given information into the proper formula.**

   That gives you $c^2 = (12)^2 + (23)^2 - 2(12)(23) \cdot \cos(39°)$.

   If you have a graphing calculator, you can plug in this formula exactly as it's written and then skip directly to Step 6. If you don't have a graphing calculator (meaning you're using a scientific calculator), be very mindful of the order of operations.

   Following the order of operations when using the Law of Cosines is extremely important. If you try to type the pieces into your calculator all in one step, without the correct use of parentheses, your results probably will be incorrect. Be sure you're comfortable with your calculator. Some scientific calculators require that you type in the degrees before you hit the trig-function button. If you try to type in an inverse cosine without parentheses to separate the top and bottom of the fraction (see Step 7), your answer will be incorrect as well. The best method is to do all the squaring separately, combine like terms in the numerator and the denominator, divide the fraction, and then take the inverse cosine, as you'll see from this point on.

4. **Square each number and multiply by the cosine separately.**

   You end up with $c^2 = 144 + 529 - 428.985$.

5. **Combine all the numbers.**

   This gives you $c^2 = 244.015$.

6. **Square root both sides.**

   You now have c, which is $c \approx 15.6$.

7. **Find the missing angles.**

   Starting with $\angle A$, you find the following:

   - $A = \cos^{-1}\left(\dfrac{b^2 + c^2 - a^2}{2bc}\right)$
   - $A = \cos^{-1}\left[\dfrac{(23)^2 + (15.6)^2 - (12)^2}{2(23)(15.6)}\right]$
   - $A = \cos^{-1}\left(\dfrac{628.36}{717.6}\right) \approx 28.9°$

When using your graphing calculator, be sure to use parentheses to separate the numerator and denominator from each other. Put parentheses around the whole numerator *and* the whole denominator.

You can find the third angle quickly by subtracting the sum of the two known angles from 180°. However, if you're not pressed for time, we recommend that you use the Law of Cosines to find the third angle because that will allow you to check your answer.

Here's how you find ∠B:

- $B = \cos^{-1}\left(\frac{a^2 + c^2 - b^2}{2ac}\right)$
- $B = \cos^{-1}\left[\frac{(12)^2 + (15.6)^2 - (23)^2}{2(12)(15.6)}\right]$
- $B = \cos^{-1}\left(\frac{-141.64}{374.4}\right) \approx 112.2°$

8. **Check to make sure that all angles add to 180°.**

   $39° + 28.9° + 112.2° = 180.1°$

   Due to rounding error, sometimes the angles won't add to 180° exactly. For most instructors, an answer within a half a degree or so is considered acceptable.

Filling in the Triangle by Calculating Area

Geometry provides a nice formula to find the area of triangles: $A = \frac{1}{2} \cdot b \cdot h$. This formula comes in handy only when you know the base and the height of the triangle. But, in an oblique triangle, where is the base? And what is the height? You can use two different methods to find the area of an oblique triangle, depending on the information you're given.

Finding area with two sides and an included angle (for SAS scenarios)

Lucky Heron (see the next section) got a formula named after him, but the SAS guy to whom this section is dedicated remains nameless. Some theorems in math are named after people, but not all of them!

You use the following formula to find the area when you know two sides of a triangle and the angle between those sides (SAS):

$$\text{Area} = (½) \cdot a \cdot b \cdot \sin C$$

In the formula, C is the angle between sides a and b.

For example, when building the sandbox — as in the earlier section "SSS: Finding angles using only sides" — you know that $a = 12$ and $b = 4$, and you find using the Law of Cosines that C = 34.62. Now you can find the area:

$$\text{Area} = (½) \cdot 12 \cdot 4 \cdot \sin 34.62 \approx 13.64$$

Heron's Formula (for SSS scenarios)

You can find the area of a triangle when given only the lengths of all three sides (no angles, in other words) by using a formula called *Heron's Formula.* It says that

$$\text{Area} = \sqrt{s(s-a)(s-b)(s-c)}, \text{ where } s = \frac{1}{2}(a+b+c)$$

The variable s is called the *semiperimeter* — or half the perimeter.

For example, you can find the area of the sandbox (see the example from the previous section) without having to solve for any angles. When you know all three sides, you can use Heron's Formula. For a triangle with sides 4, 9, and 12, follow these steps:

1. **Calculate the semiperimeter (s).**

   Follow this simple calculation: $s = \frac{1}{2}(12+4+9) = 12.5$.

2. **Plug s, a, b, and c into Heron's Formula.**

   You find that the Area $= \sqrt{12.5(12.5-12)(12.5-4)(12.5-9)} = \sqrt{12.5(0.5)(8.5)(3.5)} \approx 13.64$.

You find that the area of the sandbox is the same with both formulas we present here. In the future the triangles won't be the same, but you'll still know when to use the SAS formula and when to use Heron's Formula.

Part III
Analytic Geometry and System Solving

"Today, we'll be working with equations that calculate curves and volumes. Can everyone see this?"

In this part . . .

The term *analytic geometry* usually means drawing out a shape or an equation to study it more deeply. This part starts with the set of complex numbers and how to perform operations with them and, yes, how to graph them. It then moves on to the new system of graphing known as polar coordinates. Conic sections finish up analytic geometry as we show you how to graph and examine the parts of circles, parabolas, ellipses, and hyperbolas.

Next it's on to solving systems of equations. We cover the old favorites of graphing, substitution, and elimination. Then we introduce the idea of a matrix and explain several ways to solve a system using matrices.

After that, we move on to sequences and series: how to find the term in any sequence as well as how to find the sum of certain types of series. Lastly, we bridge to calculus with the study of limits and continuity of functions.

Chapter 11

A New Plane of Thinking: Complex Numbers and Polar Coordinates

In This Chapter

- Pitting real versus imaginary
- Exploring the complex number system
- Plotting complex numbers on a plane
- Picturing polar coordinates

Complex numbers and polar coordinates are some of the most interesting but often neglected topics in a standard pre-calculus course. Both of these concepts have very basic explanations, and they can vastly simplify a difficult problem or even allow you to solve a problem that you couldn't solve before.

In previous math courses, you were told that you can't square root a negative number. If somewhere in your calculations you stumbled upon an answer that required you to take the square root of a negative number, you simply threw that answer out the window. That changes here in *Pre-Calculus For Dummies,* however. As you advance in math, you need complex numbers to explain natural phenomena that real numbers are incapable of. In fact, there are *entire* math courses dedicated to the study of complex numbers and their applications. We won't go into that kind of depth here; we simply want to introduce you to the topics gradually.

In this chapter, we cover the concepts of complex numbers and polar coordinates. We show you where they come from and how you use them (as well as graph them).

Understanding Real versus Imaginary (According to Mathematicians)

Algebra I and II introduce the real number system to you. Pre-calculus is here to expand your horizons by adding complex numbers to your repertoire, including imaginary numbers. *Complex numbers* are numbers that include both a real *and* an imaginary part; they're widely used for complex analysis, which theorizes functions by using complex numbers as variables (see the following section for more on this number system).

You may already be familiar with *imaginary numbers,* which occur when you take the square root of a negative number. Or perhaps you were taught to disregard negative roots whenever you found them. In case you fall in the latter category, here's a quick explanation. Myth: It's impossible to take the square root of a negative number. Although the square root of a negative number isn't a real number, it does exist! It operates in the form of an imaginary number.

Imaginary numbers have the form Bi, where B is a real number and i is an imaginary number — defined as $i = \sqrt{-1}$.

Luckily, you're already familiar with the x-y coordinate plane, which you use to graph functions (such as in Chapter 3). You also can use a complex coordinate plane to graph imaginary numbers. Although these two planes are constructed the same way — two axes perpendicular to one another at the origin — they're very different. For numbers graphed on the x-y plane, the coordinate pairs represent real numbers in the form of variables (x and y). You can show the relationships between these two variables as points on the plane. On the other hand, you use the complex plane simply to plot complex numbers. If you want to graph a real number, all you really need is a real number line. However, if you want to graph a complex number, you need an entire plane so that you can graph both the real and imaginary part.

Enter the *Gauss* or *Argand coordinate plane.* In this plane, pure real numbers in the form $a + 0i$ exist completely on the real axis (the horizontal axis), and pure imaginary numbers in the form $0 + Bi$ exist completely on the imaginary axis (the vertical axis). Figure 11-1a shows the graph of a real number, and Figure 11-1b shows that of an imaginary number.

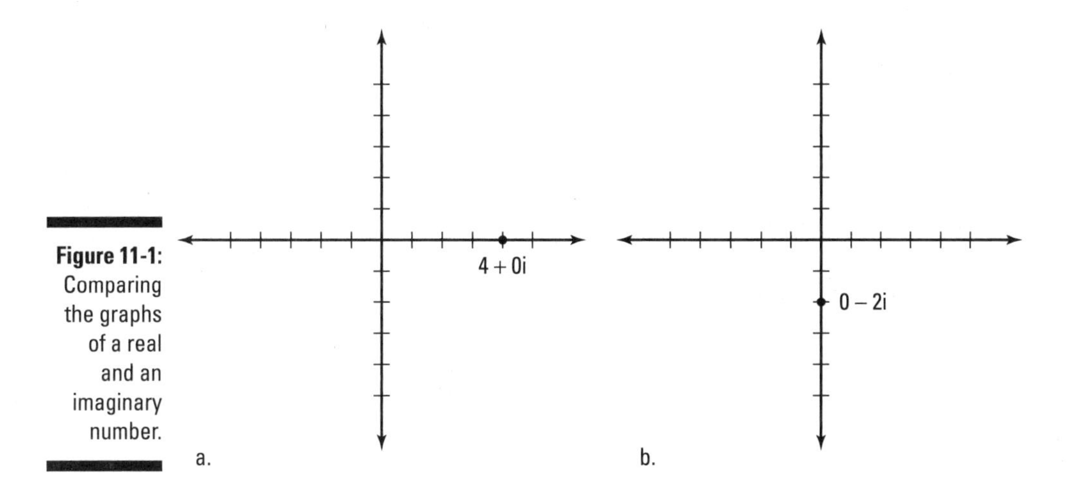

Figure 11-1: Comparing the graphs of a real and an imaginary number.

Combining Real and Imaginary: The Complex Number System

The *complex number system* is more complete than the real number system or the pure imaginary numbers in their separate forms. You can use this system to represent real numbers, imaginary numbers, and numbers that have both a real and an imaginary part. In fact, the complex number system is the most comprehensive set of numbers you deal with in pre-calculus.

Grasping the usefulness of complex numbers

You may be asking two important questions right now: When are complex numbers useful, and where will you stumble across them? Imaginary numbers are as important to the real world as real numbers, but their applications are hidden amongst some pretty heavy concepts, such as chaos theory and quantum mechanics. In addition, forms of mathematical art, called *fractals,* use complex numbers. Perhaps the most famous fractal is called the Mandelbrot Set. However, you don't have to worry about that stuff in pre-calculus. For you, imaginary numbers can be used as solutions to equations that don't have real solutions (such as quadratic equations).

For example, consider the quadratic equation $x^2 + x + 1 = 0$. This equation isn't factorable (see Chapter 4). Using the quadratic formula, you get the following:

$$\frac{-b \pm \sqrt{b^2 - 4ac}}{2a} = \frac{-1 \pm \sqrt{(-1)^2 - 4(1)(1)}}{2(1)} = \frac{-1 \pm \sqrt{-3}}{2} = \frac{-1 \pm \sqrt{3}\,i}{2}$$

Notice that the *discriminant* (the $b^2 - 4ac$ part) is a negative number, which you can't solve with only real numbers. When you first discovered the quadratic formula in algebra, you most likely used it to find real roots only. But because of complex numbers, you need not discard this solution. The previous answer is a legitimate complex solution, or a *complex root.* (Perhaps you remember encountering complex roots of quadratics in Algebra II. Check out *Algebra II For Dummies,* by Mary Jane Sterling [Wiley], for a refresher.)

Performing operations with complex numbers

Sometimes you come across situations where you need to operate on real and imaginary numbers together, so you want to write both numbers as complex numbers in order to be able to add, subtract, multiply, or divide them.

Consider the following three types of complex numbers:

- **A real number as a complex number:** $3 + 0i$

  Notice that the imaginary part of the expression is 0.

- **An imaginary number as a complex number:** $0 + 2i$

  Notice that the real portion of the expression is 0.

- **A complex number with both a real and an imaginary part:** $1 + 4i$

  This number can't be described as solely real or solely imaginary — hence the term *complex.*

You can manipulate complex numbers arithmetically just like real numbers to carry out operations. You just have to be careful to keep all the *i*'s straight. You can't combine real parts with imaginary parts by using addition or subtraction because they're not like terms, so it's important to keep them separate. Also, when multiplying complex numbers, the product of two imaginary numbers is a real number; the product of a real and an imaginary number is still imaginary; and the product of two real numbers is real. Many people get confused with this topic.

The following list presents the possible operations involving complex numbers:

- **To add and subtract complex numbers,** you simply combine like terms. For example, $(3 - 2i) - (2 - 6i) = 3 - 2i - 2 + 6i = 1 + 4i$.
- **To multiply when a complex number is involved,** you use one of three different methods, based on the situation:
  - **To multiply a complex number by a real number,** just distribute the real number to both the real and imaginary part of the complex number. For example, here's how you handle a *scalar* (a constant) multiplying a complex number in parentheses: $2(3 + 2i) = 6 + 4i$.
  - **To multiply a complex number by an imaginary number,** first realize that the real part of the complex number will become imaginary and that the imaginary part will become real. When you express your final answer, however, you still express the real part first followed by the imaginary part, in the form $A + Bi$.

    For example, here's how $2i$ multiplies into the same parenthetical number: $2i(3 + 2i) = 6i + 4i^2$. ***Note:*** You define i as $\sqrt{-1}$, so that $i^2 = -1$! This means you really have $6i + 4(-1)$, so your answer becomes $-4 + 6i$.
  - **To multiply two complex numbers,** simply follow the FOIL process (see Chapter 4). For example, $(3 - 2i)(9 + 4i) = 27 + 12i - 18i - 8i^2$, which is the same as $27 - 6i - 8(-1)$, or $35 - 6i$.
- **To divide complex numbers,** you multiply both the numerator and the denominator by the conjugate of the denominator, FOIL the numerator and denominator separately, and then combine like terms. This process is necessary because the imaginary part in the denominator is really a square root (of –1, remember?), and the denominator of the fraction must not contain an imaginary part.

  For example, say you're asked to divide $\frac{1+2i}{3-4i}$. The complex conjugate of $3 - 4i$ is $3 + 4i$. Follow these steps to finish the problem:

  1. **Multiply the numerator and the denominator by the conjugate.**

     You now have $\frac{1+2i}{3-4i} \cdot \frac{3+4i}{3+4i}$.
  2. **FOIL the numerator.**

     You go with $(1 + 2i)(3 + 4i) = 3 + 4i + 6i + 8i^2$, which simplifies to $(3 - 8) + (4i + 6i)$, or $-5 + 10i$.
  3. **FOIL the denominator.**

     You have $(3 - 4i)(3 + 4i)$, which FOILs to $9 + 12i - 12i - 16i^2$. Because $i^2 = -1$ and $12i - 12i = 0$, you're left with the real number $9 + 16 = 25$ in the denominator (which is why you multiply by $3 + 4i$ in the first place).

4. **Rewrite the numerator and the denominator.**

   You get $\frac{-5+10i}{25}$. This still isn't in the right form for a complex number, however.

5. **Separate and divide both parts by the constant denominator.**

   This gives you $\frac{-5}{25}+\frac{10i}{25}$, or $\frac{-1}{5}+\frac{2}{5}i$. Notice that the answer is finally in the form A + B*i*.

Graphing Complex Numbers

To graph complex numbers, you simply combine the ideas of the real-number coordinate plane and the Gauss or Argand coordinate plane (which we explain in the "Understanding Real versus Imaginary (According to Mathematicians)" section earlier in this chapter) to create the complex coordinate plane. In other words, you take the real portion of the complex number (A) to represent the *x* coordinate, and you take the imaginary portion (B) to represent the *y* coordinate.

Although you graph complex numbers much like any point in the real-number coordinate plane, complex numbers aren't real! The *x* coordinate is the only real part of a complex number, so you call the *x*-axis the *real axis* and the *y*-axis the *imaginary axis* when graphing in the complex coordinate plane.

Graphing complex numbers gives you a way to visualize them, but a graphed complex number doesn't have the same physical significance as a real-number coordinate pair. For an (*x*, *y*) coordinate, the position of the point on the plane is represented by two numbers. In the complex plane, the value of a single complex number is represented by the position of the point. This means that each complex number A + B*i* can be expressed as the ordered pair (A, B).

You can see several examples of graphed complex numbers in Figure 11-2:

Point A: The real part is 2 and the imaginary part is 3, so the complex coordinate is (2, 3) where 2 is on the real (or horizontal) axis and 3 is on the imaginary (or vertical) axis. This point is 2 + 3*i*.

Point B: The real part is –1 and the imaginary part is –4; you can draw the point on the complex plane as (–1, –4). This point is –1 – 4*i*.

Point C: The real part is ½ and the imaginary part is –3, so the complex coordinate is (½, –3). This point is ½ – 3*i*.

Point D: The real part is –2 and the imaginary part is 1, which means that on the complex plane, the point is (–2, 1). This coordinate is –2 + *i*.

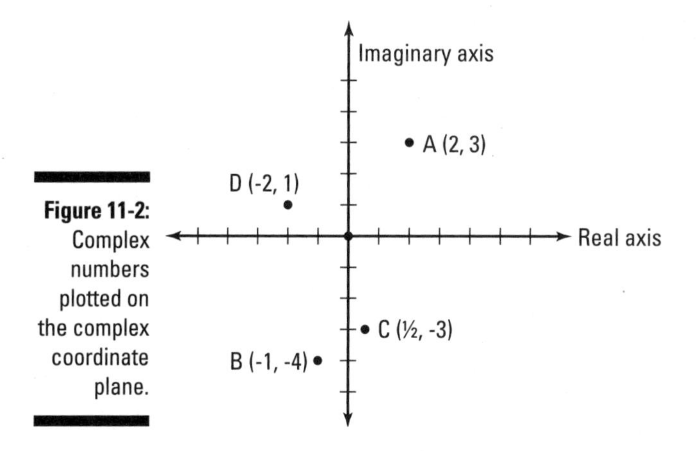

Figure 11-2: Complex numbers plotted on the complex coordinate plane.

Plotting Around a Pole: Polar Coordinates

Polar coordinates are an extremely useful addition to your mathematics toolkit because they allow you to solve problems that would be extremely ugly if you were to rely on standard x and y coordinates (for example, problems when the relationship between two quantities is most easily described in terms of the angle and distance between them, such as navigation or antenna signals). Instead of relying on the x- and y-axes as reference points, polar coordinates use only the positive x-axis (the line starting at the origin and continuing in the positive horizontal direction forever). From this line, you measure an angle (which you call theta, or θ) and a length (or radius) along the terminal side of the angle (which you call r). These coordinates replace x and y coordinates.

In polar coordinates, you always write the ordered pair as (r, θ). For instance, a polar coordinate could be $(5, \frac{\pi}{6})$ or $(-3, \pi)$.

In the following sections, we show you how to graph points in polar coordinates and how to graph equations as well. You also discover how to change back and forth between Cartesian coordinates and polar coordinates.

Wrapping your brain around the polar coordinate plane

In order to fully grasp how to plot polar coordinates, you need to see what a polar coordinate plane looks like. In Figure 11-3, you can see that the plane is no longer a grid of rectangular coordinates; instead, it's a series of concentric circles around a central point, called the *pole*. The plane appears this way because the polar coordinates are a given radius and a given angle in standard position from the pole. Each circle represents one radius unit and each line represents the special angles from the unit circle (to make finding the angles easier; see Chapter 6).

Although θ and r may seem strange as plotting points at first, they're really no more or less strange or useful than x and y. In fact, when you consider a sphere such as Earth, it becomes clear that polar coordinates make describing points on, above, or below its surface much more straightforward. Because the shape of Earth is spherical, using a coordinate plane that's similar in shape (round) makes representing things in Earth's atmosphere easier.

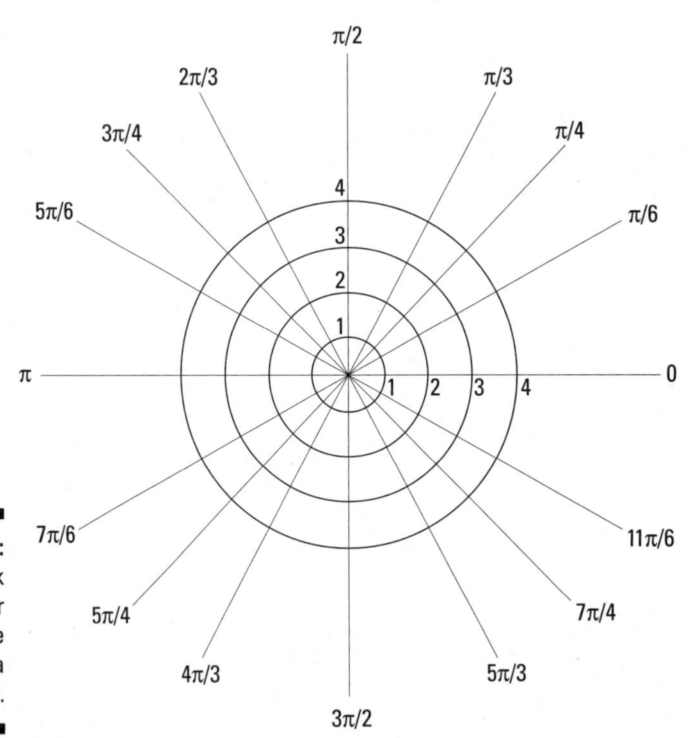

Figure 11-3: A blank polar coordinate plane (not a dartboard).

Because you write all points on the plane as (r, θ), in order to graph a point on the polar plane, we recommend that you find θ first and then locate r on that line. This allows you to narrow the location of a point to somewhere on one of the lines representing the angles. From there, you can simply count out from the pole the radial distance. If you go the other way, you could find yourself in a pickle when the problems get more complicated.

For example, to plot point E at $(2, \frac{\pi}{3})$ — which has a positive value for both the radius and the angle — you simply move from the pole counterclockwise until you reach the appropriate angle (θ). You start there in the following list:

1. **Locate the angle on the polar coordinate plane.**

   Refer to Figure 11-3 to find the angle: $\theta = \frac{\pi}{3}$.

2. **Determine where the radius intersects the angle.**

   Because the radius is 2 ($r = 2$), you start at the pole and move out 2 spots in the direction of the angle.

3. **Plot the given point.**

   At the intersection of the radius and the angle on the polar coordinate plane, plot a dot and call it a day! Figure 11-4 shows point E on the plane.

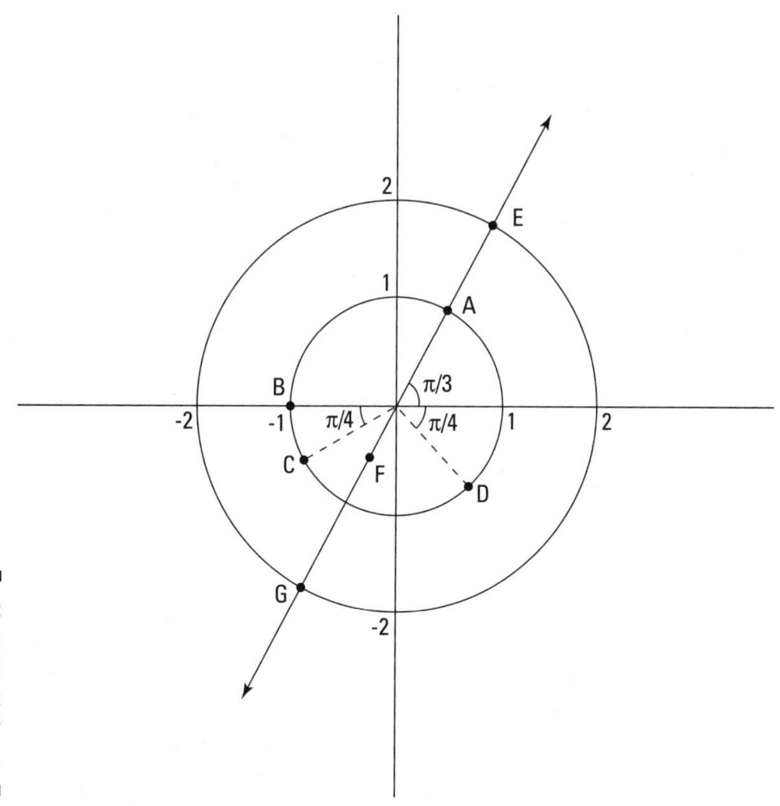

Figure 11-4: Visualizing simple and complex polar coordinates.

Polar coordinate pairs can have positive angles or negative angles for values of θ. In addition, they can have positive and negative radii. This is a new concept, because you've always heard that a radius must be positive! When graphing polar coordinates, though, the radius can be negative, which means that you move in the *opposite* direction of the angle from the pole.

Because polar coordinates are based on angles, unlike Cartesian coordinates, polar coordinates have many different ordered pairs. Because infinitely many values of θ have the same angle in standard position (see Chapter 6), there are infinitely many coordinate pairs that describe the same point. Also, a positive and a negative co-terminal angle can describe the same point for the same radius, and because the radius can be both positive or negative, there are many ways to express the point with polar coordinates. Usually, providing four different representations of the same point is sufficient.

Graphing polar coordinates with negative values

Simple polar coordinates are points where both the radius and the angle are positive. You work on graphing these in the previous section. But you also must prepare yourself for when teachers spice it up a tiny bit with *complicated polar coordinates* — points with negative angles and/or radii. The following list shows you how to plot in three situations — when the angle is negative, when the radius is negative, and when both are negative:

- **When the angle is negative:** Negative angles move in a clockwise direction (see Chapter 6 for more on these angles). Check out Figure 11-4 to see an example point, D. To locate the polar coordinate point D at $(1, -\frac{\pi}{4})$, first locate the angle $-\frac{\pi}{4}$ and then find the location of the radius, 1, on that line.
- **When the radius is negative:** When graphing a polar coordinate with a negative radius (essentially the x value), you move from the pole in the direction opposite the given positive angle (on the same line as the given angle but in the direction opposite to the angle from the pole). For example, check out point F at $(-\frac{1}{2}, \frac{\pi}{3})$ in Figure 11-4.

  Some teachers prefer to teach their students to move right along the x- (polar) axis for positive numbers (radii) and left for negative. Then you do the rotation for the angle in a positive direction. Try it; you'll get to the same spot.

  For example, take a look point F $(-\frac{1}{2}, \frac{\pi}{3})$ in Figure 11-4. Because the radius is negative, move along the left x-axis $\frac{1}{2}$ of a unit. Then rotate the angle in the positive direction (counterclockwise) $\frac{\pi}{3}$ radians. You should arrive at your destination, point F.

- **When both the angle and radius are negative:** To express a polar coordinate with a negative radius and a negative angle, locate the terminal side of the negative angle first and then move in the opposite direction to locate the radius. For example, point G in Figure 11-4 has these characteristics at $(-2, -\frac{5\pi}{3})$.

These representations should give you the location of the same point:

- Positive radius, positive angle
- Positive radius, negative angle
- Negative radius, positive angle
- Negative radius, negative angle

For example, for point E in Figure 11-4 $(2, \frac{\pi}{3})$, the three other coordinates could be

$(2, -\frac{5\pi}{3})$

$(-2, \frac{4\pi}{3})$

$(-2, -\frac{2\pi}{3})$

This is helpful with polar graphing because you can change the coordinate of any point you're given into polar coordinates that are easy to deal with (such as positive radius, positive angle).

Changing to and from polar coordinates

You can use both polar coordinates and *x-y* coordinates at any time to describe the same location on the coordinate plane. Sometimes it will be more convenient to use one form, and for this reason we teach you how to navigate between the two. Cartesian coordinates are much better suited for graphs of straight lines or simple curves. Polar coordinates can yield you a variety of pretty, very complex graphs that you couldn't plot with Cartesian coordinates.

When changing to and from polar coordinates, it often will make your work easier if you have all your angle measures in radians. You can make the change by using the conversion factor $180° = \pi$ radians. You may choose, however, to leave your angle measures in degrees, which is fine so long as your calculator is in the right mode.

Devising the changing equations

Examine the point in Figure 11-5, which illustrates a point mapped out in both (x, y) and (r, θ) coordinates, allowing you to see the relationship between them.

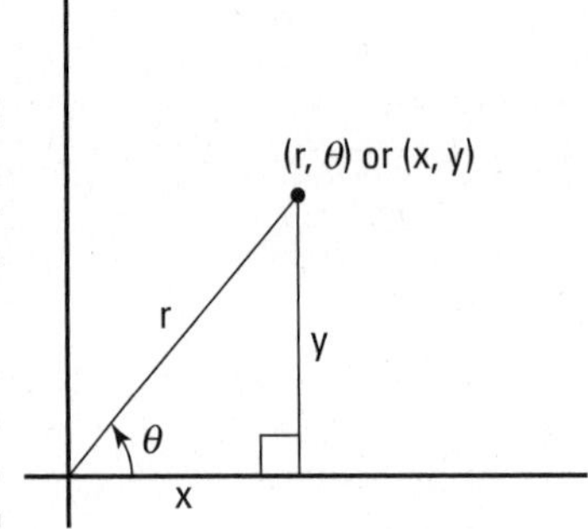

Figure 11-5: A polar and *x-y* coordinate mapped in the same plane.

What exactly is the geometric relationship between r, θ, x, and y? Look at how they're labeled on the graph — all parts of the same triangle!

Using right-triangle trigonometry (see Chapter 6), you know that $\sin\theta = \frac{y}{r}$ and $\cos\theta = \frac{x}{r}$. These simplify into two very important expressions for x and y in terms of r and θ:

$$y = r\sin\theta$$

$$x = r\cos\theta$$

Furthermore, you can use the Pythagorean Theorem in the right triangle to find the radius of the triangle if given x and y:

$$x^2 + y^2 = r^2$$

One final equation allows you to find the angle θ; it derives from the tangent of the angle:

$$\tan\theta = y/x$$

So, if you solve this equation for θ, you get the following expression:

$$\theta = \tan^{-1}(y/x)$$

With respect to the final equation, keep in mind that your calculator will always return a value of tangent that puts θ in the first or fourth quadrant. You need to look at your x and y coordinates and decide whether this is actually the case for the problem at hand. Your calculator doesn't look for tangent possibilities in the second and third quadrants, but that doesn't mean you don't have to!

As with degrees (see the section "Wrapping your brain around the polar coordinate plane"), you can add or subtract 2π to any angle to get a co-terminal angle so you have more than one way to name every point in polar coordinates. In fact, there are infinite ways of naming the same point. For instance, $(2, \pi/3)$, $(2, -5\pi/3)$, $(-2, 4\pi/3)$, $(2, -2\pi/3)$, and $(2, 7\pi/3)$ are several ways of naming the same point.

Putting the equations into action

Together, the four equations for r, θ, x, and y allow you to change x-y coordinates into polar (r, θ) coordinates and back again anytime. For example, to change the polar coordinate $(2, \pi/6)$ to a rectangular coordinate, follow these steps:

1. **Find the x value.**

   If $x = r\cos\theta$, substitute what you know — $r = 2$ and $\theta = \pi/6$ — to get $x = 2\cos\pi/6$. Use the unit circle (see the Cheat Sheet) to get $x = 2\frac{\sqrt{3}}{2}$, which means that $x = \sqrt{3}$.

2. **Find the y value.**

   If $y = r\sin\theta$, substitute what you know to get $y = 2\sin\pi/6 = 2 \cdot 1/2$, which means that $y = 1$.

3. **Express the values from Steps 1 and 2 as a coordinate point.**

   You find that $(\sqrt{3}, 1)$ is the answer as a point.

Time for an example in reverse. Given the point $(-4, -4)$, find the equivalent polar coordinate:

1. **Plot the (x, y) point first.**

   Figure 11-6 shows the location of the point in quadrant III.

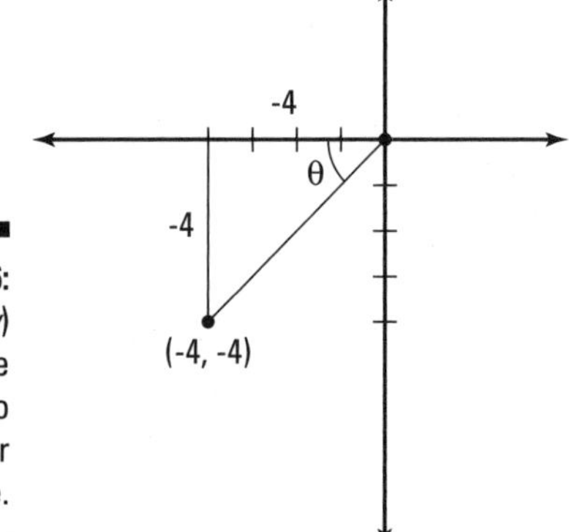

Figure 11-6: An (x, y) coordinate changed to a polar coordinate.

2. **Find the r value.**

   For this step, you use the Pythagorean Theorem for polar coordinates: $x^2 + y^2 = r^2$. Plug in what you know — $x = -4$ and $y = -4$ — to get $(-4)^2 + (-4)^2 = r^2$, or $r = 4\sqrt{2}$.

3. **Find the value of θ.**

   Use the tangent ratio for polar coordinates: $\tan\theta = \frac{-4}{-4}$, or $\tan\theta = 1$. The reference angle for this value is $\theta' = \frac{\pi}{4}$ (see Chapter 6). You know from Figure 11-6 that the point is in the third quadrant, so $\theta = \frac{5\pi}{4}$.

4. **Express the values of Steps 2 and 3 as a polar coordinate.**

   You can say that $(-4, -4) = (4\sqrt{2}, \frac{5\pi}{4})$.

Picturing polar equations

Polar coordinates allow for the graphing of some strange and remarkable equations. We give some of the most common equations and their conditions and shapes in Table 11-1.

Table 11-1 The Graphs of Common Polar Functions

Name	*Equation*	*Condition*	*Shape*
Archimedes Spiral	$r = a\theta$	$\theta \leq 0$	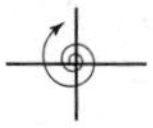
		$\theta \geq 0$	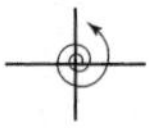
Cardiod	$r = a(1 \pm \sin\theta)$	$+\sin\theta$	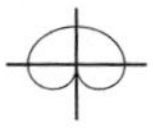
		$-\sin\theta$	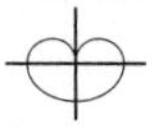
	$r = a(1 \pm \cos\theta)$	$+\cos\theta$	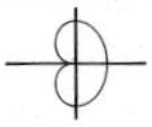
		$-\cos\theta$	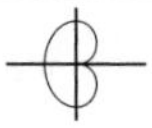
Circle	$r = a\sin\theta$		
	$r = a\cos\theta$		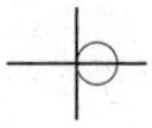

Name	*Equation*	*Condition*	*Shape*
Lemniscate	$r=\sqrt{a^2\sin 2\theta}$	$a^2\sin 2\theta > 0$	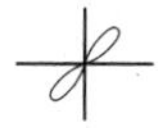
	$r=\sqrt{a^2\cos 2\theta}$	$a^2\cos 2\theta > 0$	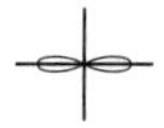
Rose	$r = a\sin b\theta$	b is odd $\rightarrow$ b petals	
		b is even $\rightarrow$ $2b$ petals	
	$r = a\cos b\theta$	b is odd $\rightarrow$ b petals	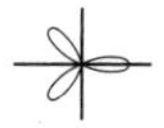
		b is even $\rightarrow$ $2b$ petals	
Limaçon	$r = a \pm b\sin\theta$	$a < b$	
		$b < a < 2b$	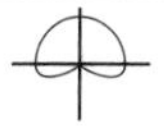
		$a \geq 2b$	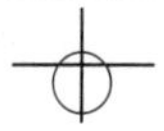
	$r = a \pm b\cos\theta$	$a < b$	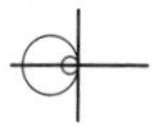
		$b < a < 2b$	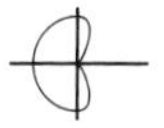
		$a \geq 2b$	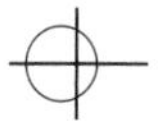

A graphing calculator is fully capable of plotting all these strange, new polar functions. However, it will spit out nothing but nonsense if you don't change two things before entering a function into your graphing utility:

1. **Make sure your calculator is in radians mode, not degrees.**
2. **Change your graphing mode to "polar."**

After you take these steps, your calculator's graphing menu should change. Instead of displaying "y =", it will display "r =". It also should give you a fairly simple way of entering θ rather than x as your variable.

Make sure that you enter a maximum and minimum θ value for the calculator to graph to (found in your graph's "window" settings), because the standard window usually is from 0 to 2π. This consideration is especially important for working polar functions like Archimedan spirals, which you want to follow out to large values of θ.

Note: Although the polar functions from Table 11-1 are much different from the types of functions you've seen before, and the graphing menu on your calculator has changed, you're still graphing in the Cartesian plane.

Chapter 12

Cutting It Up with Conics

In This Chapter

- Surveying the four conic sections
- Spinning around with circles
- Dissecting the parts and graphs of parabolas
- Exploring the ellipse
- Boxing around with hyperbolas
- Writing and graphing conics in two distinct forms

Astronomers have been looking out into space for a very long time — longer than you've been staring at the ceiling during class. Some of the things that are happening out there are mysteries; others have shown their true colors to curious observers. One phenomenon that astronomers have discovered and proven is the movement of bodies in space. They know that the paths of objects moving in space are shaped like one of four *conic sections* (shapes made from cones): the circle, the parabola, the ellipse, or the hyperbola. Conic sections have evolved into popular ways to describe motion, light, and other natural occurrences in the physical world.

In astronomical terms, an ellipse, for example, describes the path of a planet around the sun. A comet may travel so close to a planet's gravity that its path is affected and it gets swung back out into the galaxy. If you were to attach a gigantic pen to the comet, its path would trace out one huge parabola. The movement of objects as they are affected by gravity can often be described using conic sections. For example, you can describe the movement of a ball being thrown up into the air using a conic section. As you can see, the conic sections you study in pre-calculus have plenty of applications — especially for what may be the newest study of outer space . . . rocket science!

Conic sections are so named because they're made from two right circular cones (imagine two sugar cones from your favorite ice cream store). Basically, you see two right circular cones, touching pointy end to pointy end (the pointy end of a cone is called the *element*). The conic sections are formed by the intersection of a plane and the ice cream cones. When you slice through the cones with a plane, the intersection of that plane with the cones yields a variety of different curves. The plane is completely arbitrary, and where it cuts the conic and at what angle is what gives you all the different conic sections that we discuss in this chapter.

In this chapter, we break down each of the conic sections, front to back. We discuss the similarities and the differences between the four conic sections and their applications in pre-calculus. We also graph each section and look at its properties. Conic sections are the final frontier when it comes to graphing in mathematics, so sit back, relax, and enjoy the ride!

Cone to Cone: Identifying the Four Conic Sections

Each conic section has its own standard form of an equation with x and y variables that you can graph on the coordinate plane. You can write the equation of a conic section if you are given key points on the graph, or you can graph the conic section from the equation. There are various ways that you can alter the shape of each of these graphs, but the general graph shapes still remain true to the type of curve that they are.

It is important to be able to identify which conic section is which by just the equation because sometimes that's all you will be given (you won't always be told what type of curve you are graphing). Certain key points are common to all conics (vertices, foci, and axes to name a few), so you start by plotting these key points and then identifying what kind of curve they form.

In picture (graph form)

The whole point of this chapter is to be able to graph conic sections accurately, with all the necessary information. Figure 12-1 illustrates how a plane intersects the cones to create the conic sections, and the following list explains the figure:

- **Circle:** A circle is the set of all points a given distance (the radius, r) from a given point (the center). To get a circle from the right cones, the plane slice occurs parallel to the base of either cone, but does not slice through the element of the cones.
- **Parabola:** A parabola is a curve where every point on the curve is equidistant from one point (the focus) and a line (the directrix). It looks a lot like the letter U, although it may be upside down or sideways. To form a parabola, the plane slices through parallel to the side of the cones (any side works, but the bottom and top are forbidden).

- **Ellipse:** An ellipse is the set of all points where the sum of the distances from two points (the foci) is constant, and you may be more familiar with the term *oval*. In order to get an ellipse from the two right cones, the plane must cut through one cone, not parallel to the base, and not through the element.
- **Hyperbola:** A hyperbola is the set of points where the difference of the distances between two points is constant. The shape of the hyperbola is difficult to describe without a picture, but it looks visually like two parabolas (although they are very different mathematically) mirroring one another with some space between the vertices. To get a hyperbola, the slice cuts the cones perpendicular to their bases (straight up and down), but not through the element.

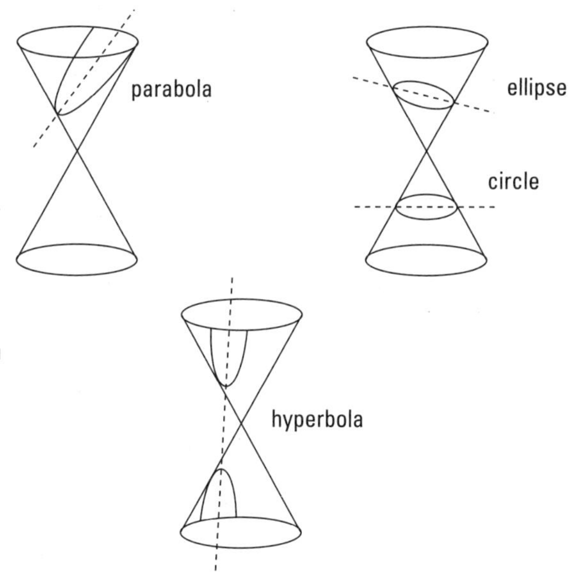

Figure 12-1: Cutting cones with a plane to get conic sections.

Most of the time, sketching a conic is not enough. Each conic section has its own set of information that you usually have to give to supplement the graph. You have to indicate where the center, vertices, major and minor axes, and the foci are located. Oftentimes, this information is more important than the graph itself. Besides, knowing all this valuable info will help you sketch the graph more accurately than you could without it.

In print (equation form)

The equations of conic sections are very important because they not only tell you which conic section you should be graphing, but they tell you what the graph should look like. There are trends in the appearance of each conic section based on the values of the constants in the equation. Usually these constants are referred to as *a, b, h, v, f,* and *d.* Not every conic will have all of these constants, but conics that do have them will be affected in the same way by changes in the same constant. Conic sections can come in all different shapes and sizes: big, small, fat, skinny, vertical, horizontal, and more. The constants listed above are the culprits of these changes.

An equation has to have x^2 and/or y^2 to create a conic. If neither x nor y is squared, then the equation will be of a line (not considered a conic section for our purposes in this book). None of the variables of a conic section may be raised to any power higher than two.

As briefly mentioned, there are certain characteristics you will find unique to each type of conic that hint to you which of the conic sections you are graphing. In order to recognize these characteristics the way we wrote them, it is important that the x^2 term and the y^2 term are on the same side of the equal sign. If they are, then these characteristics are as follows:

- **Circle: When x and y are both squared, and the coefficients on them are the same — including the sign.**

  For example, take a look at $3x^2 - 12x + 3y^2 = 2$. Notice that the x^2 and y^2 have the same coefficient (positive 3). That's all the info you need to recognize that you're working with a circle.

- **Parabola: When either x or y is squared — not both.**

  The equations $y = x^2 - 4$ and $x = 2y^2 - 3y + 10$ are both parabolas. In the first equation, you see an x^2 but no y^2, and in the second equation, you see a y^2 but no x^2. Nothing else matters — sign and coefficients will change the physical appearance of the parabola (which way it opens or how fat it is) but won't change the fact that it's a parabola.

- **Ellipse: When x and y are both squared and the coefficients are positive but different.**

  The equation $3x^2 - 9x + 2y^2 + 10y - 6 = 0$ is one example of an ellipse. The coefficients on x^2 and y^2 are different, but both are positive.

- **Hyperbola: When x and y are both squared and exactly one of the coefficients is negative (coefficients may be the same or different).**

  The equation $4y^2 - 10y - 3x^2 = 12$ is an example of a hyperbola. This time, the coefficients on x^2 and y^2 are different, but one of them is negative, which is a requirement to get the graph of a hyperbola.

The equations for the four conic sections look very similar to one another, with subtle differences (a plus sign instead of a minus sign, for instance, will give you an entirely different type of conic section). If you get the forms of equations mixed up, you'll end up graphing the wrong shape, so be forewarned!

Going Round and Round with Circles

Circles are simple to work with in pre-calculus. A circle has one center, one radius, and a whole lot of points. In this section, we show you how to graph circles on the coordinate plane and figure out from both the graph and the circle's equation where the center lies and what the radius is.

Graphing a circle

The first thing you need to know in order to graph the equation of a circle is where on a plane the center is located. The equation of a circle appears as $(x - h)^2 + (y - v)^2 = r^2$. We call this form the *center-radius* form (or standard form) because it gives you both pieces of information at the same time. The h and v represent the center of the circle at point (h, v), and r names the radius. Specifically, h represents the horizontal displacement — how far to the left or to the right the center of the circle falls from the y-axis. The variable v represents the vertical displacement — how far above or below the center falls from the x-axis. From the center, you can count from the center r units (the radius) horizontally in both directions and vertically in both directions. This will give you four different points, all equidistant from the center. Connect these four points with the best curve that you can sketch to get the graph of the circle.

At the origin

The simplest circle to graph has its center at the origin (0, 0). Because both h and v are zero, they can disappear and you can simplify the standard circle equation to look like $x^2 + y^2 = r^2$. For instance, to graph the circle $x^2 + y^2 = 16$, follow these steps:

1. **Realize that the circle is centered at the origin (no h and v) and place this point there.**
2. **Calculate the radius by solving for r.**

   Set $r^2 = 16$. In this case, you get $r = 4$.
3. **Plot the radius points on the coordinate plane.**

   You count out 4 in every direction from the center (0, 0): left, right, up, and down.

4. **Connect the dots to graph the circle using a smooth, round curve.**

   Figure 12-2 shows this circle on the plane.

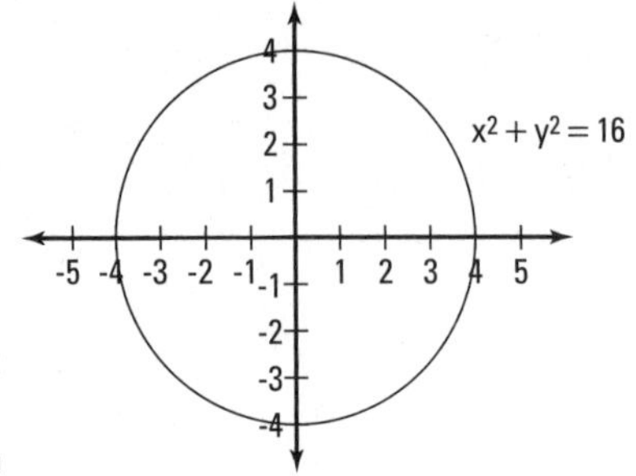

Figure 12-2: Graphing a circle centered at the origin.

Away from the origin

Graphing a circle anywhere on the coordinate plane is pretty easy when its equation appears in center-radius form. All you do is plot the center of the circle at (h, k), and then count out from the center r units in the four directions (up, down, left, right). Then, connect those four points with a nice, round circle. Unfortunately, while it is much easier to graph circles at the origin, very few are as straightforward and simple as those. In pre-calc, you work with transforming graphs of all different shapes and sizes (this is nothing new to you, right?). Fortunately, these graphs all follow the same pattern for horizontal and vertical shifts, so you don't have to remember many rules.

Don't forget to switch the sign of the h and v from inside the parentheses in the equation. This is necessary because the h and v are inside the grouping symbols, which means that the shift happens opposite from what you would think (see Chapter 3 for more information on shifting graphs).

For example, to graph the equation $(x - 3)^2 + (y + 1)^2 = 25$:

1. **Locate the center of the circle from the equation (h, v).**

   $(x - 3)^2$ means that the x-coordinate of the center is positive 3.

   $(y + 1)^2$ means that the y-coordinate of the center is negative 1.

   Place the center of the circle at (3, –1).

2. **Calculate the radius by solving for r.**

   Set $r^2 = 25$ and square root both sides to get $r = 5$.

3. **Plot the radius points on the coordinate plane.**

   Count 5 units up, down, left, and right from the center at (3, –1). This means that you should have points at (8, –1), (–2, –1), (3, –6), and (3, 4).

4. Connect the dots to the graph of the circle with a round, smooth curve.

See Figure 12-3 for a visual representation of this circle.

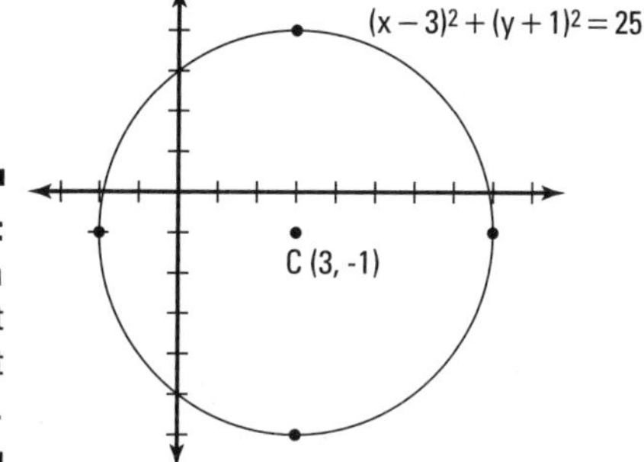

Figure 12-3: Picturing a circle not centered at the origin.

Some books use *h* and *k* to represent the horizontal and vertical displacement of circles. However, we refer to the shifts as *h* for the horizontal shift, and *v* for the vertical shifts. Who knows why everyone else chose *h* and *k*, but we think *h* and *v* are much easier to remember!

Sometimes the equation is in center-radius form (and then graphing is a piece of cake), and sometimes you have to manipulate the equation a bit to get it into a form that is easy for you to work with. When a circle doesn't appear in center-radius form, you have to complete the square in order to find the center. (We talk about this process in Chapter 4, so if you're unfamiliar with it, head back there for a refresher.)

Riding the Ups and Downs with Parabolas

Although parabolas look like simple U-shaped curves, there are actually very complicated variables at work that make them look the way they do. Because they involve squaring one value (and one value only) they become a mirror image over the axis of symmetry, just like the quadratic functions from Chapter 3. Because a positive number squared is positive, and the opposite of that number (a negative one) is also positive, you get a U-shaped graph.

The parabolas we discuss in Chapter 3 are all quadratic *functions,* which means that they passed the vertical line test. The purpose of parabolas in this chapter, however, is to discuss them not as functions, but rather as conic sections. The difference, you ask? Quadratic functions must fit the definition of a function, whereas if we discuss parabolas as conic sections, they can be vertical (like the functions) or horizontal (like a sideways U, which does not fit the definition of passing the vertical line test).

In this section, we introduce you to different parabolas that you'll encounter on your journey through conic sections.

Labeling the parts

Each of these parabolas has the same general shape; however, the width of the parabolas, their location on the coordinate plane, and which direction they open can vary greatly from one another. One thing that's true of all parabolas is their symmetry, meaning that you can fold a parabola in half over itself. The line that divides a parabola in half is called the *axis of symmetry.* The *focus* is a point inside (not on) the parabola that lies on the axis of symmetry, and the *directrix* is a line that runs outside the parabola perpendicular to the axis of symmetry. The *vertex* of the parabola is exactly halfway between the focus and the directrix. Recall from geometry that the distance from any line to a point not on that line is a line segment from the point perpendicular to the line. Therefore, the parabola is formed by all of the points equidistant from the focus and the directrix. The distance between the vertex and the focus, then, dictates how skinny or how fat the parabola is.

The first thing you must find in order to graph a parabola is where the vertex is located. From there, you can find out whether it should be up and down (a vertical parabola) or sideways (a horizontal parabola). The coefficients of the parabola will also tell you which way the parabola opens (toward the positive numbers or the negative numbers).

If you are graphing a vertical parabola, then the vertex is also the maximum or the minimum value of the curve. This has tons of real world applications for you to dive into. Bigger is usually better, and maximum area is no different. Parabolas are very useful in telling you the maximum (or sometimes minimum) area for rectangles. For example, if you are building a dog run with a preset amount of fencing, you can use parabolas to find the dimensions of the dog run that would yield the maximum area for your dog to run. Woof!

Understanding the characteristics of a standard parabola

The squaring of the variables in the equation of the parabola determines where it opens:

- **When the x is squared and y is not,** the axis of symmetry is vertical and the parabola opens up or down. For instance, $y = x^2$ is a vertical parabola; its graph is shown in Figure 12-4a.
- **When y is squared and x is not,** the axis of symmetry is horizontal and the parabola opens left or right. For example, $x = y^2$ is a horizontal parabola; it's shown in Figure 12-4b.

Both of these parabolas have the vertex located at the origin.

Be aware of negative coefficients in parabolas. If the parabola is vertical, a negative coefficient will make the parabola open downward. If the parabola is horizontal, a negative coefficient will make the parabola open to the left.

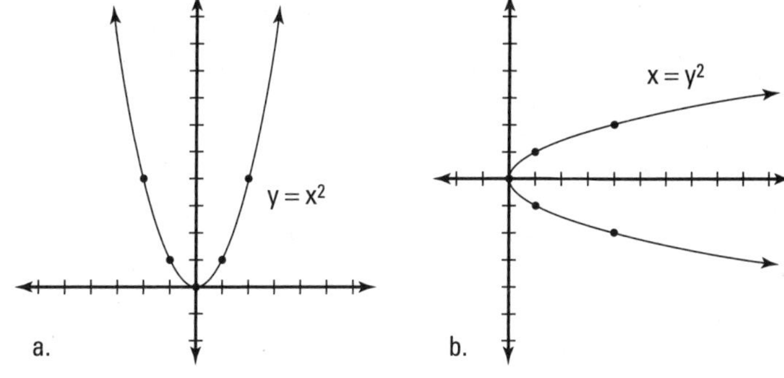

Figure 12-4: A vertical and a horizontal parabola based at the origin.

Plotting the variations: Parabolas all over the plane (not at the origin)

Just like with circles, the vertex of the parabola will not always be at the origin. You need to be comfortable with shifting parabolas around the coordinate plane too. Certain motions, especially the motion of falling ojects, move in a parabolic shape with respect to time. For example, the height of a ball launched up in the air at time t can be described by the equation $h(t)= -16t^2 + 32t$. Finding the vertex of this equation can tell you the maximum height of the ball, and also when it reached that height. Finding the x-intercepts can also tell you when the ball will hit the ground again.

A vertical parabola written in the form $y = a(x - h)^2 + v$ gives you the following information:

- **A vertical transformation (designated by the variable *a*).** For instance, for $y = 2(x - 1)^2 - 3$, every point is stretched vertically by a factor of two (see Figure 12-5a for the graph). This means that every time you plot a point on the graph, the original height of $y = x^2$ is multiplied by two.
- **The horizontal shift of the graph (designated by the variable *h*).** In this example, the vertex is shifted to the right of the origin one unit ($- h$; don't forget to switch the sign inside the parentheses).
- **The vertical shift of the graph (designated by the variable *v*).** In this example, the vertex is shifted down three ($+ v$).

All vertical parabolas with a vertical transformation of one move in the following pattern after you graph the vertex:

1. Right 1, up 1^2
2. Right 2, up 2^2
3. Right 3, up 3^2

This continues in the same pattern. Usually, just a couple of points will give you a good graph. You plot the same points on the other side of the vertex to create the mirror image over the axis of symmetry.

The transformations of horizontal parabolas on the coordinate plane are different from the transformations of vertical parabolas, because instead of moving right 1, up 1^2; right 2, up 2^2; right 3, up 3^2 and so on, the parabola is sideways. Therefore, the movement goes:

1. Up 1, right 1^2
2. Up 2, right 2^2
3. Up 3, right 3^2

A horizontal parabola appears in the form $x = a(y - v)^2 + h$. In these parabolas, the vertical shift comes with the *y* variable inside the parentheses ($- v$), and the horizontal shift is outside the parentheses ($+ h$). For instance, $x = ½(y - 1)^2 + 3$ has the following characteristics:

A vertical transformation of ½ at every point.

The vertex is moved up one (you switch the sign because it's inside the parentheses).

The vertex is moved right three.

You can see this parabola's graph in Figure 12-5b.

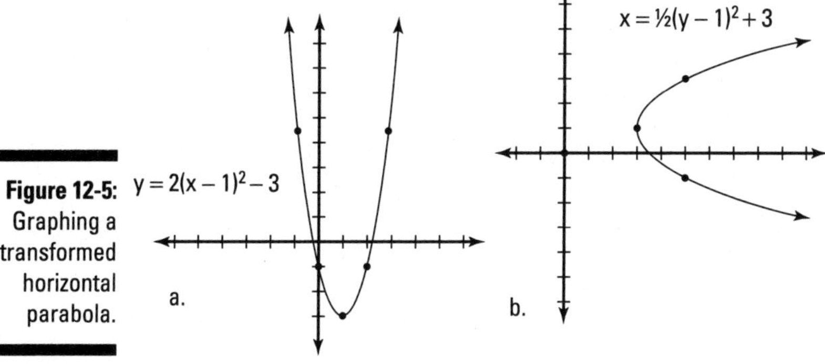

Figure 12-5: Graphing a transformed horizontal parabola.

Finding the vertex, axis of symmetry, focus, and directrix

In order to graph a parabola correctly, it is important to note whether it is a horizontal or a vertical parabola. This is because while the variables and constants in the equations for both curves serve the same purpose, their effect on the graphs in the end is slightly different. Adding a constant inside the parentheses of the vertical parabola will move the entire thing horizontally, where adding a constant inside the parentheses of a horizontal parabola will move it vertically (see the preceding section for more info). It is important to note these differences before you start graphing so that you don't accidentally move your graph in the wrong direction. In the following sections, we show you how to find all this information for both vertical and horizontal parabolas.

Of a vertical parabola

A vertical parabola has its axis of symmetry at $x = h$, and the vertex is (h, v). With this information, you can find the following parts of the parabola:

- **Focus:** The distance from the vertex to the focus is $\frac{1}{4a}$, where a can be found in the equation of the parabola (it is the scalar in front of the parentheses).The focus, as a point, is $(h, v + \frac{1}{4a})$; it should be directly above or directly below the vertex. It always appears inside the parabola.
- **Directrix:** The equation of the directrix is $y = v - \frac{1}{4a}$. It should be the same distance from the vertex along the axis of symmetry as the focus, in the opposite direction.

The directrix appears outside the parabola and is perpendicular to the axis of symmetry. Because the axis of symmetry is vertical, the directrix is a horizontal line; thus, it has an equation of the form y = a constant, which is $v - \frac{1}{4a}$.

Figure 12-6 is something we refer to as the "martini" of parabolas. The graph looks like a martini glass: The axis of symmetry is the glass stem, the directrix is the base of the glass, and the focus is the olive. You need all those parts to make a good martini *and* a parabola.

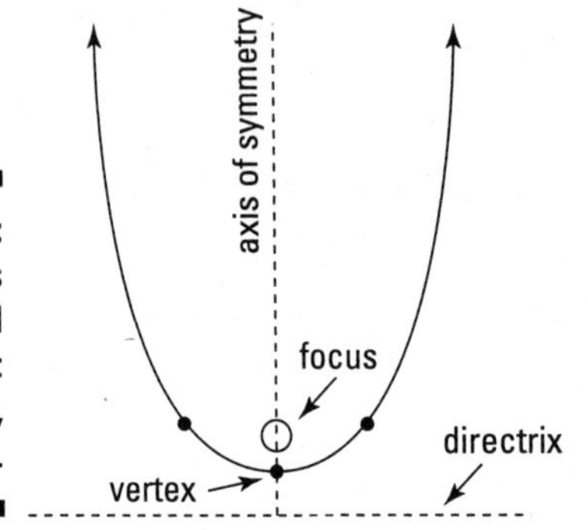

Figure 12-6: All the parts of a vertical parabola: Shaken, not stirred.

For example, the equation $y = 2(x - 1)^2 - 3$ has its vertex at (1, –3). This means that $a = 2$, $h = 1$, and $v = -3$. With this information, you can identify all the parts of a parabola (axis of symmetry, focus, and directrix) as points or equations:

1. **Find the axis of symmetry.**

   The axis of symmetry is at $x = h$, which means that $x = 1$.

2. **Determine the focal distance and write the focus as a point.**

   You can find the focal distance by using the formula $\frac{1}{4a}$. Because $a = 2$, the focal distance for this parabola is $\frac{1}{8}$. With this distance, you can write the focus as the point $(h, v + \frac{1}{4a})$, or $(1, -2\frac{7}{8})$.

3. **Find the directrix.**

   You can use the equation of the directrix: $y = v - \frac{1}{4a}$, or $y = -3\frac{1}{8}$.

4. **Graph the parabola and label all its parts.**

   You can see the graph, with all its parts, in Figure 12-7. It is always a good idea to plot at least two other points besides the vertex so that you can show that your vertical transformation is correct. Because the vertical transformation in this equation is a factor of 2, the two points on both sides of the vertex will be stretched by a factor of two. So, from the vertex, you plot a point that is to the right one, and up two (instead of up one). Then you can draw the same point on the other side of the axis of symmetry; the two other points on the graph are at (2, –1) and (0, –1).

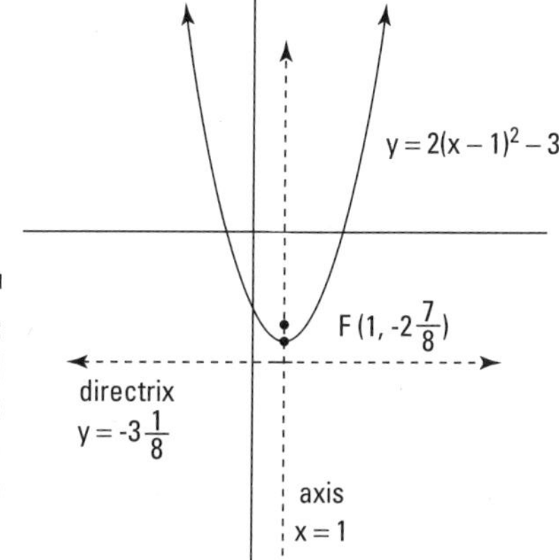

Figure 12-7: Finding all the parts of the parabola $y = 2(x - 1)^2 - 3$.

Of a horizontal parabola

A horizontal parabola features its own equations to find its parts; these are just a bit different when compared to a vertical parabola. The distance to the focus and directrix from the vertex in this case is horizontal, because they move along the axis of symmetry, which is a horizontal line. So, $\frac{1}{4a}$ is added to and subtracted from h. Here's the breakdown:

- The axis of symmetry is at $y = v$, and the vertex is still at (h, v).
- The focus is directly to the left or right of the vertex, at the point $(h + \frac{1}{4a}, v)$.
- The directrix is the same distance from the vertex as the focus in the opposite direction, at $x = h - \frac{1}{4a}$.

For example, work with the equation $x = \frac{1}{8}(y - 1)^2 + 3$:

1. **Find the axis of symmetry.**

   The vertex of this parabola is (3, 1). The axis of symmetry is at $y = v$, so for this example, it is at $y = 1$.

2. **Determine the focal distance and write this as a point.**

   For the equation above, $a = \frac{1}{2}$, and so the focal distance is 2. Add this value to h to find the focus: (3 + 2, 1) or (5, 1).

3. **Find the directrix.**

   Subtract the focal distance from Step 2 from h to find the equation of the directrix. Because this is a horizontal parabola and the axis of symmetry is horizontal, the directrix will be vertical. The equation of the directrix is $x = 3 - 2$ or $x = 1$.

4. **Graph the parabola and label its parts.**

   Figure 12-8 shows you the graph and has all of the parts labeled for you.

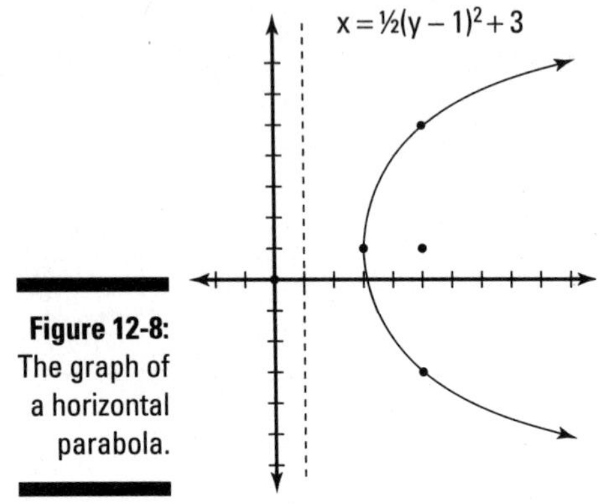

Figure 12-8: The graph of a horizontal parabola.

The focus lies inside the parabola, and the directrix is a vertical line 2 units from the vertex.

Identifying the min and max on vertical parabolas

Vertical parabolas give an important piece of information: When the parabola opens up, the vertex is the lowest point on the graph — called the *minimum.* When the parabola opens down, the vertex is the highest point on the graph — called the *maximum.* Only vertical parabolas can have minimum or maximum values, because horizontal parabolas have no limit on how high or how low they can go. Finding the maximum of a parabola can tell you the maximum height of a ball thrown into the air, the maximum area of a rectangle, the maximum or minimum value of a company's profit, and so on.

For example, say that a problem asks you to find two numbers whose sum is 10 and whose product is a maximum. You can identify two different equations hidden in this one sentence:

$$x + y = 10$$

$$x \cdot y = \text{MAX}$$

If you're like us, you don't like to mix variables when you don't have to, so we suggest that you solve one equation for one variable to substitute into the other one. This is easiest if you solve the equation that doesn't include min or max at all. So, if $x + y = 10$, you can say $y = 10 - x$. You can plug this value into the other equation to get the following:

$$(10 - x) \cdot x = \text{MAX}$$

If you distribute the x on the outside, you get $10x - x^2 = \text{MAX}$. This is a quadratic equation for which you need to find the vertex by completing the square (which will put the equation into the form you're used to seeing that identifies the vertex) — which will give you the maximum value. To do that, follow these steps:

1. **Rearrange the terms in descending order.**

   This gives you $-x^2 + 10x = \text{MAX}$.

2. **Factor out the leading term.**

   That leaves you with $-1(x^2 - 10x) = \text{MAX}$.

3. **Complete the square (see Chapter 4 for a reference).**

   This expands the equation to $-1(x^2 - 10x + 25) = \text{MAX} - 25$. Notice that -1 in front of the parentheses turned the 25 into -25, which is why you must add -25 to the right side as well.

4. **Factor the information inside the parentheses.**

   You go down to $-1(x - 5)^2 = \text{MAX} - 25$.

5. **Move the constant to the other side of the equation.**

   You end up with $-1(x - 5)^2 + 25 = \text{MAX}$.

The vertex of the parabola is (5, 25) (see the earlier section "Plotting the variations: Parabolas all over the plane (not at the origin)"). This means the number you're looking for (x) is 5, and the maximum product is 25. You can plug 5 in for x to get y in either equation: $5 + y = 10$, or $y = 5$.

Figure 12-9 shows the graph of the maximum function to illustrate that the vertex, in this case, is the maximum point.

By the way, a graphing calculator can easily find the vertex for this type of question. Even in the table form, you can see from the symmetry of the parabola that the vertex is the highest (or the lowest) point.

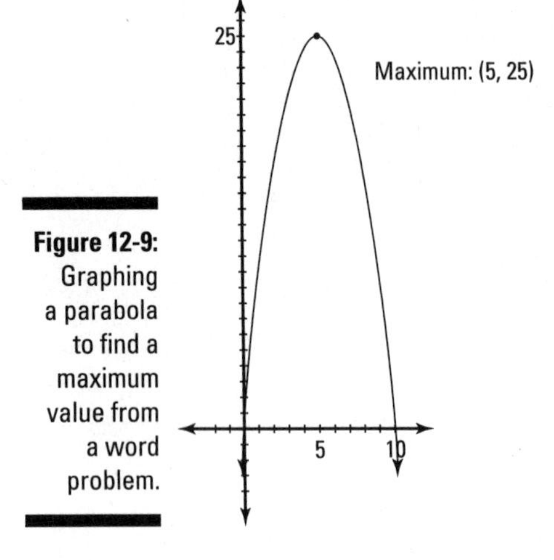

Figure 12-9: Graphing a parabola to find a maximum value from a word problem.

The Fat and the Skinny on the Ellipse (A Fancy Word for Oval)

An *ellipse* is a set of points on plane, creating an oval, curved shape, such that the sum of the distances from any point on the curve to two fixed points (the *foci*) is a constant (always the same). An ellipse is basically a circle that has been squished either horizontally or vertically.

Are you more of a visual learner? Here's how you can picture an ellipse: Take a piece of paper and pin it to a corkboard with two pins. Tie a piece of string around the two pins with a little bit of slack. Using a pencil, pull the string taut and then trace a shape around the pins — keeping the string tight the entire time. The shape you draw with this technique is an ellipse. The sums of the distances to the pins, then, is the string. The length of the string is always the same (and the different lengths of string is part of what gives you all of the different ellipses).

This definition that refers to the sums of distances can give even the best of mathematicians a headache because the idea of adding distances together can be difficult to visualize, so here's Figure 12-10 to show you what we mean. The total distance on the solid line is equal to the total distance on the dotted line.

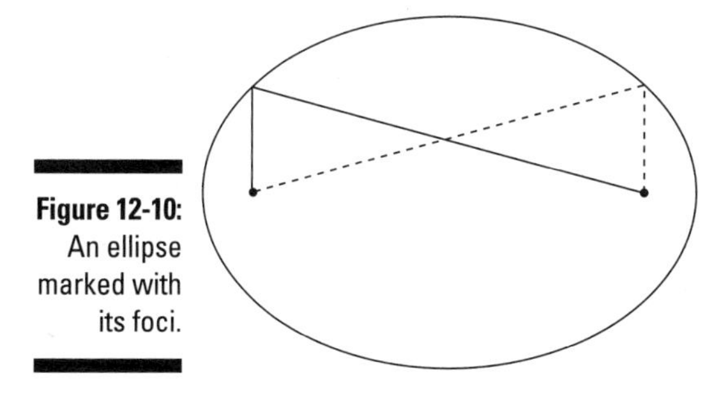

Figure 12-10: An ellipse marked with its foci.

Labeling ellipses and expressing them with algebra

Graphically speaking, you must know two different types of ellipses: horizontal and vertical. A horizontal ellipse is short and fat; a vertical one is tall and skinny. Each type of ellipse has these main parts:

- The point in the middle of the ellipse is called the *center* and is named (h, v) just like the vertex of a parabola and the center of a circle.
- The *major axis* is the line that runs through the center of the ellipse the long way. The variable a is the letter used to name the distance from the center to the ellipse on the major axis. The endpoints of the major axis are on the ellipse and are called *vertices.*
- The *minor axis* is perpendicular to the major axis and runs through the center the short way. The variable b is the letter used to name the distance to the ellipse from the center on the minor axis. Because the major axis is always longer than the minor one, $a > b$. The endpoints on the minor axis are called *co-vertices.*
- The *foci* are the two points that dictate how fat or how skinny the ellipse is. They are always located on the major axis, and can be found by the following equation: $a^2 - b^2 = F^2$ where a and b are mentioned as in the preceding bullets, and F is the distance from the center to each focus.

Figure 12-11a shows a horizontal ellipse with its parts labeled; Figure 12-11b shows a vertical one. Notice that the length of the major axis is $2a$, and the length of the minor axis is $2b$.

Figure 12-11 also shows the correct placement of the foci — always on the major axis.

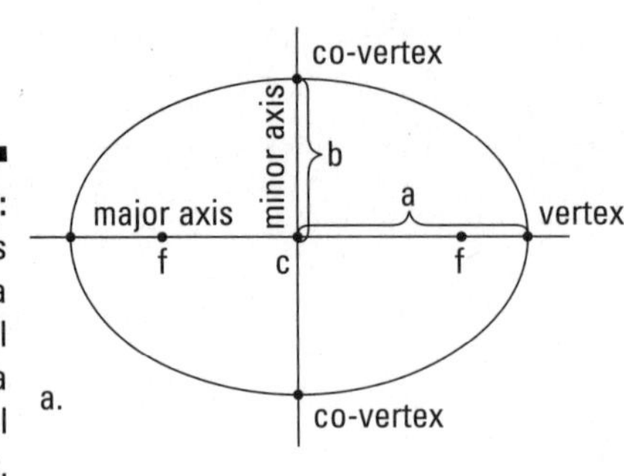

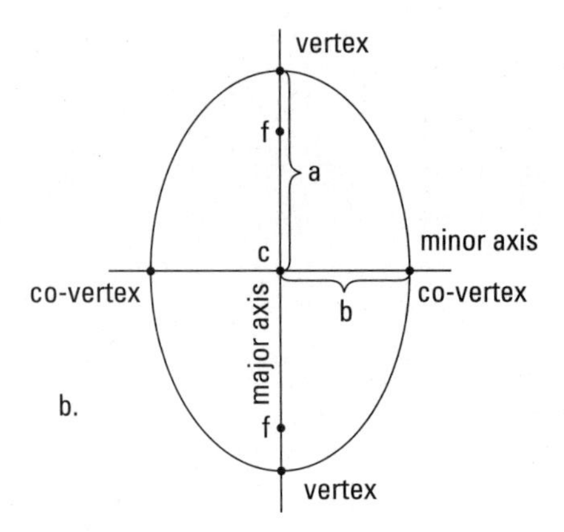

Figure 12-11: The labels of a horizontal ellipse and a vertical ellipse.

Two types of equations apply to ellipses, depending on whether they're horizontal or vertical:

The horizontal equation is $\frac{(x-h)^2}{a^2}+\frac{(y-v)^2}{b^2}=1$, with the center at (h, v), major axis of $2a$, and minor axis of $2b$.

The vertical equation is $\frac{(x-h)^2}{b^2}+\frac{(y-v)^2}{a^2}=1$, with the same parts — although a and b have switched places.

When the bigger number a is under x, the ellipse is horizontal; when the bigger number is under y, it's vertical.

Identifying the parts of the oval: Vertices, co-vertices, axes, and foci

You have to be prepared to not only graph ellipses, but also to name all their parts. If a problem asks you to calculate the parts of an ellipse, you have to be ready to deal with some ugly square roots and/or decimals. Table 12-1 presents the parts in a handy, at-a-glance format. This section prepares you to graph and to find all the parts of an ellipse.

Table 12-1	Ellipse Parts	
	Horizontal Ellipse	***Vertical Ellipse***
Equation	$\frac{(x-h)^2}{a^2}+\frac{(y-v)^2}{b^2}=1$	$\frac{(x-h)^2}{b^2}+\frac{(y-v)^2}{a^2}=1$
Center	(h, v)	(h, v)
Vertices	$(h \pm a, v)$	$(h, v \pm a)$
Co-vertices	$(h, v \pm b)$	$(h \pm b, v)$
Length of Major Axis	$2a$	$2a$
Length of Minor Axis	$2b$	$2b$
Foci where $F^2 = a^2 - b^2$	$(h \pm F, v)$	$(h, v \pm F)$

Vertices and co-vertices

To find the vertices in a horizontal ellipse, use $(h \pm a, v)$; to find the co-vertices, use $(h, v \pm b)$. A vertical ellipse has vertices at $(h, v \pm a)$ and co-vertices at $(h \pm b, v)$.

For example, look at $\frac{(x-5)^2}{9}+\frac{(y+1)^2}{16}=1$, which is already in the proper form to graph. You know that $h = 5$ and $v = -1$ (switching the signs inside the parentheses). You also know that $a^2 = 16$ (because a has to be the greater number!), or $a = 4$. If $b^2 = 9$, $b = 3$.

This example is a vertical ellipse because the bigger number is under y, so be sure to use the correct formula. This equation has vertices at $(5, -1 \pm 4)$, or $(5, 3)$ and $(5, -5)$. It has co-vertices at $(5 \pm 3, -1)$, or $(8, -1)$ and $(2, -1)$.

The axes and foci

The major axis in a horizontal ellipse is given by the equation $y = v$; the minor axis is given by $x = h$. The major axis in a vertical ellipse is represented by $x = h$; the minor axis is represented by $y = v$. The length of the major axis is $2a$, and the length of the minor axis is $2b$.

You can calculate the distance from the center to the foci in an ellipse (either variety) by using the equation $a^2 - b^2 = F^2$, where F is the distance from the center to each focus. The foci always appear on the major axis at the given distance (F) from the center.

Using the example from the previous section, you can find the foci with the equation $16 - 9 = F^2$. The focal distance is $\sqrt{7}$. Because the ellipse is vertical, the foci are at $(5, -1 \pm \sqrt{7})$.

Working with an ellipse in non-standard form

What if the elliptical equation you're given isn't in standard form? Take a look at the example $3x^2 + 6x + 4y^2 - 16y - 5 = 0$. Before you do a single thing, determine that the equation is an ellipse because the coefficients on x^2 and y^2 are both positive but not equal. Follow these steps to put the equation in standard form:

1. **Add the constant to the other side.**

   This gives you $3x^2 + 6x + 4y^2 - 16y = 5$.

2. **Complete the square.**

   You need to factor out two different constants now — the different coefficients for x^2 and y^2:

   $3(x^2 + 2x + 1) + 4(y^2 - 4y + 4) = 5$

3. **Balance the equation by adding the new terms to the other side.**

   In other words, $3(x^2 + 2x + 1) + 4(y^2 - 4y + 4) = 5 + 3 + 16$.

   Note: Adding 1 and 4 inside the parentheses really means adding $3 \cdot 1$ and $4 \cdot 4$ to each side, because you must multiply by the coefficient before adding it to the right side.

4. **Factor the left side of the equation and simplify right.**

   You now have $3(x + 1)^2 + 4(y - 2)^2 = 24$.

5. **Divide the equation by the constant on the right to get 1 and then reduce the fractions.**

   You now have the form $\frac{(x+1)^2}{8} + \frac{(y-2)^2}{6} = 1$.

6. **Determine if the ellipse is horizontal or vertical.**

   Because the bigger number is under x, this ellipse is horizontal.

7. **Find the center and the length of the major and minor axes.**

   The center is located at (h, v), or $(-1, 2)$. If $a^2 = 8$, $a = 2\sqrt{2} \approx 2.83$. If $b^2 = 6$, $b = \sqrt{6} \approx 2.45$.

8. **Graph the ellipse to determine the vertices and co-vertices.**

   Go to the center first and mark the point. Because this ellipse is horizontal, a will move to the left and right $2\sqrt{2}$ units (about 2.83) from the center, and $\sqrt{6}$ units (about 2.45) up and down from the center. Plotting these points will locate the vertices of the ellipse.

   Its vertices are at $(-1 \pm 2\sqrt{2}, 2)$ and its co-vertices are at $(-1, 2 \pm \sqrt{6})$. The major axis is at $y = 2$ and the minor axis is at $x = -1$. The length of the major axis is $2 \cdot 2\sqrt{2}$, or $4\sqrt{2}$, and the length of the minor axis is $2 \cdot \sqrt{6}$.

9. **Plot the foci of the ellipse.**

   You determine the focal distance from the center to the foci in this ellipse with the equation $8 - 6 = F^2$, so $2 = F^2$. Therefore, $F = \sqrt{2} \approx 1.41$. The foci, expressed as points, are located at $(-1 \pm \sqrt{2}, 2)$.

Figure 12-12 shows all the parts of this ellipse in its fat glory.

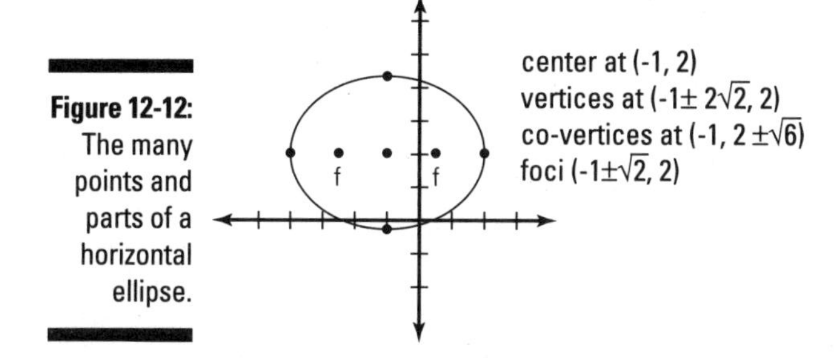

Figure 12-12: The many points and parts of a horizontal ellipse.

Pair Two Parabolas and What Do You Get? Hyperbolas

Hyperbola literally means "overshooting" in Greek, so it's a fitting name: A *hyperbola* is basically "more than a parabola." Think of a hyperbola as a mix of two parabolas — each one a perfect mirror image of the other each opening away from one another. The vertices of these parabolas are a given distance apart, and they either both open vertically or horizontally.

The mathematical definition of a hyperbola is the set of all points where the difference in the distance from two fixed points (called the *foci*) is constant. In this section, you discover the ins and outs of the hyperbola, including how to name its parts and graph it.

Visualizing the two types of hyperbolas and their bits and pieces

Similar to ellipses (see the previous section), there are two kinds of hyperbolas: horizontal and vertical.

The equation for a horizontal hyperbola is $\frac{(x-h)^2}{a^2} - \frac{(y-v)^2}{b^2} = 1$. The equation for a vertical hyperbola is $\frac{(y-v)^2}{a^2} - \frac{(x-h)^2}{b^2} = 1$. Notice that x and y switch places (as well as the h and v with them) to name horizontal versus vertical, compared to ellipses, but a and b stay put. So, for hyperbolas, a^2 should always come first, but it isn't necessarily greater. More accurately, a is always squared under the positive term (either x^2 or y^2). Basically, to get a hyperbola into standard form, you need to be sure that the positive squared term is first.

The center of a hyperbola is not actually on the curve itself, but exactly in between the two vertices of the hyperbola. Always plot the center first, and then count out from the center to find the vertices, axes, and asymptotes. A hyperbola has two axes of symmetry. The one that passes through the center and the two foci is called the *transverse axis;* the one that's perpendicular to the transverse axis through the center is called the *conjugate axis.* A horizontal hyperbola has its transverse axis at $y = v$ and its conjugate axis at $x = h$; a vertical hyperbola has its transverse axis at $x = h$ and its conjugate axis at $y = v$.

You can see the two types of hyperbolas in Figure 12-13. Figure 12-13a is a horizontal hyperbola and Figure 12-13b is a vertical one.

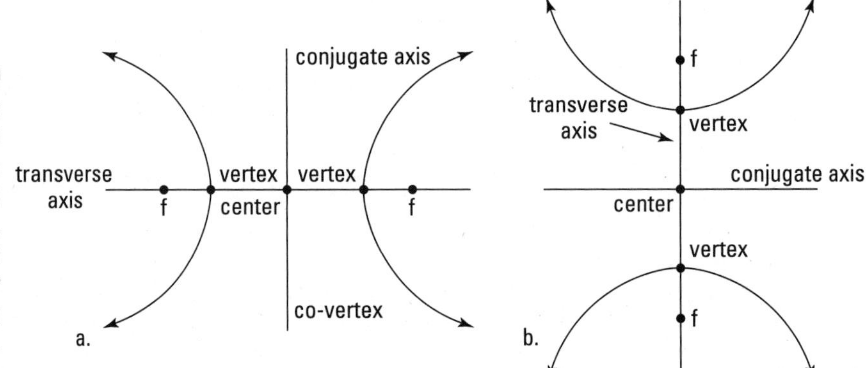

Figure 12-13: A horizontal and a vertical hyperbola, dissected for your viewing pleasure.

If the hyperbola that you are trying to graph is not in standard form, then you need to complete the square to get it into standard form. For the steps of completing the square with conic sections, check out the "Identifying the min and max on vertical parabolas" section, earlier in the chapter.

For example, the equation $\frac{(y-3)^2}{16} - \frac{(x+1)^2}{9} = 1$ is a vertical hyperbola. The center (h, v) is $(-1, 3)$. If $a^2 = 16$, $a = 4$, which means that you count vertical direction (because it is the number under the y variable); and if $b^2 = 9$, $b = 3$ (which means that you count horizontally 3 units from the center both to the left and to the right). The distance from the center to the edge of the rectangle marked "a" determines half the length of the transverse axis, and the distance to the edge of the rectangle marked "b" determines the conjugate axis. In a hyperbola, a could be greater than, less than, or equal to b. If you count out a units from the center along the transverse axis, and b units from the center in both directions along the conjugate axis, these four points will be the midpoints of the sides of a very important rectangle. This rectangle has sides that are parallel to the x- and y-axis (in other words, don't just connect the four points because they are the midpoints of the sides, not the corners of the rectangle). This rectangle will be a useful guide when it is time to graph the hyperbola.

But as you can see in Figure 12-13, hyperbolas contain other important parts that you must consider. For instance, a hyperbola has two vertices. There are two different equations — one for horizontal and one for vertical hyperbolas:

- A horizontal hyperbola has vertices at $(h \pm a, v)$.
- A vertical hyperbola has vertices at $(h, v \pm a)$.

The vertices for the previous example are at $(-1, 3 \pm 4)$, or $(-1, 7)$ and $(-1, -1)$.

You find the foci of any hyperbola by using the equation $a^2 + b^2 = F^2$, where F is the distance from the center to the foci along the transverse axis, the same axis that the vertices are on. The distance F moves in the same direction as a. Continuing our example, $16 + 9 = F^2$, or $25 = F^2$. Taking the root of both sides gives you $5 = F$.

To name the foci as points in a horizontal hyperbola, you use $(h \pm F, v)$; to name them in a vertical hyperbola, you use $(h, v \pm F)$. The foci in the example would be $(-1, 3 \pm 5)$, or $(-1, 8)$ and $(-1, -2)$. Note that this places them inside the hyperbola.

Through the center of the hyperbola and through the corners of the rectangle mentioned previously run the asymptotes of the hyperbola. These asymptotes help guide your sketch of the curves because the curves cannot cross them at any point on the graph. The slopes of these asymptotes are $m = \pm a/b$ for a vertical parabola, or $m = \pm b/a$ for a horizontal parabola.

Graphing a hyperbola from an equation

To graph a hyperbola, you take all the information from the previous section and put it to work. Follow these simple steps:

1. **Mark the center.**

   Sticking with our example hyperbola $\frac{(y-3)^2}{16} - \frac{(x+1)^2}{9} = 1$ from the previous section, you find that the center of this hyperbola is $(-1, 3)$. Remember to switch the signs of the numbers inside the parentheses, and also remember that h is inside the parentheses with x, and v is inside the parentheses with y. For this example, the quantity with y squared comes first, but that does not mean that h and v switch places. The h and v always remain true to their respective variables, x and y.

2. **From the center in Step 1, find the transverse and conjugate axes.**

   Go up and down the transverse axis a distance of 4 (because 4 is under y), and then go right and left 3 (because 3 is under x). But don't connect the dots to get an ellipse! Up until now, the steps of drawing a hyperbola were exactly the same as when you drew an ellipse, but here is where things get different. The points you marked as a (on the transverse axis) are your vertices.

3. **Use these points to draw a rectangle that will help guide the shape of your hyperbola.**

   Because you went up and down 4, the height of your rectangle is 8; going left and right 3 gives you a width of 6.

4. **Draw diagonal lines through the center and the corners of the rectangle that extend beyond the rectangle.**

   This gives you two lines that will be your asymptotes.

5. **Sketch the curves.**

   Draw the curves, beginning at each vertex separately, that hug the asymptotes the farther away from the vertices the curve gets.

   The graph approaches the asymptotes but never actually touches them.

Figure 12-14 shows the finished hyperbola.

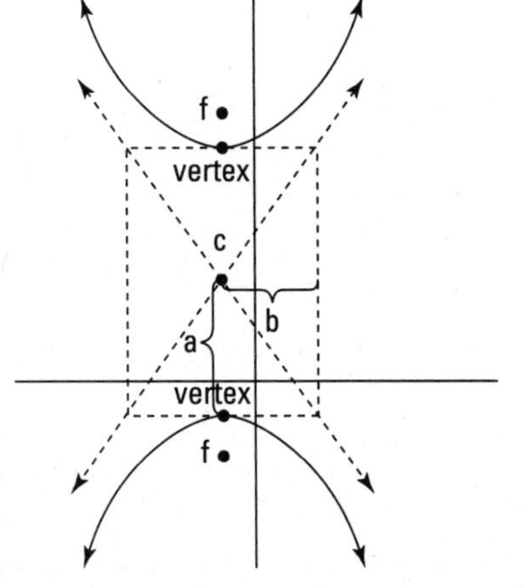

Figure 12-14: Creating a rectangle to graph a hyperbola with asymptotes.

Finding the equation of asymptotes

Because hyperbolas are formed by a curve where the difference of the distances between two points is constant, the curves will behave differently than the other conic sections in this chapter. Because distances can't be negative, this causes the graph to have asymptotes that the curve cannot cross over.

Hyperbolas are the only conic sections with asymptotes. Even though parabolas and hyperbolas look very similar, parabolas are formed by distance from a point and the distance to a line being the same. Therefore, parabolas don't have asymptotes.

Some problems ask you to find not only the graph of the hyperbola, but also the equation of the lines that determine the asymptotes. When asked to find the equation of the asymptotes, your answer depends on whether the hyperbola is horizontal or vertical.

If the hyperbola is horizontal, the asymptotes are given by the line with the equation $y = \pm\frac{b}{a}(x - h) + v$. If the hyperbola is vertical, the asymptotes have the equation $y = \pm\frac{a}{b}(x - h) + v$.

The fractions $\frac{b}{a}$ and $\frac{a}{b}$ are the slopes of the lines. You get familiar with point-slope form in Algebra II. Now that you know the slope of your line and a point (which is the center of the hyperbola), you can always write the equations without having to memorize the two asymptote formulas.

Once again, using our example, the hyperbola is vertical so the slope of the asymptotes is $m = \pm\frac{a}{b}$.

1. **Find the slope of the asymptotes.**

   Because this hyperbola is vertical, the slopes of the asymptotes are $\pm\frac{4}{3}$.

2. **Use the slope from Step 1 and the center of the hyperbola as the point to find the point–slope form of the equation.**

   Remember that the equation of a line with slope m through point (x_1, y_1) is $y - y_1 = m(x - x_1)$. Therefore if the slope is $\pm\frac{4}{3}$ and the point is $(-1, 3)$, then the equation of the line is $y - 3 = \pm\frac{4}{3}(x + 1)$.

3. **Solve for y to find the equation in slope-intercept form.**

   You have to do each asymptote separately here.

   - Distribute $\frac{4}{3}$ on the right to get $y + 3 = \frac{4}{3}x + \frac{4}{3}$, and then subtract 3 from both sides to get $y = \frac{4}{3}x + \frac{13}{3}$.

   - Distribute $-\frac{4}{3}$ to the right side to get $y + 3 = -\frac{4}{3}x - \frac{4}{3}$. Then subtract 3 from both sides to get $y = -\frac{4}{3}x + \frac{5}{3}$.

Expressing Conics Outside the Realm of Cartesian Coordinates

Up to this point in this chapter, we've been graphing conics by using rectangular coordinates (x, y). You may also be asked to graph them in two other ways:

- In parametric form, which is a fancy way of saying that in this form, you can deal with conics that aren't easily expressed as the graph of a function $y = f(x)$. Parametric equations are usually used to describe the motion or velocity of an object with respect to time. Using parametric equations will allow you to evaluate both x and y as dependent variables, as opposed to x being independent, and y dependent on x.
- In polar form (which you've seen in Chapter 11), where every point is (r, θ).

The following sections show you how to graph conics in these forms.

Graphing conic sections in parametric form

Parametric form defines both the x and the y variables of conic sections in terms of a third, arbitrary variable, called the *parameter,* which is usually represented by t, and you can find both x and y by plugging in t to the parametric equations. As t changes, so do x and y, which means that y is no longer dependent upon x, but is dependent upon t. Why switch to this form? Consider, for example, an object moving in a plane during a specific time interval. If a problem asks you to describe the path of the object and its location at any certain time, you need three variables:

- Time t, which usually is the parameter
- The coordinates (x, y) of the object at time t.

The x_t equation gives the horizontal movement of an object as t changes; the y_t equation gives the vertical movement of an object over time.

For example, one set of equations defines both x and y for the same parameter — t — and defines the parameter in a set interval:

$$x = 2t - 1$$
$$y = t^2 - 3t + 1$$
$$1 < t \leq 5$$

Time t exists only between 1 and 5 seconds for this problem.

If you're asked to graph this equation, you can do it in one of two ways. The first method is the plug and chug: set up a chart and pick t values from the given interval in order to figure out what x and y should be, and then graph these points like normal. Table 12-2 shows the results of this process. ***Note:*** We include $t = 1$ in the chart, even though the parameter isn't defined there. You need to see what it would've been, because you graph the point where $t = 1$ with an open circle to show what happens to the function arbitrarily close to 1. Be sure to make that point an open circle on your graph.

Table 12-2 Plug and Chug *t* Values from the Interval

Variable	*Interval Time*				
***t* value**	**1**	**2**	**3**	**4**	**5**
***x* value**	1	3	5	7	9
***y* value**	–1	–1	1	5	11

The other way to graph a parametric curve is to solve one equation for the parameter and then substitute that equation into the other equation. You should pick the simplest equation to solve and start there.

Sticking with the same example, we'll solve the linear equation $x = 2t - 1$ for t:

1. **Solve the simplest equation.**

   For our chosen equation, we get $t = \frac{x-1}{2}$.

2. **Plug the solved equation into the other equation.**

   For this, we get $y = \left(\frac{x-1}{2}\right)^2 - 3\left(\frac{x-1}{2}\right) + 1$.

3. **Simplify this equation if necessary.**

   We now have $y = \frac{1}{4}x^2 - 2x + \frac{11}{4}$.

   Because this gives you an equation in terms of x and y, you can graph the points on the coordinate plane just like you always do. The only problem is that you don't draw the entire graph, because you have to look at a specific interval of t.

4. **Substitute the endpoints of the *t* interval into the *x* function to know where the graph starts and stops.**

   We do this in Table 12-2. When $t = 1$, $x = 1$, and when $t = 5$, $x = 9$.

Figure 12-15 shows the parametric curve from this example (for both methods). You end up with a parabola, but it's also possible to write parametric equations for ellipses, circles, and hyperbolas.

If you have a graphing calculator, you can set it to parametric mode to graph. When you go into your graphing utility, you'll get two equations — one is "$x =$" and the other is "$y =$." Input both equations exactly as they're given, and the calculator will do the work for you!

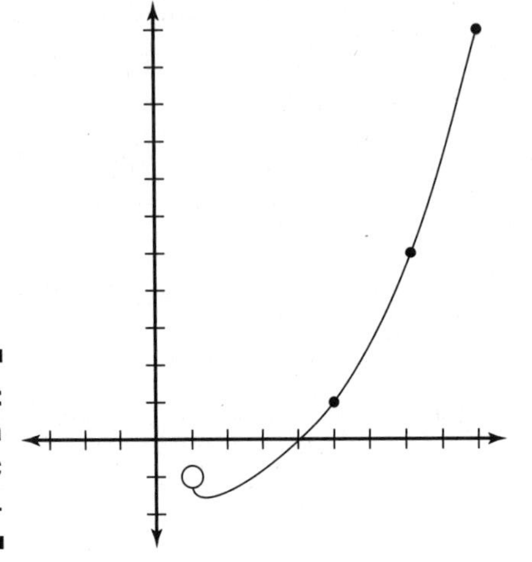

Figure 12-15: Graphing a parametric curve.

The equations of conic sections on the polar coordinate plane

Graphing conic sections on the polar plane (see Chapter 11) is based on equations that depend on a special value known as *eccentricity,* which describes the overall shape of a conic section. The value of a conics eccentricity can tell you what type of conic section the equation describes, as well as how fat or skinny it is. When graphing equations in polar coordinates, it can be difficult to tell which conic section you should be graphing based solely on the equation (unlike graphing in Cartesian coordinates, where each conic section has its own unique equation). Therefore, you can use the eccentricity of a conic section to find out exactly which type of curve you should be graphing.

Here are the two equations that allow you to put conic sections in polar coordinate form, where (r, θ) is the coordinate of a point on the curve in polar form. Recall from Chapter 11 that r is the radius, and θ is the angle in standard position on the polar coordinate plane.

$$r = \frac{ke}{1 - e\cos\theta} \text{ or } \frac{ke}{1 - e\sin\theta}$$

$$r = \frac{-ke}{1 + e\cos\theta} \text{ or } \frac{-ke}{1 - e\sin\theta}$$

When graphing conic sections in polar form, you can plug in various values of θ to get the graph of the curve. In each equation above, k is a constant value, θ takes the place of time, and e is the eccentricity. The variable e determines the conic section:

If $e = 0$, the conic section is a circle.

If $0 < e < 1$, the conic section is an ellipse.

If $e = 1$, the conic section is a parabola.

If $e > 1$, the conic section is a hyperbola.

For example, to graph $r = \frac{2}{4 - \cos\theta}$, first realize that as shown, it does not fit the form of any of the equations we have introduced you to for the conic sections. This is because all the denominators of the conic sections begin with 1, and this equation begins with 4. Have no fear, you can factor out that 4, which will in turn tell you what k is!

Factoring out the 4 from the denominator gives you $1 - e \cdot \cos\theta$. In order to keep the equation as close to the standard form for polar conics, multiply the numerator and denominator by ¼. This gives you $r = \frac{2}{4\left(1 - \frac{1}{4}\cos\theta\right)}$, which is the same as $r = \frac{2 \cdot \frac{1}{4}}{1 - \frac{1}{4}\cos\theta}$. Therefore, the constant k is 2 and the eccentricity, e, is ¼, which tells you that you have an ellipse, because e is between 0 and 1.

In order to graph the polar function of the ellipse mentioned above, you can plug in values of θ and solve for r. Then plot the coordinates of (r, θ) on the polar coordinate plane to get the graph. For the graph of the previous equation, $r = \frac{2}{4 - \cos\theta}$, you can plug in 0, $\pi/2$, π, and $3\pi/2$ and find r:

$r(0)$: The cosine of 0 is 1, so $r(0) = 2/3$.

$r(\pi/2)$: The cosine of $\pi/2$ is 0, so $r(\pi/2) = 1/2$.

$r(\pi)$: The cosine of π is -1, so $r(\pi) = 2/5$.

$r(3\pi/2)$: The cosine of $3\pi/2$ is 0, so $r(3\pi/2) = 1/2$.

These four points should be enough to give you a rough sketch of the graph. You can see the graph of the example ellipse in Figure 12-16.

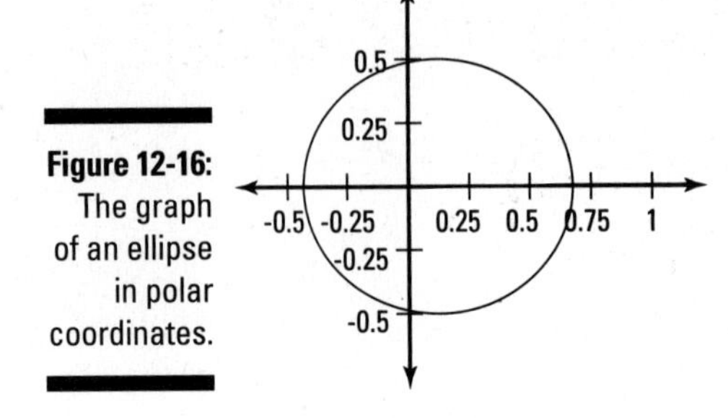

Figure 12-16: The graph of an ellipse in polar coordinates.

Chapter 13

Solving Systems and Mingling with Matrices

In This Chapter

- Taking down two-equation systems with substitution and elimination
- Breaking down systems with more than two equations
- Graphing systems of inequalities
- Forming and operating on matrices
- Putting matrices into simpler forms
- Solving systems of equations using matrices

When you have one variable and one equation, you can almost always solve the equation. Finding a solution may take you some time, but it is *usually* possible. When a problem has two variables, however, you need at least two equations to solve; this is called a *system*. When you have three variables, you need at least three equations in the system. Basically, for every variable present, you need a separate unique equation if you want to solve for it.

For an equation with three variables, an infinite number of values for two variables would work for that particular equation. Why? Because you can pick any two numbers to plug in for two of the variables in order to find the value of the third one that makes the equation true. If you add another equation into the mix, the solutions to the first equation now have to *also* work in the second equation, which makes for fewer solutions that will work. The set of solutions (usually x, y, z, and so on) must work when plugged into each and every equation in the system. More equations mean fewer values will work as solutions.

Of course, the bigger a system of equations becomes, the longer it will take you to solve it algebraically. Therefore, it's easier to solve certain systems in certain ways, which is why math books usually show each way. In this book, we follow suit, showing you when each method is preferable to all the others. It is a good idea to be comfortable with as many of these methods as possible so that you can choose the best route possible, which is the one with the least amount of steps (and that means fewer places to get mixed up).

As if systems of equations weren't enough on their own, we introduce you to systems of inequalities. These systems of inequalities require you to graph the solution, which is actually easier than finding the solutions to systems of equations algebraically because you don't need to find the exact values of the solutions, because there aren't any. The solution to a system of inequalities is shown as a shaded region on a graph.

A Primer on Your System-Solving Options

Your system-solving options are as follows:

- If the system has only two or three variables, you can use substitution or elimination (which you have seen before in Algebra I and Algebra II).
- If the system has four or more variables, you can use matrices, which are rectangular tables that display either numbers or variables, called *elements*. With matrices, you have your choice of the following, all of which we discuss later in this chapter:
    - Gaussian elimination
    - Inverse matrices
    - Cramer's rule

A note to the calculator savvy: Some instructors teach the material contained in this chapter and tell students to just plug the numbers into a calculator, don't ask any questions, and move on. If you're lucky enough to have a graphing calculator and a calculator-happy teacher, you could let the calculator do the math for you. However, because we are mathematicians, we always recommend that you buckle down and learn it anyway; then pat yourself on the back for going the extra mile!

No matter which system-solving method you use, check the answers you get, because even the best mathematicians make mistakes sometimes. The more variables and equations you have in a system, the more likely you are to make a mistake. And if you make a mistake in calculations somewhere, it can affect more than one answer, because one variable usually is dependent on another. Always verify!

Finding Solutions of Two-Equation Systems Algebraically

When you solve systems with two variables and therefore two equations, the equations can be linear or nonlinear. Linear systems are usually expressed in the form $Ax + By = C$, where A, B, and C are real numbers.

Nonlinear equations can include circles, other conics, polynomials, and exponential, logarithmic, or rational functions. Elimination won't work in nonlinear systems if x appears in one equation of the system but x^2 appears in another equation, because the terms aren't alike and, therefore, can't be added together. In that case, you're stuck with substitution to solve the nonlinear system.

Like most algebra problems, systems can have a number of possible solutions. If a system has one or more unique solutions that can be expressed as coordinate pairs, it's called *consistent and independent.* If it has no solution, it's called an *inconsistent system.* If there are infinite solutions, this is called a *dependent system.* It can be difficult to tell which of these categories your system of equations falls into just by looking at the problem. A linear system can only have no solution, one solution, or infinite solutions because it's impossible for two different straight lines to intersect in more than one place (if you don't believe us, draw a picture). A line and a conic section can intersect no more than twice, and two conic sections can intersect a maximum of four times.

Solving linear systems

When solving linear systems, you have two methods at your disposal, and which one you choose depends on the problem:

- If the coefficient of any variable is 1, which means you can easily solve for it in terms of the other variable, then substitution is a very good bet. If you use this method, then it doesn't matter how each equation is set up.
- If all the coefficients are anything other than 1, then you can use elimination, but only if the equations can be added together to make one of the variables disappear. However, if you use this method, be sure that all the variables and the equal sign line up with one another before you add the equations together.

With the substitution method

In the *substitution method,* you use one equation to solve for one variable and then substitute that expression into the other equation to solve for the other variable. Look for a variable with a coefficient of 1 . . . that's how you'll know where to begin. If the coefficient on a variable is 1, then that is the variable you should solve for because solving for that variable will solely entail adding or subtracting terms in order to move everything to the other side of the equal sign, just like you've been doing to solve for variables since Algebra I. That way, you won't have to divide by the coefficient when you're solving, which means you won't have any fractions.

For example, suppose you're managing a theater, and you need to know how many adults and children are in attendance at a show. The auditorium is sold out and contains a mixture of adults and children. The tickets cost $23.00 per adult and $15.00 per child. If the auditorium has 250 seats and the total ticket revenue for the event is $4,846.00, how many adults and children are in attendance?

To solve the problem with the substitution method, follow these steps:

1. **Express the word problem as a system of equations.**

   You can use the information given in the word problem to set up two different equations. You want to solve for how many adult tickets (a) and child tickets (c) you sold. If the auditorium has 250 seats and was sold out, the sum of the adult tickets and child tickets must be 250.

   The ticket prices also lead you to the revenue (or money made) from the event. The adult ticket price times the number of adults present lets you know how much money you made from the adults. You can do the same calculation with the child tickets. The sum of these two calculations must be the total ticket revenue for the event.

   Here's how you write this system of equations:

   $$\begin{cases} a + c = 250 \\ 23a + 15c = 4{,}846 \end{cases}$$

2. **Solve for one of the variables.**

   Pick the variable with a coefficient of 1 if you can, because solving for this variable will be easy. For this example, you can choose to solve for a in the first equation. To do this, subtract c from both sides: $a = 250 - c$.

   You can always move things from one side of an equation to the other, but don't fall prey to the trap that $250 - c$ is $249c$, like some people do. Those are not like terms, so you can't combine them.

3. **Substitute the solved variable into the other equation.**

   In this example, you solve for a in the first equation. You take this value ($250 - c$) and substitute it into the other equation for a. (Make sure that you don't substitute into the equation you used in Step 1; otherwise, you'll be going in circles.)

The second equation now says $23(250 - c) + 15c = 4{,}846$.

4. **Solve for the unknown variable.**

   When you distribute the number 23, you get $5{,}750 - 23c + 15c = 4{,}846$. When you simplify this, you get $5{,}750 - 8c = 4{,}846$, or $-8c = -904$. So, $c = 113$. A total of 113 children attended the event.

5. **Substitute the value of the unknown variable into one of the original equations to solve for the other unknown variable.**

   You don't have to substitute into one of the original equations, but your answers tend to be more accurate if you do.

   When you plug 113 into the first equation for *c*, you get $a + 113 = 250$. Solving this equation, you get $a = 137$. You sold a total of 137 adult tickets.

6. **Check your solution.**

   When you plug *a* and *c* into the original equations, you should get two true statements. Does $137 + 113 = 250$? Yes. Does $23(137) + 15(113) = 4{,}846$? Indeed.

By using the process of elimination

If solving a system of two equations with the substitution method proves difficult or the system involves fractions, the elimination method is your next best option. (Who wants to deal with fractions anyway?) In the *elimination method,* you make one of the variables cancel itself out by adding the two equations.

Sometimes, you have to multiply one or both equations by constants in order to add the equations; this situation occurs when you can't eliminate one of the variables by just adding the two equations together. (Remember that in order for one variable to be eliminated, the coefficients of one variable must be opposites.)

For example, the following steps show you how to solve the system

$$\begin{cases} 20x + 24y = 10 \\ \frac{1}{3}x + \frac{4}{5}y = \frac{5}{6} \end{cases}$$

by using the process of elimination:

1. **Rewrite the equations, if necessary, to make like variables line up underneath each other.**

   The order of the variables doesn't matter; just make sure that like terms line up with like terms from top to bottom. The equations in this system have the variables *x* and *y* lined up already:

$$\begin{cases} 20x + 24y = 10 \\ \frac{1}{3}x + \frac{4}{5}y = \frac{5}{6} \end{cases}$$

2. **Multiply the equations by constants to make one set of variables match coefficients.**

   Decide which variable you want to eliminate.

   Say you decide to eliminate the *x* variables; first, you have to find their least common multiple. What number do 20 and $\frac{1}{3}$ both go into? The answer is 60. But one of them has to be negative so that when you add the equations, the terms cancel out (that's why it's called elimination!). Multiply the top equation by –3 and the bottom equation by 180. (Be sure to distribute this number to each term — even on the other side of the equal sign.) Doing this gives you the following:

   $$\begin{cases} -60x - 72y = -30 \\ 60x + 144y = 150 \end{cases}$$

3. **Add the two equations.**

   You now have $72y = 120$.

4. **Solve for the unknown variable that remains.**

   Dividing by 72 gives you $y = \frac{5}{3}$.

5. **Substitute the value of the found variable into either equation.**

   We chose the first equation: $20x + 24(\frac{5}{3}) = 10$.

6. **Solve for the final unknown variable.**

   You end up with $x = -\frac{3}{2}$.

7. **Check your solutions.**

   Always verify your answer by plugging the solutions back into the original system. These check out!

   $20(-\frac{3}{2}) + 24(\frac{5}{3})$

   $= -30 + 40 = 10$

   It works! Now check the other equation:

   $\frac{1}{3} \cdot (-\frac{3}{2}) + \frac{4}{5} \cdot (\frac{5}{3})$

   $-\frac{1}{2} + \frac{4}{3} = -\frac{3}{6} + \frac{8}{6} = \frac{5}{6}$. Because both values are solutions to both equations, the solution to the system is correct.

Working nonlinear systems

In a *nonlinear system,* at least one equation will have a graph that isn't a straight line. You can always write a linear equation in the form $Ax + By = C$ (where A, B, and C are real numbers); a nonlinear system is represented by any other form. Examples of nonlinear equations include, but are not limited to, any conic section, polynomial, rational function, exponential, or logarithm (all of which we cover in other parts of this book). The nonlinear systems you'll see in pre-calc will have two equations with two variables, as the three-dimensional systems are extremely difficult to solve (trust us on that one!). Because you're really working with a system with two equations and two variables (even though one or both equations are nonlinear), you have the same two methods at your disposal: substitution and elimination.

The method of solving nonlinear systems is different from that of linear systems in that these systems are much more complicated and therefore require much more work (those exponents really screw things up). Unlike before (with linear systems), nonlinear systems are less forgiving than the systems we cover earlier in the chapter. Usually, substitution is your best bet. Unless the variable you want to eliminate is raised to the same power in both equations, elimination won't get you anywhere.

When one system equation is nonlinear

If one equation in a system is nonlinear, your first thought before solving should be, "Bingo! Substitution method!" (or something to that effect). In this situation, you can solve for one variable in the linear equation and substitute this expression into the nonlinear equation, because solving for a variable in a linear equation is a piece of cake! And any time you can solve for one variable easily, you can substitute that expression into the other equation to solve for the other one.

For example, follow these steps to solve the system $\begin{cases} x - 4y = 3 \\ xy = 6 \end{cases}$:

1. **Solve the linear equation for one variable.**

   In the example system, the top equation is linear. If you solve for x, you get $x = 3 + 4y$.

2. **Substitute the value of the variable into the nonlinear equation.**

   When you plug $3 + 4y$ into the second equation for x, you get $(3 + 4y)y = 6$.

3. **Solve the nonlinear equation for the variable.**

   When you distribute the y, you get $4y^2 + 3y = 6$. Because this is a quadratic equation (refer to Chapter 4), you must get 0 on one side, so subtract the 6 from both sides to get $4y^2 + 3y - 6 = 0$. You have to use the quadratic formula to solve this equation for y:

   $$y = \frac{-3 \pm \sqrt{9 - 4(4)(-6)}}{2(4)} = \frac{-3 \pm \sqrt{9 + 96}}{8} = \frac{-3 \pm \sqrt{105}}{8}$$

   When you square root something, you get a positive and a negative answer, which means you have two different answers in this situation.

4. **Substitute the solution(s) into either equation to solve for the other variable.**

   Because you find two solutions for y, you have to substitute them both to get two different coordinate pairs. Here's what happens when you do:

- $x = 3 + 4\left(\frac{-3 \pm \sqrt{105}}{8}\right) = 3 + \frac{-3 \pm \sqrt{105}}{2}$
- $3 + \frac{-3 + \sqrt{105}}{2} = \frac{6}{2} + \frac{-3 + \sqrt{105}}{2} = \frac{3 + \sqrt{105}}{2}$
- $3 + \frac{-3 - \sqrt{105}}{2} = \frac{6}{2} + \frac{-3 - \sqrt{105}}{2} = \frac{3 - \sqrt{105}}{2}$

This gives you the solutions to the system: $\left(\frac{3 + \sqrt{105}}{2}, \frac{-3 + \sqrt{105}}{8}\right)$ and $\left(\frac{3 - \sqrt{105}}{2}, \frac{-3 - \sqrt{105}}{8}\right)$. These solutions represent the intersection of the line $x - 4y = 3$ and the rational function $xy = 6$.

When both system equations are nonlinear

If both of the equations in a system are nonlinear, well, you just have to get more creative to find the solutions. Unless one variable is raised to the same power in both equations, elimination is out of the question. Solving for one of the variables in either equation will not necessarily be easy, but it can usually be done. Then, plug this expression into the other equation and solve for the other variable just as you did before. Unlike linear systems, there may be many operations involved in the simplification or solving of these equations. Just remember to keep your order of operations in mind at each step of the way.

When both equations in a system are conic sections, you'll never find more than four solutions (unless the two equations describe the same conic section, in which case the system has an infinite number of solutions — and therefore is a dependent system). This is because conic sections are all very smooth curves with no sharp corners or crazy bends, so there is no way for two different conic sections to intersect more than four times.

For example, suppose a problem asks you to solve the following system:

$$\begin{cases} x^2 + y^2 = 9 \\ y = x^2 - 9 \end{cases}$$

Doesn't that just make your skin crawl? Don't break out the calamine lotion just yet, though. Follow these steps to find the solutions:

1. **Solve for x^2 or y^2 in one of the given equations.**

   The second equation is attractive because all you have to do is add 9 to both sides to get $y + 9 = x^2$.

2. **Substitute the value from Step 1 into the other equation.**

   You now have $y + 9 + y^2 = 9$. A-ha! This is a quadratic equation, and you know how to solve that (from Chapter 4).

3. Solve the quadratic equation.

Subtract 9 from both sides to get $y + y^2 = 0$.

Remember that you're not allowed, ever, to divide by a variable.

You must factor out the greatest common factor (GCF) instead to get $y(1 + y) = 0$. Use the zero product property to solve for $y = 0$ and $y = -1$. (Chapter 4 covers the basics of how to complete these tasks.)

4. Substitute the value(s) from Step 3 into either equation to solve for the other variable.

We chose to use the equation solved for in Step 1. When y is 0, $9 = x^2$, so $x = \pm 3$. When y is -1, $8 = x^2$, so $x = \pm 2\sqrt{2}$.

Be sure to keep track of which solution goes with which variable, because you have to express these solutions as points on a coordinate pair. Your answers are $(-3, 0)$, $(3, 0)$, $\left(-2\sqrt{2}, -1\right)$, and $\left(2\sqrt{2}, -1\right)$.

This solution set represents the intersections of the circle and the parabola given by the equations in the system.

Solving Systems with More than Two Equations

Larger systems of linear equations involve more than two equations that go along with more than two variables. These larger systems can be written in the form $Ax + By + Cz + \ldots = K$ where all coefficients (and K) are constants. These linear systems can have many variables, and you can solve those systems as long as you have one unique equation per variable. In other words, while three variables need three equations to find a unique solution, four variables need four equations, and ten variables would have to have ten equations, and so on. You do not need to concern yourself with larger systems of non-linear equations. That would be far too complicated for pre-calc, and larger linear systems are complicated enough. For these types of systems, the solutions you can find vary widely:

- You may find no solution.
- You may find one unique solution.
- You may come across infinitely many solutions.

The number of solutions you find depends on how the equations interact with one another. Because linear systems of three variables describe equations of planes, not lines (as two-variable equations do), the solution to the system depends on how the planes lie in three-dimensional space relative to one another. Unfortunately, just like in the systems of equations with two variables, you can't tell how many solutions the system has without doing the problem. Treat each problem as if it has one solution, and if it doesn't, you will either arrive at a statement that is never true (no solutions) or is always true (which means there are infinite solutions).

Typically, you must use the elimination method more than once to solve systems with more than two variables and two equations (see the earlier section "By using the process of elimination").

For example, suppose a problem asks you to solve the following system:

$$\begin{cases} x + 2y + 3z = -7 \\ 2x - 3y - 5z = 9 \\ -6x - 8y + z = -22 \end{cases}$$

To find the solution(s), follow these steps:

1. **Look at the coefficients of all the variables and decide which variable is easiest to eliminate.**

   With elimination, you want to find the least common multiple (LCM) for one of the variables, so go with the one that's the easiest. In this case, we recommend that you eliminate the *x* variable.

2. **Set apart two of the equations and eliminate one variable.**

   Looking at the first two equations, you have to multiply the top by –2 and add it to the second equation. Doing this, you get the following:

   $$\begin{array}{r} \cancel{-2x} - 4y - 6z = 14 \\ \cancel{2x} - 3y - 5z = 9 \\ \hline -7y - 11z = 23 \end{array}$$

3. **Set apart another two equations and eliminate the *same variable.***

   The first and the third equations allow you to easily eliminate *x* again. Multiply the top equation by 6 and add it to the third equation to get the following:

   $$\begin{array}{r} \cancel{6x} + 12y + 18z = -42 \\ \cancel{-6x} - 8y + z = -22 \\ \hline 4y + 19z = -64 \end{array}$$

4. **Repeat the elimination process with your two new equations.**

   You should now have two equations with two variables:

   $$-7y - 11z = 23$$
   $$4y + 19z = -64$$

You need to eliminate one of these variables. We've chosen to eliminate the *y* variable by multiplying the top equation by 4 and the bottom by 7 and then adding the equations. Here's what that gives you:

$$\begin{array}{r} \cancel{-28y} - 44z = 92 \\ \cancel{28y} + 133z = -448 \\ \hline 89z = -356 \end{array}$$

5. **Solve the final equation for the variable that remains.**

   If $89z = -356$, $z = -4$.

6. **Substitute the value of the solved variable into one of the equations that has two variables to solve for another one.**

   We chose the equation $-7y - 11z = 23$. Substituting, you have $-7y - 11(-4) = 23$, which simplifies to $-7y + 44 = 23$. Now finish the job:

   $$-7y = -21$$
   $$y = 3$$

7. **Substitute the two values you now have into one of the original equations to solve for the last variable.**

   We chose the first equation in the original system, which now becomes $x + 2(3) + 3(-4) = -7$. Simplify to get your final answer:

   $$x + 6 - 12 = -7$$
   $$x - 6 = -7$$
   $$x = -1$$

   The solutions to this equation are $x = -1$, $y = 3$, and $z = -4$.

This process is called back-substitution because you literally solve for one variable and then work your way backwards to solve for the others (you see this again later when solving matrices). In this last example, we went from the solution for one variable in one equation to two variables in two equations to the last step with three variables in three equations . . . always move from the more simple to the more complicated.

Decomposing Partial Fractions

A process called *partial fractions* takes one fraction and expresses it as the sum or difference of two other fractions. We can think of many reasons why you'd need to do this. In calculus, this process is useful before you integrate a function. Because integration is so much easier when the degree of a rational function is 1 in the denominator, partial fraction decomposition is a useful tool for you.

The process of decomposing partial fractions requires you to separate the fraction into two (or sometimes more) disjointed fractions with variables (usually *A*, *B*, *C*, and so on) standing in as placeholders in the numerator. Then you can set up a system of equations to solve for these variables. For instance, you must follow these steps to write the partial fraction decomposition of $\frac{11x+21}{2x^2+9x-18}$:

1. **Factor the denominator (see Chapter 4) and rewrite it as *A* over one factor and *B* over the other.**

   You do this because you want to break the fraction into two. The process unfolds as follows:

   $$\frac{11x+21}{2x^2+9x-18}=\frac{11x+21}{(2x-3)(x+6)}=\frac{A}{2x-3}+\frac{B}{x+6}$$

2. **Multiply every term you've created by the factored denominator and then cancel.**

   You'll multiply a total of three times in this example:

   $$\frac{11x+21}{\cancel{(2x-3)}\cancel{(x+6)}}\cdot\cancel{(2x-3)}\cancel{(x+6)}=$$

   $$\frac{A}{\cancel{(2x-3)}}\cdot\cancel{(2x-3)}(x+6)+\frac{B}{\cancel{(x+6)}}\cdot(2x-3)\cancel{(x+6)}$$

   This equals $11x + 21 = A(x + 6) + B(2x - 3)$.

3. **Distribute *A* and *B*.**

   This gives you $11x + 21 = Ax + 6A + 2Bx - 3B$.

4. **On the right side of the equation only, put all terms with an *x* together and all terms without it together.**

   Rearranging gives you $11x + 21 = Ax + 2Bx + 6A - 3B$.

5. **Factor out the *x* from the terms on the right side.**

   You now have $11x + 21 = (A + 2B)x + 6A - 3B$.

6. **Create a system out of this equation by pairing up terms.**

   For an equation to work, everything must be in balance. Because of this fact, the coefficients of x must be equal and the constants must be equal. If the coefficient of x is 11 on the left and $A + 2B$ on the right, you can say that $11 = A + 2B$ is one equation. Constants are the terms with no variable, and in this case, the constant on the left is 21. On the right side, $6A - 3B$ is the constant (because there is no variable attached) and so $21 = 6A - 3B$.

7. **Solve the system, using either substitution or elimination (see the earlier sections of this chapter).**

   We use elimination in this system. If $\begin{cases} A+2B=11 \\ 6A-3B=21 \end{cases}$, you can multiply the top equation by –6 and then add to eliminate and solve. You find that $A = 5$ and $B = 3$.

8. Write the solution as the sum of two fractions.

The partial fraction of $\frac{11x+21}{2x^2+9x-18}$ is $\frac{5}{2x-3}+\frac{3}{x+6}$.

Surveying Systems of Inequalities

In a *system of inequalities,* you see more than one inequality with more than one variable. Before pre-calculus, teachers tend to focus mostly on systems of linear inequalities. Those are inequalities whose graphs are straight lines. In pre-calc, though, you expand your study to systems of nonlinear inequalities because they are more thorough in the types of equations they cover (straight lines are so boring!).

In these systems of inequalities, at least one inequality isn't linear. The only way to solve a system of inequalities is to graph the solution. Fortunately, these graphs will look very similar to the graphs you have been graphing throughout your entire pre-calc course, and beyond. You may be required to graph inequalities that you haven't seen since pre-algebra. But for the most part, these inequalities will probably resemble the parent functions from Chapter 3 and conic sections from Chapter 12. The only difference between then and now is that the line that you graph will either be solid or dashed, depending on the problem, and you will get to color (or shade) where the solutions lie!

For example, consider the following nonlinear system of inequalities:

$$\begin{cases} x^2+y^2\leq 25 \\ y\geq -x^2+5 \end{cases}$$

To solve this system of equations, first graph the system. The fact that these are inequalities, and not equations, doesn't change the general shape of the graph at all. Therefore, you can graph these inequalities just as you would graph them if they were equations. The top equation of this example is a circle (for a quick refresher on graphing circles, refer to Chapter 12 about conic sections). This circle is centered at the origin, and the radius is 5. The second equation is an upside-down parabola (there's those pesky conic sections again!). It is shifted vertically 5 units, and flipped upside down. Because both of the inequality signs in this example include the equality line underneath (the first one is "less than or equal to" and the second is "greater than or equal to"), both lines should be solid.

If the inequality symbol says "strictly greater than: >" or "strictly less than: <" then the boundary line for the curve (or line) should be dashed.

After graphing, pick one test point that isn't on a boundary and plug it into the equations to see if you get true or false statements. The point(s) that you pick as a solution must work in every equation.

For example, our test point is (0, 4). If you plug this into the inequality for the circle (see Chapter 12), you get $0^2 + 4^2 \leq 25$. This is a true statement, because $16 \leq 25$, so you shade inside the circle. Now plug the same point into the parabola to get $4 \geq -0^2 + 5$, but because 4 isn't greater than 5, this is a false statement. You shade outside the parabola.

The solution of this system of inequalities is where the shading overlaps.

See Figure 13-1 for the final graph.

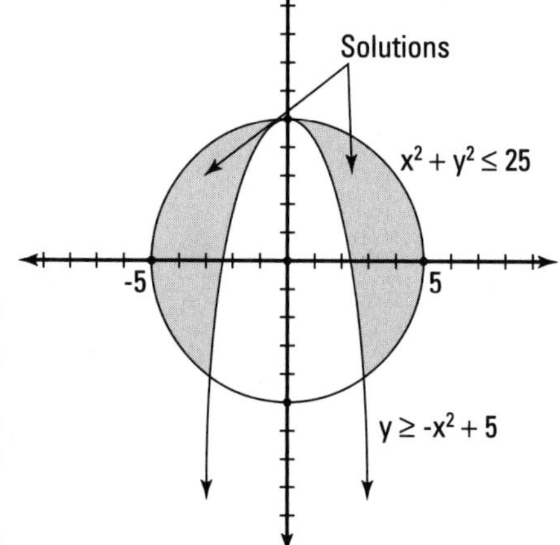

Figure 13-1: Graphing a nonlinear system of inequalities.

Introducing Matrices: The Basics

In the previous sections of this chapter, we cover how to solve systems of two or more equations by using substitution or elimination. But these methods will get very messy when the size of a system rises above three equations. Not to worry; whenever you have four or more equations to solve simultaneously, matrices are your best bet.

A *matrix* is a rectangle of numbers arranged in rows and columns. You use matrices to organize complicated data — say, for example, you want to keep track of sales records in your store. Matrices help you do that, because they can separate the sales by day in columns while different types of sales are organized by row.

After you get comfortable with what matrices are and how they are important, you can start adding, subtracting, and multiplying them by scalars and each other. Operating on matrices is useful when you need to add, subtract, or multiply large groups of data in an organized fashion. (***Note:*** There's no such thing as matrix division, so don't spend time worrying about it.) This section shows you how to perform all the above operations.

One thing to always remember when working with matrices is the order of operations, which is the same across all math applications: First do any multiplication, and then do the addition/subtraction.

You express the *dimensions,* sometimes called *order,* of a matrix as the number of rows by the number of columns. For example, if matrix M is 3×2, it has three rows and two columns.

To remember that rows come first, think of how you read in English — from left to right and then down, so the horizontal comes first.

Applying basic operations to matrices

Operating on matrices works very much like operating on multiple terms within parentheses; you just have more terms in the "parentheses" to work with. Just like with operations on numbers, there is a certain order involved with operating on matrices. Multiplication comes before addition and/or subtraction. When multiplying by a scalar, each and every element of the matrix gets multiplied. A *scalar* is a constant that multiplies a quantity (which changes its size, or "scale"). The following sections show you how to compute some of the more basic operations on matrices: addition, subtraction, and multiplication.

When adding or subtracting matrices, you just add or subtract their corresponding terms. It's as simple as that. Figure 13-2 shows how to add and subtract two matrices.

$$A = \begin{bmatrix} -5 & 1 & -3 \\ 6 & 0 & 2 \\ 2 & 6 & 1 \end{bmatrix} \qquad B = \begin{bmatrix} 2 & 4 & 5 \\ -8 & 10 & 3 \\ -2 & -3 & -9 \end{bmatrix}$$

$$A + B = \begin{bmatrix} -3 & 5 & 2 \\ -2 & 10 & 5 \\ 0 & 3 & -8 \end{bmatrix} \qquad A - B = \begin{bmatrix} -7 & -3 & -8 \\ 14 & -10 & -1 \\ 4 & 9 & 10 \end{bmatrix}$$

Figure 13-2: Addition and subtraction of matrices.

Note, however, that you can add or subtract matrices only if their dimensions are exactly the same. To add or subtract matrices, you add or subtract their corresponding terms; if the dimensions aren't exactly the same, then the terms won't line up. This will present a problem because you can't add or subtract terms that aren't there! For this reason, it's a general math rule that in order to add or subtract matrices, the dimensions must be the same.

When you multiply a matrix by a scalar, you're just multiplying by a constant. To do that, you multiply each term inside the matrix by the constant on the outside. Using the same matrix A from the previous example, you can find 3A by multiplying each term of matrix A by 3. Figure 13-3 shows this example:

$$3A = 3\begin{bmatrix} -5 & 1 & -3 \\ 6 & 0 & 2 \\ 2 & 6 & 1 \end{bmatrix} = \begin{bmatrix} -15 & 3 & -9 \\ 18 & 0 & 6 \\ 6 & 18 & 3 \end{bmatrix}$$

Figure 13-3: Multiplying matrix A by 3.

Suppose a problem asks you to combine operations. You simply multiply each matrix by the scalar separately, and then add or subtract them. For example, if $A = \begin{bmatrix} 3 & -4 \\ 2 & 6 \end{bmatrix}$ and $B = \begin{bmatrix} 8 & -10 \\ -5 & 4 \end{bmatrix}$, you find 3A – 2B as follows:

1. **Insert the matrices into the problem.**

   The setup is $3\begin{bmatrix} 3 & -4 \\ 2 & 6 \end{bmatrix} - 2\begin{bmatrix} 8 & -10 \\ -5 & 4 \end{bmatrix}$.

2. **Multiply the scalars into the matrices.**

   You now have $\begin{bmatrix} 9 & -12 \\ 6 & 18 \end{bmatrix} - \begin{bmatrix} 16 & -20 \\ -10 & 8 \end{bmatrix}$.

3. **Complete the problem by adding or subtracting the matrices.**

   Your final answer is $\begin{bmatrix} -7 & 8 \\ 16 & 10 \end{bmatrix}$.

Multiplying matrices by each other

Multiplying matrices is very useful when solving systems of equations, because you can multiply a matrix by its inverse (don't worry, we'll tell you how to find that) on both sides of the equal sign to eventually get the variable matrix on one side, and the solution to the system on the other.

Multiplying two matrices together is not as simple as multiplying the corresponding terms (although we wish it were!). Each element of each matrix gets multiplied by each term of the other at some point. In fact, multiplication of matrices and dot products of vectors are actually quite similar. There is a very methodical way of multiplying certain terms and then adding them together. The difference is that with vectors, this only needs to be done once. With matrices, you could be here doing this all day, depending on the size of the matrices.

In order to multiply two matrices, say for instance AB (for matrix multiplication, the matrices are written right next to each other with no symbol in between), the number of columns in A must match the number of rows in B. This is because to multiply A times B, each element in the first row of A gets multiplied by each corresponding element from the first column of B, and then all of these products get added together to give you the element in the [first row, first column] of AB. To find the value in the [first row, second column] position, multiply each element in the first row of A by each element in the second column of B and then add them all together. In the end, after all the multiplication and addition are finished, your new matrix should have the same number of rows as A, and the same number of columns as B.

For example, to multiply a matrix with 3 rows and 2 columns by a matrix with 2 rows and 4 columns, you would multiply the first row with each of the columns to 4 terms in the new row. Multiplying the second row with the columns produces a row of another 4 terms. And the same goes for the last row. You end up with a matrix of 3 rows and 4 columns.

Note: If matrix A has dimensions $m \times n$ and matrix B has dimensions $n \times p$, AB is an $m \times p$ matrix. See Figure 13-4 for a visual representation of matrix multiplication.

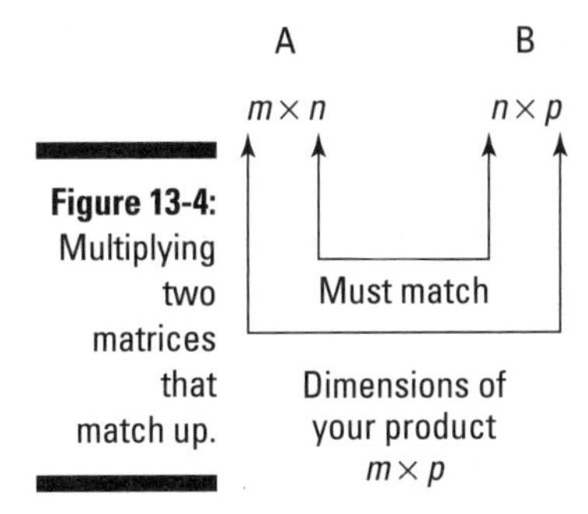

Figure 13-4: Multiplying two matrices that match up.

When you multiply matrices, you don't multiply the corresponding parts like when you add or subtract. You don't multiply the [first row, first column] term of the first matrix by the [first row, first column] term of the second matrix. Matrix multiplication follows many of the same ideas as dot products in vectors where you multiply many things and then add them together. Also, in matrix multiplication, AB doesn't equal BA. In fact, just because you can multiply A by B doesn't even mean you *can* multiply B by A. This is because if the columns in A are equal to the rows in B, it is not necessarily true that the columns in B will equal the rows in A. Take the example above into consideration. You can multiply a matrix with 3 rows and 2 columns by a matrix with 2 rows and 4 columns. However, you can't do the multiplication the other way. There is no way to multiply the matrix with 2 rows and 4 columns by the matrix with 3 rows and 2 columns. If you tried to multiply the correct terms together and then add them, somewhere along the way you would run out!

Also note that AB isn't the same as A × B when it comes to matrices. When two matrices are written right next to each other without any symbols in between, this stands for matrix multiplication. The multiplication dot (·) is reserved for scalar multiplication. The symbol × is used to symbolize the *cross product,* and it represents something completely different. Cross products are only used in 3 × 3 matrices, and have certain applications to physics. Because of their specific nature, we do not cover them in this book.

Time for an example. Say you have matrix $A = \begin{bmatrix} 5 & -6 \\ -3 & 9 \\ 2 & 4 \end{bmatrix}$ and matrix $B = \begin{bmatrix} -2 & 4 & 8 & -5 \\ 1 & 3 & -4 & -2 \end{bmatrix}$ and a problem asks you to multiply them. First, check to make sure that you can multiply the two matrices. Matrix A is 3 × 2 and B is 2 × 4, so you can multiply them to get a 3 × 4 matrix as an answer. Now you can proceed to multiply every row of the first matrix times every column of the second.

We lay out this process for you in Figure 13-5. You can start by multiplying each term in the first row of A by the sequential terms in the columns of matrix B. Note that multiplying row one by column one and adding them together gives you [row one, column one]'s answer. Similarly, multiplying row two by column three gives you [row two, column three]'s answer.

5 -6 ×-2 1 -10 + -6 = -16	5 -6 × 4 3 20 + -18 = 2	5 -6 × 8 -4 40 + 24 = 64	5 -6 ×-5 -2 -25 + 12 = -13
-3 9 -2 1 6 + 9 = 15	-3 9 × 4 3 -12 + 27 = 15	-3 9 × 8 -4 -24 + -36 = -60	-3 9 ×-5 -2 15 + -18 = -3
2 4 ×-2 1 -4 + 4 = 0	2 4 × 4 3 8 + 12 = 20	2 4 × 8 -4 16 + -16 = 0	2 4 ×-5 -2 -10 + -8 = -18

Figure 13-5: The process of multiplying AB.

Taking out all the fluff, $\begin{bmatrix} -16 & 2 & 64 & -13 \\ 15 & 15 & -60 & -3 \\ 0 & 20 & 0 & -18 \end{bmatrix}$ is the answer matrix.

Simplifying Matrices to Ease the Solving Process

In a system of linear equations, where each equation is in the form $Ax + By + Cz + \ldots = K$, the coefficients of this system can be represented in a matrix, called the *coefficient matrix*. If all the variables line up with one another vertically, then the first column of the coefficient matrix is dedicated to all the coefficients of the first variable, the second row is for the second variable, and so on. Each row then represents the coefficients to each variable in order as they appear in the system of equations. Through a couple of different processes, you can manipulate the coefficient matrix in order to make the solutions easier to find. Solving a system of equations using a matrix is a great method, especially for larger systems (with more variables and more equations). But don't get us wrong, these methods work for systems of all sizes, so its up to you which method you choose for which problem. The following sections break down the available simplifying processes.

Writing a system in matrix form

You can write any system of equations as a matrix.

Take a look at the following system:

$$\begin{cases} x + 2y + 3z = -7 \\ 2x - 3y - 5z = 9 \\ -6x - 8y + z = -22 \end{cases}$$

To express this system in matrix form, you follow three simple steps:

1. **Write all the coefficients in one matrix first (this is called a *coefficient matrix*).**
2. **Multiply this matrix with the variables of the system set up in another matrix (some books call this the *variable matrix*).**
3. **Insert the answers on the other side of the equal sign in another matrix (some books call this the *answer matrix*).**

The setup appears as follows:

$$\begin{bmatrix} 1 & 2 & 3 \\ 2 & -3 & -5 \\ -6 & -8 & 1 \end{bmatrix} \begin{bmatrix} x \\ y \\ z \end{bmatrix} = \begin{bmatrix} -7 \\ 9 \\ -22 \end{bmatrix}$$

Notice that the coefficients in the matrix go in order — you see a column for x, y, and z.

Reduced row echelon form

You can find the *reduced row echelon form* of a matrix to find the solutions to a system of equations. This, however, is a complicated process and we don't really recommend it. However, just like with completing the square in Chapter 4, sometimes you'll be required to solve problems in a specific manner. If you are asked to find the reduced row echelon form of a matrix, don't worry. We'll help you through it. It's beneficial to put a matrix into reduced row echelon form because this form of a matrix is unique to each matrix (and that unique matrix could give you the solutions to your system of equations). There can't be two different matrices with the same reduced row echelon form. There can, however, be infinitely many matrices in row echelon form (not reduced) that are not unique; they will all be scalar multiples of one another.

Reduced row echelon form shows a matrix with a very specific set of requirements. These requirements pertain to where any rows of all 0s lie as well as what the first number in any row is. ***Note:*** The first number in a row of a matrix that is not 0 is called the *leading coefficient.* If any of the following requirements are not met, then the matrix is *not* considered to be in reduced row echelon form:

- All rows containing all 0s are at the bottom of the matrix.
- All leading coefficients are 1.
- Any element above or below a leading coefficient is 0.
- The leading coefficient of any row is always to the left of the leading coefficient below it.

Figure 13-6a shows you a matrix in reduced row echelon form, and Figure 13-6b is not in reduced row echelon form because the 7 is directly above the leading coefficient of the last row, and the 2 is above the leading coefficient in row two.

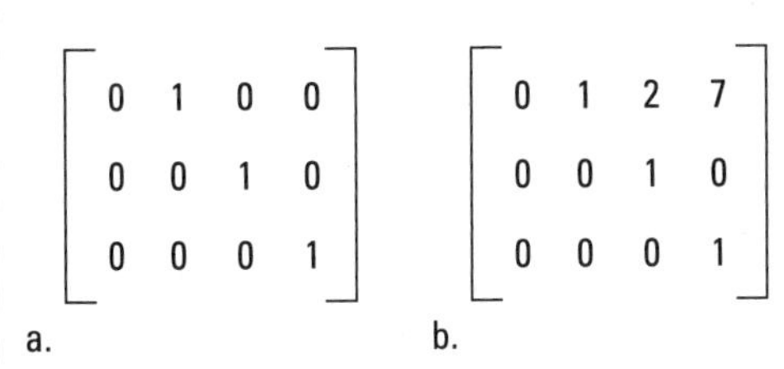

Figure 13-6: A matrix (a) in reduced row echelon form and (b) not in reduced row echelon form.

The row echelon form of a matrix comes in handy for solving systems of equations that are 4×4 or larger, because the method of elimination would entail an enormous amount of work on your part. Here, we show you how to get a matrix into row echelon form using *elementary row operations.* These operations are different from the operations on matrices discussed in the previous section because these operations are only carried out on *one* row of a matrix at a time. These are the operations that you can use on a row of a matrix to get it into row echelon form:

- Multiply each element in a single row by a constant.
- Interchange two rows.
- Add two rows together.

Using these elementary row operations, you can rewrite any matrix so that the solutions to the system that the matrix represents become apparent. We'll show you how in a bit in the section called "Conquering Matrices."

Use the reduced row echelon form *only* if you're specifically told to do so by a pre-calc teacher or textbook (and there will be some that do, so we've included it to help you). Otherwise, use any of the other methods we tell you about in this chapter (Cramer's rule is a *great* one!). Reduced row echelon form takes a lot of time, energy, and precision. It can take a ton of steps, which means that there are a ton of places to get mixed up. If you have the choice, we recommend opting for a less rigorous tactic (unless, of course, you're trying to show off).

Perhaps the most famous (and useful) matrix in pre-calculus is called the identity matrix, which has 1s along the diagonal from the upper-left corner to the lower-right, and 0s everywhere else. It is a square matrix in reduced row echelon form, and stands for the identity element of multiplication in the world of matrices (remember the identity property of multiplication from Algebra II?). This means that multiplying a matrix by the identity will result in the same matrix.

This is an important idea in solving systems because if you can manipulate the coefficient matrix to look like the identity matrix (using legal matrix operations, which we told you about in the previous section), then the solution to the system will be on the other side of the equal sign.

$$\begin{bmatrix} 1 & 0 & 0 \\ 0 & 1 & 0 \\ 0 & 0 & 1 \end{bmatrix} \begin{bmatrix} x \\ y \\ z \end{bmatrix} = \begin{bmatrix} -1 \\ 3 \\ -4 \end{bmatrix}$$

Rewriting this matrix as a system produces the values $x = -1$, $y = 3$, and $z = -4$. But you don't have to take the coefficient matrix this far just to get a solution. You can write it in row echelon form, as follows:

$$\begin{bmatrix} 1 & 2 & 3 \\ 0 & 1 & \frac{11}{7} \\ 0 & 0 & 1 \end{bmatrix} \begin{bmatrix} x \\ y \\ z \end{bmatrix} = \begin{bmatrix} -7 \\ -\frac{23}{7} \\ -4 \end{bmatrix}$$

This is different from reduced row echelon form because row echelon form allows for there to be numbers above the leading coefficients, but not below. Rewriting this system gives you the following from the rows:

$$\begin{aligned} x + 2y + 3z &= -7 \\ y + \frac{11}{7}z &= \frac{-23}{7} \\ z &= -4 \end{aligned}$$

How do you get to the solution from there? The answer to that question is *back solving,* also known as *back substitution.* If a matrix is written in row echelon form, then the variable on the bottom row should be solved for. You can then plug this value into the equation above to solve for another variable. You should be able to continue this process, moving your way up (or backwards) until you have solved for all the variables. This is the same as a system of equations where you move from the simplest equation to the most complicated.

Here's how you execute the back solving: Now that you know $z = -4$, you can substitute that value into the second equation to get y:

$$\begin{aligned} y + \frac{11}{7}(-4) &= \frac{-23}{7} \\ y - \frac{44}{7} &= \frac{-23}{7} \\ y &= 21/7 = 3 \end{aligned}$$

And now that you know z and y, you can go back further into the first equation to get x:

$$\begin{aligned} x + 2(3) + 3(-4) &= -7 \\ x + 6 - 12 &= -7 \\ x - 6 &= -7 \\ x &= -1 \end{aligned}$$

Augmented form

You also can write a matrix in what's known as *augmented form,* where the coefficient matrix and the solution matrix are written in the same matrix, separated in each row by colons. This makes using elementary row operations to solve a matrix much simpler because you only have one matrix on your plate at a time (as opposed to three!) You can do this because you know that your main job is to solve for the variables, and each column in the coefficient matrix represents a different variable. Mathematicians are a very lazy bunch, and they like to write as little as possible. Using augmented form will cut down on the amount that you have to write. And when you're attempting to solve a system of equations that requires many steps, you'll be thankful to be writing less! Then, you can use elementary row operations just as before to get the solution to your system.

Consider this matrix equation:

$$\begin{bmatrix} 1 & 2 & 3 \\ 2 & -3 & -5 \\ -6 & -8 & 1 \end{bmatrix} \begin{bmatrix} x \\ y \\ z \end{bmatrix} = \begin{bmatrix} -7 \\ 9 \\ -22 \end{bmatrix}$$

Written in augmented form, it looks like this:

$$\begin{bmatrix} 1 & 2 & 3 & : & -7 \\ 2 & -3 & -5 & : & 9 \\ -6 & -8 & 1 & : & -22 \end{bmatrix}$$

Conquering Matrices

When you are comfortable with changing the appearances of matrices (to get them into augmented, and then reduced row echelon, form, for instance), you are ready to tackle matrices and really start solving difficult systems. Hopefully, for really large systems (four or more variables) you'll have the aid of a graphing calculator. Computer programs can also be very helpful with matrices, and can solve systems of equations in a variety of ways. The three ways we introduce to you in this section are Gaussian elimination, matrix inverses, and Cramer's rule. Gaussian elimination is probably the best method to use if you don't have a graphing calculator or computer program to help you. If you do have these tools, then you can use either of them to find the inverse of any matrix, and then the inverse operation is the best plan. If the system only has two or three variables, and you don't have a graphing calculator to help you, then Cramer's rule is a good way to go.

The previous section is all about getting the matrices into the form that would make it easy for you to solve. Now we take you one step further to getting into the form and then actually solving these messy things. By the time we're finished with you, you'll be an expert at solving complicated systems of equations.

Using Gaussian elimination to solve systems

Gaussian elimination requires the use of the elementary row operations from the section "Reduced row echelon form." We'll be using an augmented matrix, because that is most often how you're asked to solve (and the easiest way to get to the solutions as well).

The goals of Gaussian elimination are to make the upper-left corner element a 1; use elementary row operations to get 0s in all positions underneath that first 1; get 1s for leading coefficients in every row diagonally from the upper-left to lower-right corner; and get 0s beneath all leading coefficients. Basically, you eliminate all variables except for one in the last row, all variables except for two in the equation above that one, and so on and so forth to the top equation, which will have all the variables. Then you can use back-substitution to solve for one variable at a time by plugging the values you know into the equations from the bottom up.

You accomplish this elimination by eliminating the x variable (or whatever comes first) in all equations except for the first one. Then eliminate the second variable in all equations except for the first two. This process continues, eliminating one more variable per line until there is only one variable left in the last line. Then solve for that variable.

Elementary operations for Gaussian elimination are the same as the elementary row operations used on matrices in the previous section. We have stated them again here so that you don't have to look back.

You can perform three operations on matrices in order to eliminate variables in a system of linear equations:

- **You can switch any two rows:** $r_1 \leftrightarrow r_2$ would swap rows one and two.
- **You can multiply any row by a constant:** $-2r_3 \rightarrow r_3$ would multiply row three by –2 to give you a new row three.
- **You can add two rows together:** $r_1 + r_2 \rightarrow r_2$ adds rows 1 and 2 and writes it in row 2.

Then, this means you can multiply a row by a constant and then add it to another row to change that row: $3r_1 + r_2 \rightarrow r_2$ would multiply row one by 3 and then add that to row two to create a new row two.

Consider the following augmented matrix:

$$\begin{bmatrix} 1 & 2 & 3 & : & -7 \\ 2 & -3 & -5 & : & 9 \\ -6 & -8 & 1 & : & -22 \end{bmatrix}$$

Now take a look at the goals of Gaussian elimination in order to complete the following steps to solve this matrix:

1. **Complete the first goal: to get 1 in the upper-left corner.**

   You already have it!

2. **Complete the second goal: to get 0s underneath the 1 in the first column.**

   You need to use the combo of the second and third matrix operations together here. Here's what you should ask: "What do I need to add to row two to make a 2 become a 0?" The answer is –2. So, you perform the following operation: $-2r_1 + r_2 \rightarrow r_2$. Completing the math, you now have this matrix:

   $$\begin{bmatrix} 1 & 2 & 3 & : & -7 \\ 0 & -7 & -11 & : & 23 \\ -6 & -8 & 1 & : & -22 \end{bmatrix}$$

3. **Get a 0 under the 1 in the third row.**

   To do this, you need the following operation: $6r_1 + r_3 \rightarrow r_3$. With this calculation, you should now have the following matrix:

   $$\begin{bmatrix} 1 & 2 & 3 & : & -7 \\ 0 & -7 & -11 & : & 23 \\ 0 & 4 & 19 & : & -64 \end{bmatrix}$$

4. **Get a 1 in the second row, second column.**

   To do this, you need to use the second row operation; in other words, multiply row two by the appropriate reciprocal: $\frac{-1}{7}r_2 \rightarrow r_2$. This produces a new second row:

   $$\begin{bmatrix} 1 & 2 & 3 & : & -7 \\ 0 & 1 & \frac{11}{7} & : & -\frac{23}{7} \\ 0 & 4 & 19 & : & -64 \end{bmatrix}$$

5. **Get a 0 under the 1 you created in row two.**

   Back to the good old combo operation for the third row: $-4r_2 + r_3 \rightarrow r_3$. Here's yet another version of the matrix:

   $$\begin{bmatrix} 1 & 2 & 3 & : & -7 \\ 0 & 1 & \frac{11}{7} & : & \frac{-23}{7} \\ 0 & 0 & \frac{89}{7} & : & \frac{-356}{7} \end{bmatrix}$$

6. **Get another 1, this time in the third row, third column.**

   Multiply the third row by the reciprocal of the coefficient to get a 1: $\frac{7}{89}r_3 \rightarrow r_3$. You've completed the main diagonal after doing the math:

   $$\begin{bmatrix} 1 & 2 & 3 & : & -7 \\ 0 & 1 & \frac{11}{7} & : & \frac{-23}{7} \\ 0 & 0 & 1 & : & -4 \end{bmatrix}$$

You now have a matrix in row echelon form, which gives you the solutions when you use back substitution (referring to this matrix in the section "Reduced row echelon form," you know that $z = -4$). However, if you want to know how to get this matrix into reduced row echelon form to find the solutions, follow these steps:

1. **Get a 0 in row two, column three.**

   The operation $\frac{-11}{7}r_3 + r_2 \rightarrow r_2$ gives you the following:

   $$\begin{bmatrix} 1 & 2 & 3 & : & -7 \\ 0 & 1 & 0 & : & 3 \\ 0 & 0 & 1 & : & -4 \end{bmatrix}$$

2. **Get a 0 in row one, column three.**

   The operation $-3r_3 + r_1 \rightarrow r_1$ gives you the following:

   $$\begin{bmatrix} 1 & 2 & 0 & : & 5 \\ 0 & 1 & 0 & : & 3 \\ 0 & 0 & 1 & : & -4 \end{bmatrix}$$

3. Get a 0 in row one, column two.

Finally, the operation $-2r_2 + r_1 \rightarrow r_1$ gives you the following:

$$\begin{bmatrix} 1 & 0 & 0 & : & -1 \\ 0 & 1 & 0 & : & 3 \\ 0 & 0 & 1 & : & -4 \end{bmatrix}$$

This matrix, in reduced row echelon form, is actually the solution to the system. If you multiply the two matrices on the left side, and change the colons back into equal signs, you get a matrix that looks like $\begin{bmatrix} x \\ y \\ z \end{bmatrix} = \begin{bmatrix} -1 \\ 3 \\ -4 \end{bmatrix}$.

Multiplying a matrix by its inverse

You can add another way to solve a system of equations by using matrices to your arsenal; it's based on the simple idea that if you have a coefficient tied to a variable on one side of an equation, you can multiply by the coefficient's inverse to make that coefficient go away and leave you with just the variable. For example, if $3x = 12$, how would you solve the equation? You'd divide both sides by 3, which is the same thing as multiplying by ⅓, to get $x = 4$. So it goes with matrices.

In variable form, an inverse function is written as $f^{-1}(x)$, where f^{-1} is the inverse of the function f. You name an inverse matrix similarly; the inverse of matrix A is A^{-1}. If A, B, and C are matrices in the matrix equation AB = C, and you want to solve for B, how do you do that? Just multiply by the inverse of matrix A, which you write like this:

$$A^{-1}[AB] = A^{-1}C$$

So, the simplified version is $B = A^{-1}C$.

Now that you've simplified the basic equation, you need to calculate the inverse matrix in order to calculate the answer to the problem.

Finding a matrix's inverse

First off, we must establish that only square matrices have inverses — in other words, the number of rows must be equal to the number of columns. And even then, not every square matrix has an inverse. If the determinant of a matrix is not 0, then the matrix will have an inverse. See the following section on Cramer's rule for more on determinants.

When a matrix has an inverse, you have several ways to find it, depending on how big the matrix is. If the matrix is a 2×2 matrix, then there is a simple formula to find the inverse. However, for anything larger than 2×2, we recommend that you use a graphing calculator or computer program (there are many Web sites that will find matrix inverses for you, and most teachers and textbooks will give you the inverse matrix for any system that is 3×3 or bigger).

If you don't use a graphing calculator, you can augment your original, invertible matrix with the identity matrix and use elementary row operations to get the identity matrix where your original matrix once was. This leaves the inverse matrix where you had the identity originally. This, however, is extremely difficult, and we don't really recommend it.

With that said, here's how you find an inverse of a 2×2 matrix:

If matrix A is the 2×2 matrix $\begin{bmatrix} a & b \\ c & d \end{bmatrix}$, its inverse is as follows:

$$\frac{1}{ad - bc}\begin{bmatrix} d & -b \\ -c & a \end{bmatrix}$$

Simply follow this format with any 2×2 matrix you're asked to find.

Using an inverse to solve a system

Armed with a system of equations and the knowledge of how to use inverse matrices (see the previous section), you can follow a series of simple steps to arrive at a solution to the system, again using the trusty old matrix. For instance, you can solve the system that follows by using inverse matrices:

$$\begin{cases} 4x + 3y = -13 \\ -10x - 2y = 5 \end{cases}$$

These steps show you the way:

1. **Write the system as a matrix equation.**

   When written as a matrix equation (see the earlier section "Writing a system in matrix form"), you get

   $$\begin{bmatrix} 4 & 3 \\ -10 & -2 \end{bmatrix}\begin{bmatrix} x \\ y \end{bmatrix} = \begin{bmatrix} -13 \\ 5 \end{bmatrix}$$

2. **Create the inverse matrix out of the matrix equation.**

   The inverse matrix is

   $$\frac{1}{22}\begin{bmatrix} -2 & -3 \\ 10 & 4 \end{bmatrix}$$

3. **Multiply the inverse in the front on both sides of the equation.**

   You now have the following equation:

$$\frac{1}{22}\begin{bmatrix}-2 & -3\\ 10 & 4\end{bmatrix}\cdot\begin{bmatrix}4 & 3\\ -10 & -2\end{bmatrix}\begin{bmatrix}x\\ y\end{bmatrix}=\frac{1}{22}\begin{bmatrix}-2 & -3\\ 10 & 4\end{bmatrix}\begin{bmatrix}-13\\ 5\end{bmatrix}=\frac{1}{22}\begin{bmatrix}(-2)(-13)+(-3)(5)\\ (10)(-13)+(4)(5)\end{bmatrix}$$

4. **Cancel the matrix on the left and multiply the matrices on the right (see the section "Multiplying matrices by each other").**

   An inverse matrix times a matrix cancels out. You're left with

$$\begin{bmatrix}x\\ y\end{bmatrix}=\frac{1}{22}\begin{bmatrix}11\\ -110\end{bmatrix}$$

5. **Multiply the scalar to solve the system.**

   You finish with the x and y values: $\begin{bmatrix}x\\ y\end{bmatrix}=\begin{bmatrix}\frac{1}{2}\\ 5\end{bmatrix}$

It's usually easier to multiply the scalar after you multiply the two matrices.

Using determinants: Cramer's rule

The final method that we show you for solving systems (almost home!) was thought up by Gabriel Cramer and is named after him. As with much of what this chapter covers, the graphing calculator enables you to bypass much of the legwork and has made life a ton easier for pre-calc students. However, if your teacher asks you to use Cramer's rule, and some certainly will, you can impress him or her with all the know-how you pick up in this section!

Cramer's rule says that if the determinant of a coefficient matrix $|A|$ (see the earlier section "Simplifying Matrices to Ease the Solving Process" for more info on how to find the coefficient matrix) is not 0, then the solutions to a system of linear equations can be found as follows:

If the matrix describing the system of equations looks like

$$\begin{bmatrix}a_1 & b_1 & c_1 & \cdots\\ a_2 & b_2 & c_2 & \cdots\\ a_3 & b_3 & c_3 & \cdots\\ \vdots & \vdots & \vdots & \end{bmatrix}\begin{bmatrix}x_1\\ x_2\\ x_3\\ \vdots\end{bmatrix}=\begin{bmatrix}k_1\\ k_2\\ k_3\\ \vdots\end{bmatrix}, \text{ then}$$

$$x_1 = \frac{\begin{vmatrix} k_1 & b_1 & c_1 & \cdots \\ k_2 & b_2 & c_2 & \cdots \\ k_3 & b_3 & c_3 & \cdots \\ \vdots & \vdots & \vdots & \end{vmatrix}}{\begin{vmatrix} a_1 & b_1 & c_1 & \cdots \\ a_2 & b_2 & c_2 & \cdots \\ a_3 & b_3 & c_3 & \cdots \\ \vdots & \vdots & \vdots & \end{vmatrix}}$$

$$x_2 = \frac{\begin{vmatrix} a_1 & k_1 & c_1 & \cdots \\ a_2 & k_2 & c_2 & \cdots \\ a_3 & k_3 & c_3 & \cdots \\ \vdots & \vdots & \vdots & \end{vmatrix}}{\begin{vmatrix} a_1 & b_1 & c_1 & \cdots \\ a_2 & b_2 & c_2 & \cdots \\ a_3 & b_3 & c_3 & \cdots \\ \vdots & \vdots & \vdots & \end{vmatrix}}$$

$$x_3 = \frac{\begin{vmatrix} a_1 & b_1 & k_1 & \cdots \\ a_2 & b_2 & k_2 & \cdots \\ a_3 & b_3 & k_3 & \cdots \\ \vdots & \vdots & \vdots & \end{vmatrix}}{\begin{vmatrix} a_1 & b_1 & c_1 & \cdots \\ a_2 & b_2 & c_2 & \cdots \\ a_3 & b_3 & c_3 & \cdots \\ \vdots & \vdots & \vdots & \vdots \end{vmatrix}}$$

and so on until you have solved for all the variables.

This rule is helpful when the systems are very small, or when you can use a graphing calculator to determine the determinants, because it helps you find the solutions with minimal places to get mixed up. To use it, you simply find the determinant of the coefficient matrix.

The determinant of a 2×2 matrix $\begin{bmatrix} a & b \\ c & d \end{bmatrix}$ is defined to be $ad - bc$. The determinant of a 3×3 matrix is a bit more complicated. If the matrix is $A = \begin{bmatrix} a_1 & b_1 & c_1 \\ a_2 & b_2 & c_2 \\ a_3 & b_3 & c_3 \end{bmatrix}$, then you can find the determinant as follows. Rewrite the first two columns immediately after the third column. Draw three diagonal lines from the upper left to the lower right, and three diagonal lines from the lower left to the upper right, as shown in Figure 13-7.

$$|A| = \begin{vmatrix} a_1 & b_1 & c_1 \\ a_2 & b_2 & c_2 \\ a_3 & b_3 & c_3 \end{vmatrix} \begin{matrix} a_1 & b_1 \\ a_2 & b_2 \\ a_3 & b_3 \end{matrix}$$

Figure 13-7: How to find a matrix's determinant.

Then multiply down the three diagonals from left to right, and up the other three. The determinant of the 3 × 3 matrix is:

$$(a_1b_2c_3 + b_1c_2a_3 + c_1a_2b_2) - (a_3b_2c_1 + b_3c_2a_1 + c_3a_2b_1)$$

To find the determinant of the 3 × 3 matrix $\begin{vmatrix} 1 & 2 & 3 \\ 2 & -3 & -5 \\ -6 & -8 & 1 \end{vmatrix}$, you use a process known as *using diagonals,* which you can see in Figure 13-8.

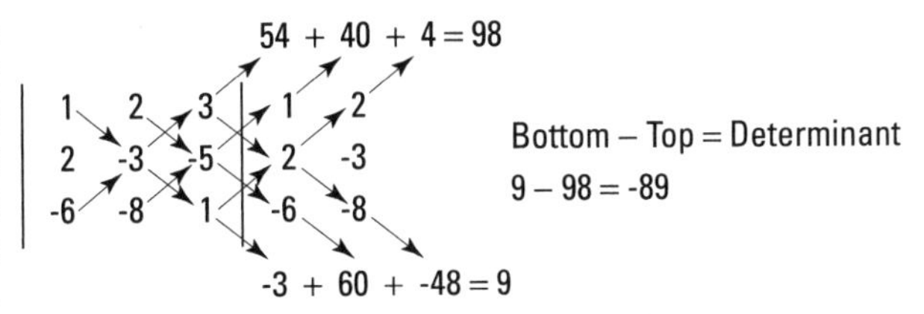

Figure 13-8: Finding the determinant of a 3 × 3 matrix is all about the diagonals.

After you find the determinant of the coefficient matrix (either by hand or with a technological device), replace the first column of the coefficient matrix with the solution matrix from the other side of the equal sign, and find the determinant of that new matrix. Then, replace the second column of the coefficient matrix with the solution matrix and find the determinant of that matrix. Continue this process until you have replaced each column and found each new determinant. The values of the respective variables are equal to the determinant of the new matrix (when you replaced the respective column) divided by the determinant of the coefficient matrix.

You can't use Cramer's rule when the matrix is not square, or when the determinant of the coefficient matrix is 0, because you can't divide by 0. Cramer's rule is most useful for a 2×2 or higher system of linear equations.

To solve a 3×3 system of equations such as

$$\begin{cases} ax + by + cz = d \\ ex + fy + gz = h \\ jx + ky + mz = n \end{cases}$$

using Cramer's rule, you set up the variables as follows:

$$x = \frac{\begin{vmatrix} d & b & c \\ h & f & g \\ n & k & m \end{vmatrix}}{\begin{vmatrix} a & b & c \\ e & f & g \\ j & k & m \end{vmatrix}}$$

$$y = \frac{\begin{vmatrix} a & d & c \\ e & h & g \\ j & n & m \end{vmatrix}}{\begin{vmatrix} a & b & c \\ e & f & g \\ j & k & m \end{vmatrix}}$$

$$z = \frac{\begin{vmatrix} a & b & d \\ e & f & h \\ j & k & n \end{vmatrix}}{\begin{vmatrix} a & b & c \\ e & f & g \\ j & k & m \end{vmatrix}}$$

Chapter 14

Sequences, Series, and Expanding Binomials

In This Chapter

- Exploring the terms and formulas of sequences
- Grasping arithmetic and geometric sequences
- Summing sequences to create a series
- Applying the binomial theorem to expand binomials

It's time to put aside your graph paper and the many complex, intangible concepts that pre-calculus presents to you, such as the unit circle, conics, and logs. This chapter is dedicated to how you can really use pre-calculus in the real world. The real-world applications from previous chapters are each useful to probably a handful of people. This chapter is different because the applications are useful to *everyone*. No matter who you are, or what you do, you should probably understand the value of your belongings. We focus on a few different topics to take math out of the classroom and into the fresh air:

- **Sequences** help you understand patterns. You can see patterns develop, for example, in how much your car depreciates, how credit-card interest builds, and how scientists estimate the growth of bacteria populations.
- **Series** help you understand the sum of a sequence of numbers, such as annuities, the height a ball bounces (if you really want to figure that out in your free time), and so on.

This chapter dives into these topics and debunks the myth that math isn't useful in the real world.

Speaking Sequentially: Grasping the General Method

A *sequence* is basically an ordered list of numbers, following some sort of pattern. This pattern can usually be described by a general rule that will allow you to find out any of the numbers in this list without having to find *all* the numbers in between. It's infinite, meaning it can continue in the same pattern forever. A sequence's mathematical definition is a function defined over the set of positive integers, usually written in the following form:

$$\{a_n\} = a_1, a_2, a_3, \ldots, a_n, \ldots$$

The $\{a_n\}$ portion represents the notation for the entire set of numbers. Each a_n is called a *term of the sequence;* a_1 is the first term, a_2 is the second term, and so on. The a_n is the nth term, meaning it can be any term you need it to be.

In the real world, sequences are helpful when describing any quantity that increases or decreases with time — financial interest, debt, sales, populations, and asset depreciation or appreciation, to name a few. Any quantity that changes with time based on a certain percentage will follow a pattern that can be described using a sequence. Depending on the rule for the sequence, you can multiply the initial value of an object by a certain percentage to find a new value after a certain length of time. Repeating this process will reveal the general pattern and the change in value for the object.

Calculating a sequence's terms by using the sequence expression

The general formula for any sequence involves the letter n, which is the number of the term (the first term would be $n = 1$, while the 20th term would be $n = 20$), as well as the rule to find each term. You can find any term of a sequence by plugging n into the general formula, which will give you specific instructions on what to do with this value n. If you are given a few terms of a sequence, you can use these terms to find the general formula for the sequence. If you are given the general formula (complete with n as the variable), you can find any term by plugging in the number of the term you want for n.

Unless otherwise noted, the first term of any sequence $\{a_n\}$ begins with $n = 1$. The next n always goes up by 1.

For example, you can use the formula to find the first three terms of $a_n = (-1)^{n-1} \cdot (n^2)$:

1. **Find a_1 first by plugging in 1 wherever you see n.**

   This gives you $a_1 = (-1)^{1-1} \cdot (1^2) = (-1)^0 \cdot 1 = 1 \cdot 1 = 1$.

2. **Continue plugging in consecutive integers for n.**

   This will give you terms two and three:

   - $a_2 = (-1)^{2-1} \cdot (2^2) = (-1)^1 \cdot 4 = -1 \cdot 4 = -4$
   - $a_3 = (-1)^{3-1} \cdot (3^2) = (-1)^2 \cdot 9 = 1 \cdot 9 = 9$

Working in reverse: Forming an expression from terms

If you know the first few terms of a sequence, you can write a general expression for the sequence to find the nth term. To write the general expression, you must look for a pattern in the first few terms of the sequence, which demonstrates logical thinking (and we all want to be logical thinkers, right?). The formula you write must work for every integer value of n, starting with $n = 1$.

Sometimes this calculation is an easy task, and sometimes it's less apparent and more complicated. Sequences involving fractions and/or exponents tend to be more complicated and less obvious in their patterns. The easy ones to write include addition, subtraction, multiplication, or division by integers.

For example, to find the general formula for the nth term of the sequence $\frac{2}{3}, \frac{3}{5}, \frac{4}{7}, \frac{5}{9}, \frac{6}{11}$, you should look at the numerator and the denominator separately:

> The numerators begin with 2 and increase by one each time. This sequence is described by $a_n = n + 1$.
>
> The denominators start with 3 and increase by two each time. This sequence is described by $a_n = 2n + 1$.

Therefore, this sequence can be expressed by the general formula $\frac{n+1}{2n+1}$.

To double-check your formula and ensure that the answers work, plug in 1, 2, 3, and so on to make sure you get the original numbers from the given sequence.

$$n = 1: a_1 = \frac{1+1}{2 \cdot 1 + 1} = 2/3$$

$$n = 2: a_2 = \frac{2+1}{2 \cdot 2 + 1} = 3/5$$

$$n = 3: a_3 = \frac{3+1}{2 \cdot 3 + 1} = 4/7$$

$$n = 4: a_4 = \frac{4+1}{2 \cdot 4 + 1} = 5/9$$

$$n = 5: a_5 = \frac{5+1}{2 \cdot 5 + 1} = 6/11$$

They all work, so we did it right!

Recursive sequences: One type of general sequence

A *recursive sequence* is a sequence where each term depends on the term before it. To find any term in a recursive sequence, you use the given term (at least one term [usually the first] will be given for the problem) and the given formula that allows you to find the other terms.

You'll recognize recursive sequences because the given formula will typically have a_n (the *n*th term of the sequence) as well as a_{n-1} (the term before the *n*th term of the sequence). You will be given a formula (a different one for each problem), and the directions for these types of problems will ask you to find the terms of the sequence.

For example, the most famous recursive sequence is the Fibonacci Sequence, where each term after the second term (the sequence begins looking like a sequence when $n > 2$) is defined as the sum of the two terms before it. The first term of this sequence is 1, and the second term is 1 also. The formula for the Fibonacci Sequence is $a_n = a_{n-2} + a_{n-1}$ for $n \geq 3$.

So if you were asked to find the next three terms of the sequence, you'd have to use the formula as follows:

$$a_3 = a_{3-2} + a_{3-1} = a_1 + a_2 = 1 + 1 = 2$$

$$a_4 = a_{4-2} + a_{4-1} = a_2 + a_3 = 1 + 2 = 3$$

$$a_5 = a_{5-2} + a_{5-1} = a_3 + a_4 = 2 + 3 = 5$$

This sequence is very famous because many things in the natural world follow the pattern of the Fibonacci Sequence. For example, lilies and iris both have three petals. Buttercups have five petals, and corn marigolds have 13 petals. Seeds of coneflowers and sunflowers have also been observed to follow the same pattern as the Fibonacci Sequence. Pine cones and cauliflower also follow this pattern.

Covering the Distance between Terms: Arithmetic Sequences

One of the most common types of sequences is called an *arithmetic sequence.* In an arithmetic sequence, each term differs from the one before it by the same number, called the *common difference.* To determine whether a sequence is arithmetic, you subtract each term by its preceding term; if the difference between each term is the same, the sequence is arithmetic.

Arithmetic sequences are very helpful to identify, because they all follow one formula, whereas every formula in the previous section would be totally different and not necessarily follow any rules at all. The formula for the nth term of an arithmetic sequence is always the same:

$$a_n = a_1 + (n - 1)d$$

where a_1 is the first term and d is the common difference.

For exercises involving arithmetic sequences, you will be asked to find a term somewhere in a given sequence. You'll recognize the sequence as arithmetic because there will be a common difference between each term. This lets you know to start off with the general formula for any arithmetic sequence. There are always three steps to finding the desired terms: finding the common difference, writing the formula for the specific given sequence using the first term and the common difference, and then finding the term you were asked to find by plugging in the number of the term for n. There are two main types of problems that you could encounter, however, as you'll discover in the next two sections: one where you're given a list of consecutive terms (which is easy), and one where you're given two terms that are not consecutive (where finding the common difference is no piece of cake).

Using consecutive terms to find another in an arithmetic sequence

If you are given two consecutive terms of an arithmetic sequence, the common difference between these terms is not too far away.

For example, an arithmetic sequence is –7, –4, –1, 2, 5. . . . If you want to find the 55th term of this arithmetic sequence, you can continue the pattern begun by the first few terms 50 more times. However, that would be very time consuming and not very effective to find terms that come later in the sequence.

Instead, you can use a general formula to find any term of an arithmetic sequence. Finding the general formula for the *n*th term of an arithmetic sequence is easy as long as you know the first term and the common difference.

1. **Find the common difference.**

   To find the common difference, simply subtract one term from the one after it: $-4 - (-7) = 3$.

2. **Plug a_1 and d into the general formula for any arithmetic sequence to write the specific formula for the given sequence.**

   Start with $a_n = a_1 + (n - 1)d$. Plug in what you know: The first term of the sequence is –7, and the common difference is 3. So $a_n = -7 + (n - 1)3 = -7 + 3n - 3 = 3n - 10$.

3. **Plug in the number of the term you are trying to find for n.**

   To find the 55th term, plug 55 in for n into the general formula for a_n:

   $a_{55} = 3(55) + 1 = 165 + 1 = 166.$

Using any two terms

At times you'll need to find the general formula for the *n*th term of an arithmetic sequence without knowing the first term or the common difference. In this case, you're given two terms (not necessarily consecutive), and you use this information to find a_1 and d. Your steps are still the same: Find the common difference, write the specific formula for the given sequence, and then find the term you're looking for (we can't say that often enough).

For instance, to find the general formula of an arithmetic sequence where $a_4 = -23$, and $a_{22} = 40$, follow these steps:

1. **Find the common difference.**

   You will have to be more creative in finding the common difference for these types of problems.

   a. **Use the formula $a_n = a_1 + (n - 1)d$ to set up two equations that use the given information.**

      For the first equation, you know that when $n = 4$, $a_n = -23$:

      $-23 = a_1 + (4 - 1)d$, or $-23 = a_1 + 3d$.

For the second equation, you know that when $n = 22$, $a_n = 40$:

$40 = a_1 + (22 - 1)d$, or $40 = a_1 + 21d$.

b. Set up a system of equations (see Chapter 13) and solve for *d*.

The system will look like this:

$$\begin{cases} -23 = a_1 + 3d + 3d \\ 40 = a_1 + 21d \end{cases}$$

You can use elimination or substitution to solve the system, like we show you in Chapter 13. Elimination works nicely because you can multiply either equation by –1 and add the two together to get $63 = 18d$. Therefore, $d = 3.5$.

2. **Write the formula for the specific sequence.**

This is also a little more work than before.

a. Plug *d* into one of the equations to solve for a_1.

You can plug 3.5 back into either equation:

$-23 = a_1 + 3(3.5)$, or $a_1 = -33.5$.

b. Use a_1 and *d* to find the general formula for a_n.

This becomes a simple three-step simplification:

$a_n = -33.5 + (n - 1)3.5$

$a_n = -33.5 + 3.5n - 3.5$

$a_n = 3.5n - 37$

3. **Find the term you were looking for.**

We didn't ask in the directions to this problem to find any specific term (always read the directions!), but if we did, you could plug that number in for n and then find the term you were looking for.

Sharing Ratios with Consecutive Paired Terms: Geometric Sequences

A *geometric sequence* is one where consecutive terms have a common ratio. In other words, if you divide each term by the term before it, the quotient should be the same, denoted by the letter r.

Certain objects, such as cars, depreciate with time. You can describe this depreciation by using a geometric sequence. The common ratio will always be the rate as a percent (sometimes called APR, which stands for Annual

Percentage Rate). Finding the value of the car at any time, as long as you know its original value, is easy to do. The following sections show you how to identify the terms and expressions of geometric sequences, which allow you to apply the sequences to real-world situations (such as trading in your car!).

Here we begin to work with geometric sequences: how to find a term in the sequence as well as how to find the formula for the specific sequence when you're not given it. But first, here are some general ideas to remember.

The first term of any sequence is denoted as a_1. To find the second term of a geometric sequence, multiply the first term by the common ratio, r. You can follow this pattern infinitely to find any term of a geometric sequence:

$$\{a_n\} = a_1, a_2, a_3, a_4, a_5, \ldots, a_n \ldots$$

$$\{a_n\} = a_1, a_1 \cdot r, a_1 \cdot r^2, a_1 \cdot r^3, a_1 \cdot r^4, \ldots, a_1 \cdot r^{n-1}, \ldots$$

More simply put, the formula for the nth term of a geometric sequence is

$$a_n = a_1 \cdot r^{n-1}$$

In the formula, a_1 is the first term and r is the common ratio.

Identifying a term when you know consecutive terms

The steps for dealing with geometric sequences are remarkably similar to those in the arithmetic sequence sections. First you find the common ratio (not the difference!), then you write the specific formula for the given sequence, and then you find the term you're looking for.

An example of a geometric sequence is 2, 4, 8, 16, 32. To find the 15th term follow these steps:

1. **Find the common ratio.**

   In this sequence, each consecutive term is twice the previous term. If you can't see the common difference by looking at the sequence, divide any term by the term before it.

2. **Find the formula for the given sequence.**

   In terms of the formula, $a_1 = 2$ and $r = 2$. The general formula for this sequence is $a_n = 2 \cdot 2^{n-1}$, which simplifies (using the rules of exponents) to $2^1 \cdot 2^{n-1} = 2^{1+(n-1)} = 2^n$.

3. **Find the term you're looking for.**

   If $a_n = 2^n$, then $a_{15} = 2^{15} = 32{,}768$.

The formula in the previous example simplifies nicely because the bases of the two exponents are the same. If the first term and r don't have the same base, you can't combine them. (For more on rules such as this, head to Chapter 5.)

Going out of order: Finding a term when the terms are nonconsecutive

If you know any two nonconsecutive terms of a geometric sequence, you can use this information to find the general formula of the sequence, as well as any specified term. For example, if the 5th term of a geometric sequence is 64 and the 10th term is 2, you can find the 15th term. Just follow these steps:

1. **Determine the value of *r*.**

   You can use the geometric formula to create a system of two formulas to find r: $a_5 = a_1 \cdot r^{5-1}$ and $a_{10} = a_1 \cdot r^{10-1}$, or

   $$\begin{cases} 64 = a_1 \cdot r^4 \\ 2 = a_1 \cdot r^9 \end{cases}$$

   You can use substitution to solve one equation for a_1 (see Chapter 13 for more on this method of solving systems): $a_1 = \frac{64}{r^4}$.

   Plug this expression in for a_1 in the other equation: $2 = \left(\frac{64}{r^4}\right) \cdot (r^9)$. Now simplify this equation:

   $$2 = 64r^5$$

   $$2/64 = 1/32 = r^5$$

   $$1/2 = r$$

2. **Find the specific formula for the given sequence.**

   a. **Plug *r* into one of the equations to find a_1.**

      This gives you $a_1 = \frac{64}{\left(\frac{1}{2}\right)^4} = 64(2)^4 = 1{,}024$.

   b. **Plug a_1 and *r* into the formula.**

      Now that you know a_1 and r, you can write the formula:

      $a_n = 1{,}024(1/2)^{(n-1)}$ before moving on.

3. **Find the term you're looking for.**

   In this case, you want to find the 15th term ($n = 15$):

   $a_{15} = 1{,}024(1/2)^{15-1} = 1{,}024(1/2)^{14} = 1{,}024(1/16{,}384) = 1/16$.

The annual depreciation of a car's value is approximately 30 percent. Every year, the car is actually worth 70 percent of its value from the year before. If a_1 represents the value of a car when it was new and n represents the number of years that have passed, $a_n = a_1 \cdot (0.7)^n$ when $n \geq 0$. Notice that this sequence starts at 0, which is okay as long as the information says that it starts at 0.

Creating a Series: Summing Terms of a Sequence

A *series* is the sum of terms in a sequence. Except for one situation where you can add the sum of an infinite series, you will be asked to find the sum of a certain number of terms (the first 12, for example). It's especially helpful in calculus when you begin discussing integration. Before some of the newer calculus concepts were discovered, mathematicians used series to find the areas under curves. Finding the area of a rectangle was easy, but curves aren't straight, so finding the area under them wasn't as easy. So they broke up the region into very small rectangles and added them together. This concept then evolved into an integral, and you will see a ton of that in calculus.

Reviewing general summation notation

The sum of the first k terms of a sequence is referred to as the *kth partial sum*. Don't let the use of a different variable here confuse you. Your book may still even use n and call it an nth partial sum. Remember that a variable just stands in for an unknown, so it really can be any variable you want — even those Greek variables that we used in the trig chapters. But we've most often seen books use k to represent the number of terms in a series and n for the number of terms in a sequence. They're called partial sums because you'll only be able to find the sum of a certain number of terms — no infinite series here! You may use partial sums when you want to find the area under a curve (graph) between two certain values of x. Although it's not usually possible to find the *entire* area under the graph (because it could be infinite if the curve goes on forever), you can find the area underneath a piece of it.

The notation of the kth partial sum of a sequence is as follows:

$$\sum_{n=1}^{k} a_n = a_1 + a_2 + a_3 + \ldots + a_k$$

You read this as "the *k*th partial sum of a_n is . . ." where $n = 1$ is the *lower limit* of the sum and *k* is the *upper limit* of the sum. To find the *k*th partial sum, you begin by plugging the lower limit into the general formula and continue in order, plugging in integers until you reach the upper limit of the sum. At that point, you simply add all the terms to find the sum.

To find the fifth partial sum of $a_n = n^3 - 4n + 2$, for example, follow these steps:

1. **Plug all values of *n* (starting with 1 and ending with *k*) into the formula.**

   Because you want to find the fifth partial sum, plug in 1, 2, 3, 4, and 5:

   - $a_1 = (1)^3 - 4(1) + 2 = 1 - 4 + 2 = -1$
   - $a_2 = (2)^3 - 4(2) + 2 = 8 - 8 + 2 = 2$
   - $a_3 = (3)^3 - 4(3) + 2 = 27 - 12 + 2 = 17$
   - $a_4 = (4)^3 - 4(4) + 2 = 64 - 16 + 2 = 50$
   - $a_5 = (5)^3 - 4(5) + 2 = 125 - 20 + 2 = 107$

2. **Add all the values from a_1 to a_k to find the sum.**

   This gives you $-1 + 2 + 17 + 50 + 107 = 175$.

3. **Rewrite the final answer, using summation notation.**

   $\sum_{n=1}^{5}(n^3 - 4n + 2) = 175$

Summing an arithmetic sequence

The *k*th partial sum of an arithmetic sequence still calls for you to add the first *k* terms. But in the arithmetic sequence, you do have a formula to use instead of plugging in each of the values for *n*. The *k*th partial sum of an arithmetic series is:

$$S_k = \sum_{n=1}^{k} a_n = \frac{k}{2}(a_1 + a_k)$$

You simply plug the lower and upper limits into the formula for a_n to find a_1 and a_k.

One real-world application of an arithmetic sum involves stadium seating. Say, for example, a stadium has 35 rows of seats; there are 20 seats in the first row, 21 seats in the second row, 22 seats in the third row, and so on. How many seats do all 35 rows contain? Follow these steps to find out:

1. **Find the first term of the sequence.**

   The first term of this sequence (or the number of seats in the first row) is given: 20.

2. **Find the *k*th term of the sequence.**

   Because the stadium has 35 rows, find a_{35}. Use the formula for the *n*th term of an arithmetic sequence (see the earlier section "Covering the Distance between Terms: Arithmetic Sequences"). The first term is 20, and each row has one more seat than the row before it, so $d = 1$. Plug these values into the formula:

   $a_{35} = a_1 + (35 - 1)d = 20 + (34) \cdot 1 = 54$

   Note: This is the number of seats in the 35th row, not the answer to how many seats the stadium contains.

3. **Use the formula for the *k*th partial sum of an arithmetic sequence to find the sum.**

   You now have $S_{35} = {}^{35}\!/\!_2(a_1 + a_{35}) = {}^{35}\!/\!_2(20 + 54) = {}^{35}\!/\!_2(74) = 1{,}295$.

Seeing how a geometric sequence adds up

Just like when you found the sum of an arithmetic sequence, you can find the sum of a geometric sequence. Also, because the formulas to find specific terms in the two types of sequences are different, so is the formula to find their sums. Here, we show you how to find the sum of two different types of geometric sequences. The first type is a finite sum (comparable to a *k*th partial sum from the previous section), and it too will have an upper limit and a lower limit. There are no specific restrictions on the common ratio of partial sums of this type. The second type of geometric sum is called an *infinite* geometric sum, and the common ratio for this type is very specific (it *must* be strictly between –1 and 1). This type of geometric sequence is very helpful if you drop a ball and count how far it travels up and down, then up and down, until it finally starts rolling.

Cars depreciate at an annual rate of 30 percent, starting the second you drive your new, shiny car off the lot. Say you originally pay $22,500 for a car; you can use the rate of depreciation and the price to figure out how much your car is worth at any given time — all by using geometric sequences. Just find the common ratio (which is the percent of the car that remains when the depreciation has been taken away) as a decimal. Using the original price as the first term, when $t = 0$ (because its brand new), you can use a geometric sequence to find out how much the car is worth after *t* years.

By definition, a geometric series continues infinitely, for as long as you want to keep plugging in values for *n*. However, in a specific type of geometric series, no matter how long you plug in values for *n,* the sum will never get larger than a certain value. This type of series has a specific formula to find the infinite sum. The sum isn't infinite, the number of terms is. In mathematical terms, you say that some geometric sequences — ones with a common ratio between –1 and 1 — have a limit to their sequence of partial sums. In other words, the partial sum comes closer and closer to (without ever actually reaching) a particular number. You call this number the *sum of the sequence,* as opposed to the *k*th partial sum you find in previous geometric sequence sections in this chapter.

Stop right there: Determining the partial sum of a finite geometric sequence

You can find a partial sum of a geometric sequence by using the following formula:

$$\sum_{n=1}^{k} a_n = a_1\left(\frac{1-r^k}{1-r}\right)$$

For example, to find $\sum_{n=1}^{7} 9\left(\frac{-1}{3}\right)^{n-1}$, follow these steps.

1. **Find a_1 by plugging in 1 for n.**

   This gives you $9(-\frac{1}{3})^{1-1} = 9(1) = 9$.

2. **Find a_2 by plugging in 2 for n.**

   For this, you have $9(-\frac{1}{3})^{2-1} = 9(-\frac{1}{3})^1 = -3$.

3. **Divide a_2 by a_1 to find r.**

   For this example, $r = -\frac{3}{9} = -\frac{1}{3}$. Notice that this value is the same as the fraction in the parentheses.

   You may have noticed that $9(-\frac{1}{3})^{n-1}$ follows the general formula for $a_n = a_1 \cdot r^{n-1}$ (the general formula for a geometric sequence) exactly, where $a_1 = 9$ and $r = -\frac{1}{3}$. However, if you didn't notice it, the method used in Steps 1–3 works to the tee.

4. **Plug a_1, r, and k into the sum formula.**

   The problem now boils down to the following simplifications:

   - $$S_7 = 9\left(\frac{1-\left(\frac{-1}{3}\right)^7}{1-\left(\frac{-1}{3}\right)}\right) = 9\left(\frac{1-\left(\frac{-1}{2{,}187}\right)}{1+\frac{1}{3}}\right)$$
   - $$S_7 = 9\left(\frac{1+\frac{1}{2{,}187}}{\frac{4}{3}}\right) = 9\left(\frac{\frac{2{,}187}{2{,}187}+\frac{1}{2{,}187}}{\frac{4}{3}}\right)$$
   - $$S_7 = 9\left(\frac{\frac{2{,}188}{2{,}187}}{\frac{4}{3}}\right) = 9\left(\frac{2{,}188}{2{,}187}\right)\left(\frac{3}{4}\right) = \frac{547}{81}$$

Geometric summation problems will take quite a bit of work with fractions, so make sure to find a common denominator, invert, and multiply when necessary. Or you can use a calculator and then reconvert to a fraction. Just be careful to use correct parentheses when entering the numbers.

To geometry and beyond: Finding the value of an infinite sum

Finding the value of an infinite sum in a geometric sequence is actually quite simple — as long as you keep your fractions and decimals straight. If r lies outside the range $-1 < r < 1$, a_n will grow without bound infinitely, so there's no limit on how large the absolute value of a_n ($|a_n|$) can get. If $|r| < 1$, for every value of n, $|r^n|$ will continue to decrease infinitely until it becomes arbitrarily close to 0. This is because when you multiply a fraction between –1 and 1 by itself, the absolute value of that fraction continues to get smaller until it becomes so small that you hardly even notice it any more. Therefore, the term r^k in the finite geometric sum formula $S_k = \sum_{n=1}^{k} a_1 \cdot r^{n-1} = a_1\left(\frac{1-r^k}{1-r}\right)$ almost disappears completely. And if the r^k disappears — or gets very small — the finite formula changes to the following and allows you to find the sum of an infinite geometric series: $\sum_{n=1}^{\infty} a_n = \frac{a_1}{1-r}$.

For example, to find the value of $\sum_{n=1}^{\infty} 4\left(\frac{2}{5}\right)^{n-1}$, follow these steps:

1. **Find the value of a_1 by plugging in 1 for n.**

   This gives you $a_1 = 4(2/5)^{1-1} = 4(2/5)^0 = 4 \cdot 1 = 4$.

2. **Calculate a_2 by plugging in 2 for n.**

   For this example, $a_2 = 4(2/5)^{2-1} = 4(2/5)^1 = 8/5$.

3. **Determine r.**

   To find r, you divide a_2 by a_1:

   $$\frac{a_2}{a_1} = \frac{\frac{8}{5}}{4} = \frac{2}{5}$$

4. **Plug a_1 and r into the formula to find the infinite sum.**

   Plug in and simplify to find the following:

   - $\sum_{n=1}^{\infty} 4\left(\frac{2}{5}\right)^{n-1} = \frac{4}{1-\frac{2}{5}}$
   - $= \frac{4}{\frac{5}{5}-\frac{2}{5}} = \frac{4}{\frac{3}{5}}$
   - $= 4 \cdot (5/3) = 20/3$

Repeating decimals also can be expressed as infinite sums. Consider the number 0.5555555. . . . You can write this number as 0.5 + 0.05 + 0.005 + . . . , and so on forever. The first term of this sequence is 0.5; to find r, 0.05 ÷ 0.5 = 0.1. Plug these values into the infinite sum formula:

$$\sum_{n=1}^{k} 0.5(0.1)^{n-1} = \frac{0.5}{1-0.1} = \frac{0.5}{0.9} = \frac{5}{9}$$

This sum is finite only if r lies strictly between –1 and 1.

Expanding with the Binomial Theorem

A *binomial* is a polynomial with exactly two terms. Expressing the multiplication of binomials without any parentheses is called *binomial expansion*. Using the binomial theorem requires you to find the coefficients of this expansion.

Expanding many binomials takes a rather extensive application of the distributive property and quite a bit of time. Multiplying two binomials is easy if you use the FOIL method (see Chapter 4), and multiplying three binomials doesn't take much more effort. Multiplying ten binomials, however, takes long enough that you may end up quitting short of the half-way point. And if you make a mistake somewhere along the line, it snowballs and affects every subsequent step.

Therefore, in the interest of saving bushels of time and energy, we present to you the binomial theorem. If you need to find the entire expansion for a binomial, this theorem is the greatest thing since sliced bread:

$$(a+b)^n =$$

$$\binom{n}{0}a^n b^0 + \binom{n}{1}a^{n-1}b^1 + \binom{n}{2}a^{n-2}b^2 + \ldots + \binom{n}{n-2}a^2 b^{n-2} +$$

$$\binom{n}{n-1}a^1 b^{n-1} + \binom{n}{n}a^0 b^n$$

This formula gives you a very abstract view of how to multiply a binomial n times. It's quite hard to read, actually. But this is the way your textbook will show it to you.

We promise, the actual use of this formula is not as hard as it looks. Each $\binom{n}{r}$ comes from a combination formula and gives you the coefficients for each term (they're sometimes called *binomial coefficients*). We tell you how to deal with $\binom{n}{r}$ in the section called "Using algebra."

For example, to find $(2y-1)^4$ you start off the binomial theorem by replacing *a* with 2*y*, *b* with –1, and *n* with 4 to get:

$$\binom{4}{0}(2y)^4(-1)^0+\binom{4}{1}(2y)^3(-1)^1+\binom{4}{2}(2y)^2(-1)^2+\binom{4}{3}(2y)^1(-1)^3+\binom{4}{4}(2y)^0(-1)^4$$

You then have to simplify this mess. We break it down in the next sections. First, a closer look at the binomial theorem, then how to find those dreaded binomial coefficients, and last (but certainly not least) how to put all the parts together to get the final answer.

Breaking down the binomial theorem

The binomial theorem looks extremely intimidating, but it becomes much simpler if you break it up into smaller steps and examine the parts. Allow us to point out a few things to be aware of so that you don't get confused somewhere along the way; after you have all this info straightened out, your task will seem much more manageable:

- The binomial coefficients $\binom{n}{r}$ won't necessarily be the coefficients in your final answer. You're raising each monomial to a power, including any coefficients attached to each of them.
- The theorem is written as the sum of two monomials, so if your task is to expand the difference of two monomials, the terms in your final answer should alternate between positive and negative numbers.
- The exponent of the first monomial begins at *n* and decreases by 1 with each sequential term, until it reaches 0 at the last term. The exponent of the second monomial begins at 0 and increases by 1 each time until it reaches *n* at the last term.
- The exponents of both monomials should add to *n* — unless the monomials themselves have powers greater than 1.

Starting at the beginning: Binomial coefficients

Depending on how many times you must multiply the same binomial — a value also known as an *exponent* — the coefficients for that particular exponent will always be the same. The binomial coefficients are found by using the $\binom{n}{r}$ combinations formula. If the exponent is relatively small, you can use a shortcut called Pascal's triangle to find these coefficients. If not, you can always rely on algebra!

Using Pascal's triangle

Pascal's triangle, named after the famous mathematician Blaise Pascal, names the coefficients for a binomial expansion. It is especially useful with lower degrees. For example, if a sadistic teacher asked you to find $(3x + 4)^{10}$, we wouldn't recommend using this shortcut; instead, you'd just use the formula as described in the next section, "Using algebra." Figure 14-1 illustrates this concept. Each row gives the coefficients to $(a + b)^n$, starting with $n = 0$, depending on the exponent. To find any row of the triangle, you always start with the beginning. The top number of the triangle is 1, as well as all the numbers on the outer sides. To get any term in the triangle, you find the sum of the two numbers above it.

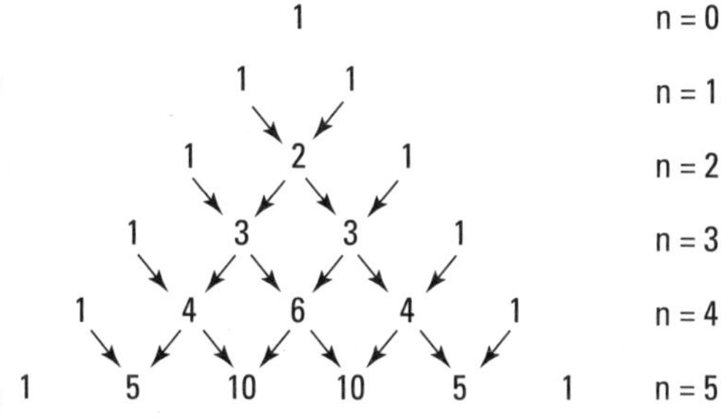

Figure 14-1: Determining coefficients with Pascal's triangle.

For instance, the binomial coefficients for $(a + b)^5$ are 1, 5, 10, 10, 5, 1 — in that order.

Using algebra

If you need to find the coefficients of binomials algebraically, we offer a formula for that as well. The rth coefficient for the nth binomial expansion is written in the following form:

$$\binom{n}{r} = \frac{n!}{r!(n-r)!}$$

You may recall the term *factorial* from your earlier math classes. If not, allow us to remind you: $n!$, read "n factorial," is defined as $1 \cdot 2 \cdot 3 \cdot \ldots \cdot (n-2) \cdot (n-1) \cdot n$. You read the expression for the binomial coefficient $\binom{n}{r}$ as "n choose r." You usually can find a button for combinations on a calculator. If not, you can use the factorial button and do each part separately.

To make things a little easier, 0! is defined as 1. Therefore, you can define $\binom{n}{0} = 1$ and $\binom{n}{n} = 1$.

For example, to find the binomial coefficient given by $\binom{5}{3}$, substitute the values into the formula:

$$= \frac{5!}{3!(5-3)!} = \frac{5!}{3!2!}$$

$$= \frac{120}{6 \cdot 2} = 10$$

Expanding by using the binomial theorem

Using the binomial theorem can save you time, but it can be dangerous (the whole "nothing in life comes easy" philosophy). Keeping each of the steps separate until the very end should help. It depends, as well, on whether the original monomial had no coefficients or exponents (other than 1) on the variables — we show you how to use the theorem in the next section, "Normal expansion problems." When the original monomial has coefficients or exponents other than 1 on the variable(s), you have to be careful you take those into account. We show you an example of that as well in the section called "Raising monomials to a power pre-expansion."

Normal expansion problems

To find the expansion of binomials with the theorem in a basic situation, follow these steps:

1. **Write out the binomial expansion by using the theorem, changing the variables where necessary.**

   For example, consider the problem $(m + 2)^4$. According to the theorem, you should replace the letter a with m, the letter b with 2, and the exponent n with 4:

   $$(m+2)^4 =$$

   $$\binom{4}{0}(m)^4(2)^0 + \binom{4}{1}(m)^3(2)^1 + \binom{4}{2}(m)^2(2)^2 + \binom{4}{3}(m)^1(2)^3 + \binom{4}{4}(m)^0(2)^4$$

   The exponents of m begin at 4 and end at 0 (see the section "Breaking down the binomial theorem"). Similarly, the exponents of 2 begin at 0 and end at 4. For each term, the sum of the exponents in the expansion is always 4.

2. **Find the binomial coefficients (see the section "Starting at the beginning: Binomial coefficients").**

   We used the combinations formula to find the five coefficients, but you could use the Pascal's triangle shortcut because the degree is so low (it wouldn't hurt you to write out 5 rows of Pascal's triangle — starting with 0 through 4).

- $\binom{4}{0} = 1$
- $\binom{4}{1} = 4$
- $\binom{4}{2} = 6$
- $\binom{4}{3} = 4$
- $\binom{4}{4} = 1$

You may have noticed that after you reach the middle of the expansion, the coefficients are a mirror image of the first half. This is another time-saving trick you can employ so you don't need to do all the calculations for $\binom{n}{r}$.

3. **Replace all $\binom{n}{r}$ with the coefficients from Step 2.**

   This gives you $1(m)^4(2)^0 + 4(m)^3(2)^1 + 6(m)^2(2)^2 + 4(m)^1(2)^3 + 1(m)^0(2)^4$.

4. **Raise the monomials to the powers specified for each term.**

   You now have $1 \cdot m^4 \cdot 1 + 4 \cdot m^3 \cdot 2 + 6 \cdot m^2 \cdot 4 + 4 \cdot m \cdot 8 + 1 \cdot 1 \cdot 16$.

5. **Combine like terms and simplify.**

   You end up with $m^4 + 8m^3 + 24m^2 + 32m + 16$.

Notice that the coefficients you get in the final answer aren't the binomial coefficients you find in Step 1. This is because you must raise each monomial to a power (Step 4), and the constant in the original binomial changed each term.

Raising monomials to a power pre-expansion

At times, monomials can have coefficients and/or be raised to a power before you begin the binomial expansion. When this is the case, you have to raise the entire monomial to the appropriate power in each step. For example, here's how you expand the expression $(3x^2 - 2y)^7$:

1. **Write out the binomial expansion by using the theorem, changing the variables where necessary.**

   Replace the letter a in the theorem with the quantity $(3x^2)$ and the letter b with $(-2y)$. Don't let those coefficients or exponents scare you — you're still substituting them into the binomial theorem. Replace n with 7. You end up with

$$(3x^2 - 2y)^7 = \binom{7}{0}(3x^2)^7(-2y)^0 + \binom{7}{1}(3x^2)^6(-2y)^1 + \binom{7}{2}(3x^2)^5(-2y)^2 + \binom{7}{3}(3x^2)^4(-2y)^3 + \binom{7}{4}(3x^2)^3(-2y)^4 + \binom{7}{5}(3x^2)^2(-2y)^5 + \binom{7}{6}(3x^2)^1(-2y)^6 + \binom{7}{7}(3x^2)^0(-2y)^7$$

2. **Find the binomial coefficients (see the section "Starting at the beginning: Binomial coefficients").**

   Using the combination formula gives you the following:

   - $\binom{7}{0} = 1$
   - $\binom{7}{1} = 7$
   - $\binom{7}{2} = 21$
   - $\binom{7}{3} = 35$ (the halfway mirror point)
   - $\binom{7}{4} = 35$
   - $\binom{7}{5} = 21$
   - $\binom{7}{6} = 7$
   - $\binom{7}{7} = 1$

3. **Replace all $\binom{n}{r}$ with the coefficients from Step 2.**

   This gives you $1(3x^2)^7(-2y)^0 + 7(3x^2)^6(-2y)^1 + 21(3x^2)^5(-2y)^2 + 35(3x^2)^4(-2y)^3 + 35(3x^2)^3(-2y)^4 + 21(3x^2)^2(-2y)^5 + 7(3x^2)^1(-2y)^6 + 1(3x^2)^1(-2y)^7$.

4. **Raise the monomials to the powers specified for each term.**

   You now have the following: $1(2{,}187x^{14})(1) + 7(729x^{12})(-2y) + 21(243x^{10})(4y^2) + 35(81x^8)(-8y^3) + 35(27x^6)(16y^4) + 21(9x^4)(-32y^5) + 7(3x^2)(64y^6) + 1(1)(-128y^7)$.

5. **Simplify.**

   You end up with the following: $2{,}187x^{14} - 10{,}206x^{12}y + 20{,}412x^{10}y^2 - 22{,}680x^8y^3 + 15{,}120x^6y^4 - 6{,}048x^4y^5 + 1{,}344x^2y^6 - 128y^7$.

Expansion with complex numbers

The most complicated type of binomial expansion involves the complex number, i (for more on complex numbers, see Chapter 11), because you are not only dealing with the binomial theorem, but you are dealing with imaginary numbers as well. When raising complex numbers to a power, note that $i^1 = i$, $i^2 = -1$, $i^3 = -i$, and $i^4 = 1$. If you run into higher powers this pattern repeats: $i^5 = i$, $i^6 = -1$, $i^7 = -i$, and so on. Because powers of the imaginary number, i, can be simplified, your final answer to the expansion should not include powers of i. Instead, use the information given here to simplify the powers of i and then combine your like terms.

For example, to expand $(1 + 2i)^8$, follow these steps:

1. **Write out the binomial expansion by using the theorem, changing the variables where necessary.**

   $(1 + 2i)^8$ expands to $\binom{8}{0}(1)^8(2i)^0 + \binom{8}{1}(1)^7(2i)^1 + \binom{8}{2}(1)^6(2i)^2 + \binom{8}{3}(1)^5(2i)^3 + \binom{8}{4}(1)^4(2i)^4 + \binom{8}{5}(1)^3(2i)^5 + \binom{8}{6}(1)^2(2i)^6 + \binom{8}{7}(1)^1(2i)^7 + \binom{8}{8}(1)^0(2i)^8$

2. **Find the binomial coefficients.**

   Using the combination formula gives you the following:

   - $\binom{8}{0} = 1$
   - $\binom{8}{1} = 8$
   - $\binom{8}{2} = 28$
   - $\binom{8}{3} = 56$
   - $\binom{8}{4} = 70$ (the halfway mirror point)

- $\binom{8}{5} = 56$
- $\binom{8}{6} = 28$
- $\binom{8}{7} = 8$
- $\binom{8}{8} = 1$

3. **Replace all $\binom{n}{r}$ with the coefficients from Step 2.**

   This gives you $1(1)^8(2i)^0 + 8(1)^7(2i)^1 + 28(1)^6(2i)^2 + 56(1)^5(2i)^3 + 70(1)^4(2i)^4 + 56(1)^3(2i)^5 + 28(1)^2(2i)^6 + 8(1)^1(2i)^7 + 1(1)^0(2i)^8$.

4. **Raise the monomials to the powers specified for each term.**

   You now have $1(1)(1) + 8(1)(2i) + 28(1)(4i^2) + 56(1)(8i^3) + 70(1)(16i^4) + 56(1)(32i^5) + 28(1)(64i^6) + 8(1)(128i^7) + 1(1)(256i^8)$.

5. **Simplify any *i*'s that you can.**

   The problem breaks down to $1(1)(1) + 8(1)(2i) + 28(1)(4 \cdot -1) + 56(1)(8 \cdot -i) + 70(1)(16 \cdot 1) + 56(1)(32 \cdot i) + 28(1)(64 \cdot -1) + 8(1)(128 \cdot -i) + 1(1)(256 \cdot 1)$.

6. **Combine like terms and simplify.**

   You end up with the following:

   $1 + 16i - 112 - 448i + 1{,}120 + 1{,}792i - 1{,}792 - 1{,}024i + 256 = -527 + 336i$

Chapter 15

Looking Forward to Calculus

In This Chapter

- Determining limits graphically, analytically, and algebraically
- Pairing limits and operations
- Identifying continuity and discontinuity in a function

Every good thing must come to an end, and for pre-calculus, the end is actually the beginning — the beginning of calculus. *Calculus* is the study of change and rates of change (not to mention a big change for you!). Before calculus, everything had to be *static* (stationary or motionless), but calculus shows you that things can be different over time. This branch of mathematics enables you to study how things move, grow, travel, expand, and shrink and helps you do so much more than any other math subject before.

This chapter helps prepare you for calculus by introducing you to the very first bits of the subject. First, we run through the difference between pre-calculus and calculus. Then we look at *limits,* which dictate that a graph can get really close to values without ever actually reaching them. Before you get to calculus, math problems always give you a function $f(x)$ and ask you to find the y value at one specific x in the domain (see Chapter 3). But when you get to calc, you look at what happens to the function the closer you get to certain values (like a really tough game of hide-and-seek).

Getting even more specific, a function can be *discontinuous* at that point. We look at those points one at a time in efforts to take a really good look at what's going on in the function at that particular value — that information comes in really handy in calculus when you start studying change. When studying limits and continuity, you're not working with the study of change specifically, which is why the limit isn't truly a calc topic — most calc textbooks consider the topics in this chapter to be review material.

The Differences between Pre-Calc and Calc

Here are a few basic distinctions between pre-calculus and calculus to illustrate the change:

- **Pre-calculus:** Study slope of a line. **Calculus:** Study slope of a tangent line to a curve.

  A straight line has the same slope all the time. No matter which point you choose to look at, the slope is the same. However, because a curve moves and changes, the slope of the tangent line will be different at different points.

- **Pre-calculus:** Study area of geometric shapes. **Calculus:** Study area under a curve.

  In pre-calc, you can rest easy knowing that a geometric shape is always basically the same, so you can find its area with a formula using certain measurements. A curve goes on forever and, depending on which section you're looking at, its area will change. No more nice, pretty formulas to find area here; instead, you use a process called *integration.*

- **Pre-calculus:** Study volume of a geometric solid. **Calculus:** Study volume of complicated shapes called *solids of revolution.*

  The geometric solids you find volume for (prisms, cylinders, and pyramids, for example) have formulas that are always the same, based on the basic shapes of the solid and their dimensions. The only way to find the volume of a solid of revolution, however, is to cut the shape into infinitely small pieces, each of which you can find the volume. However, the volume changes over time based on the section of the curve you're looking at.

- **Pre-calculus:** Study objects moving with constant velocities. **Calculus:** Study objects moving with acceleration.

  Using algebra, you can find the average rate of change of an object over a certain time interval. Using calculus, you can find the *instantaneous* rate of change for an object at an exact moment in time.

- **Pre-calculus:** Study functions in terms of x. **Calculus:** Study changes to functions in terms of x, with those changes in terms of t.

  Graphs of functions generally are referred to as $f(x)$, and you can find those graphs by plotting points. In calculus, you describe the changes to the graph $f(x)$ by using the variable t, as in $\frac{dx}{dt}$.

Our best advice: Calculus is best taken with an open mind and two aspirin! It's best to not view it as a bunch of material to memorize. Instead, try to build upon your pre-calculus knowledge and experience. Try to glean a deep understanding of *why* calculus does what it does. In this arena, concepts are key.

Understanding and Communicating about Limits

You can calculate a function's limit because not every function is defined at every value of x. Rational functions, for example, are undefined if the denominator of the function is 0. This is actually a perfect example of how you can use a limit to look at a function to see what it *would* do if it could. Take a look at the behavior of the function near the undefined value(s). Literally, you're looking at the function when it gets really close. If a function is undefined at $x = 3$, you could look at $x = 2$, $x = 2.9$, $x = 2.99$, $x = 2.999$, and so on. Now do it again from the other side: $x = 4$, $x = 3.1$, $x = 3.01$, and so on. All these values are defined, *except* for $x = 3$.

In symbols, you write $\lim_{x \to n} f(x) = L$, which is read as "the limit as x approaches n of $f(x)$ is L." L is the limit you'll be looking for. For the limit of a function to exist, the left limit and the right limit must both exist and be equivalent.

- A left limit starts at a value that's less than the number x is approaching and gets closer and closer from the left-hand side.
- A right limit is the exact opposite; it starts out greater than the number x is approaching and gets closer and closer from the right-hand side.

If, and only if, the left-hand limit equals the right-hand limit can you say that the function has a limit for that particular value of x.

Mathematically, you'd write let f be a function and let c and L be real numbers. Then $\lim_{x \to c} f(x) = L$ exactly when $\lim_{x \to c^-} f(x) = L$ and $\lim_{x \to c^+} f(x) = L$. In real-world language, this means that if you took two pencils, one in each hand, and started tracing along the graph of the function in equal measures, the two pencils would have to meet in one spot in order for the limit to exist. (Figure 15-1 shows that even though the function isn't defined at $x = 3$, the limit exists.)

Finding the Limit of a Function

You can look for the limit of a function in three ways for a certain value of x: graphically, analytically, and algebraically. We save any discussion for *how* to do it for the sections that follow. However, you may not always be able to reach a conclusion (the function doesn't approach just one y value at the particular x value you're looking at). In these cases, the graph will jump and you say it's not continuous (we cover continuity later in this chapter).

If you're ever asked to find the limit of a function, and plugging the x value actually works in the function, you've also found the limit. It's literally that easy!

We recommend using the graphing method only when you've been given the graph and asked to find a limit (because graph reading can be very inaccurate, especially if you're doing the graphing). The analytical method always works for any function, but it's slow. If you can use the algebraic method, it will save you time. We expand on each method in the sections that follow.

Graphically

When you're given the graph of a function and the problem asks you to find the limit, you read values from the graph — something you've been doing ever since you learned what a graph was! If you're looking for a limit from the left, you follow that function from the left-hand side toward the x value in question. Repeat this process from the right to find the right-hand limit. If the y value there is the same from the left as it is from the right (did the pencils meet?), that y value is the limit. Because the process of graphing a function can be long and complicated, we don't recommend using the graphing approach unless you've been given the graph.

For example, in Figure 15-1, find $\lim_{x \to -1} f(x)$, $\lim_{x \to 3} f(x)$, and $\lim_{x \to -5} f(x)$:

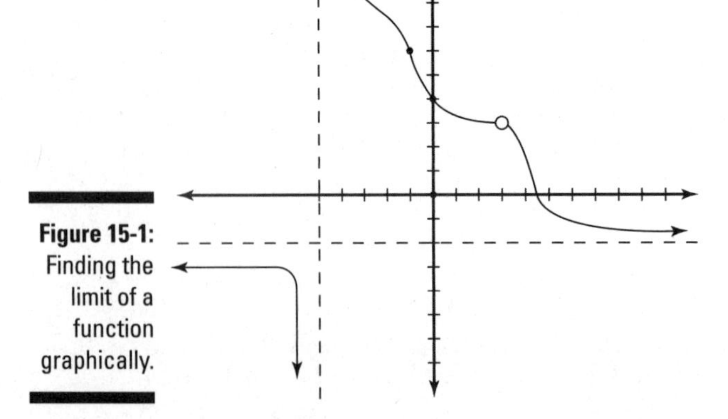

Figure 15-1: Finding the limit of a function graphically.

- **$\lim\limits_{x \to -1} f(x)$:** Because the function is defined at $x = -1$ — which you can see on the graph because the graph actually has a value (a dot) — its limit is the value $f(x) = 6$ (the y value when $x = -1$).
- **$\lim\limits_{x \to 3} f(x)$:** In the graph, you can see a hole in the function at $x = 3$, which means that the function is undefined — but that doesn't mean you can't state a limit. If you look at the function's values from the left — $\lim\limits_{x \to 3^-} f(x)$ — and from the right — $\lim\limits_{x \to 3^+} f(x)$ — you see the y value getting really close to 3. So, you say that the limit of the function as x approaches 3 is 3.
- **$\lim\limits_{x \to -5} f(x)$:** You can see that the function has a vertical asymptote at $x = -5$ (for more on asymptotes, see Chapter 3). From the left, the function approaches $-\infty$ as it nears $x = -5$. You can express this mathematically as $\lim\limits_{x \to -5^-} f(x) = -\infty$. From the right, the function approaches ∞ as it nears $x = -5$. You write this as $\lim\limits_{x \to -5^+} f(x) = \infty$. Therefore, the limit doesn't exist at this value, because one side is $-\infty$ and the other side is ∞.

For a function to have a limit, the left and right values must be the same. You can have a function with a hole in the graph, like $\lim\limits_{x \to 3} f(x)$, that has a limit, but the function can't jump over an asymptote at a value and have a limit (like $\lim\limits_{x \to -5} f(x)$).

Analytically

To find a limit analytically, basically you set up a chart and put the number that x is approaching smack dab in the middle of it. Then, coming in from the left in the same row, randomly choose numbers that get closer to the number. Do the same thing coming in from the right. In the next row, you compute the y values that correspond to these values that x is approaching.

Solving analytically is the long way of finding a limit, but sometimes you'll come across a function (or teacher) that requires this technique, so it's good for you to know. Basically, if you can use the algebraic technique we describe in the next section, you should. When you can't, you're stuck with this method. In typical mathematical style, instructors always teach you the long way before showing you a shortcut. This is why we, too, include the analytic method before the algebraic one. We don't like to go against the grain; it gives us splinters!

For example, the function $f(x) = \frac{x^2 - 6x + 8}{x - 4}$ is undefined at $x = 4$ because that value makes the denominator 0. But you can find the limit of the function as x approaches 4 by using a chart. Table 15-1 shows how to set it up.

Table 15-1 Finding a Limit Analytically

x	3.0	3.9	3.99	3.999	4.0	4.001	4.01	4.1	5.0
$f(x)$ (or the y value)	1.0	1.9	1.99	1.999	???	2.001	2.01	2.1	3.0

The values that you pick for x are completely arbitrary — they can be anything you want. Just make sure they get closer and closer to the value you're looking for from both directions. The closer you get to the actual x value, though, the closer your limit will be as well. If you look at the y values in the chart, you'll notice that they get closer and closer to 2 from both sides; so, 2 is the limit of the function, determined analytically.

It's very easy to make this chart with a calculator and its table feature. Look in the manual for your particular calculator to discover how.

Algebraically

The last way to find a limit is to do it algebraically. When you can use one of the techniques we describe in this section, you should. You have four techniques you have to know to find a limit algebraically. The best place to start is the first technique; if you plug in the value that x is approaching and the answer is undefined, you must move on to the other techniques to simplify so that you can plug in the approached value for x. The following sections break them all down.

Plugging in

The first technique for solving for a limit algebraically is to plug the number that x is approaching into the function. If you get an undefined value (0 in the denominator), you must move on to another technique. But when you do get a value, you're done; you've found your limit! For example, you can find the $\lim_{x \to 5} \frac{x^2 - 6x + 8}{x - 4}$ with this method. The limit is 3, because $f(5) = 3$.

Factoring

Factoring is the method to try when plugging in fails — especially when any part of the given function is a polynomial expression. (If you've forgotten how to factor a polynomial, refer to Chapter 4.)

If you're asked to find $\lim_{x \to 4} \frac{x^2 - 6x + 8}{x - 4}$, you first try to plug 4 into the function, and you get 0 in the numerator *and* the denominator, which tells you to move on to the next technique. The quadratic expression in the numerator screams for you to try factoring it. Notice that the numerator of the previous function factors to $(x-4)(x-2)$. The $x-4$ cancels on the top and the bottom of the fraction. This leaves you with $f(x) = x - 2$. You can plug 4 into this function to get 2.

Now, if you graph this function, it looks like the straight line $f(x) = x - 2$, but with a hole when $x = 4$, because the original function is still undefined there (this creates 0 in the denominator). See Figure 15-2 for an illustration of what we mean.

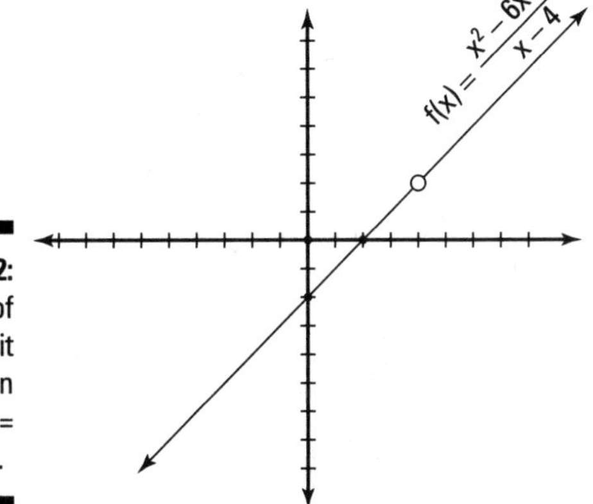

Figure 15-2: The graph of the limit function $f(x) = \frac{x^2-6x+8}{x-4}$.

If, after you've factored the top and bottom of the fraction, a term in the denominator didn't cancel and the value that you're looking for is undefined, the limit of the function at that value of x doesn't exist (sometimes you write this as DNE).

For example, $f(x) = \frac{x^2-3x-28}{x^2-6x-7}$ factors to $\frac{(x-7)(x+4)}{(x-7)(x+1)}$, and the $x-7$s on the top and bottom cancel. So, if you're asked to find the limit of the function as x approaches 7, you could plug it into the cancelled version and get $^{11}/_{8}$. But if you're looking at the $\lim_{x \to -1}$, the limit DNE, because you'd get 0 on the denominator. This function, therefore, has a limit anywhere except as x approaches –1.

Rationalizing the numerator

The third technique you need to know to find limits algebraically requires you to rationalize the numerator. Functions that require this method have a square root in the numerator and a polynomial expression in the denominator. For example, if you're asked to find the limit of $g(x) = \frac{\sqrt{x-4}-3}{x-13}$ as

x approaches 13, plugging in fails when you get 0 in the denominator of the fraction. Factoring fails because there's no polynomial to factor. In this situation, if you multiply the top by its conjugate, the term in the denominator that was a problem should cancel, and you'll be able to find the limit:

1. **Multiply the top and bottom of the fraction by the conjugate. (See Chapter 2 for more info.)**

   The conjugate here is $\sqrt{x-4}+3$. Multiplying through, the setup you get is $\frac{(\sqrt{x-4}-3)}{(x-13)} \cdot \frac{(\sqrt{x-4}+3)}{(\sqrt{x-4}+3)}$. FOIL the tops to get $(x-4)+3\sqrt{x-4}-3\sqrt{x-4}-9$, which simplifies to $x-13$ (the middle two terms cancel and you combine like terms from the FOIL).

2. **Cancel factors.**

   This gives you $\frac{(x-13)}{(x-13)(\sqrt{x-4}+3)}$. The $(x-13)$ terms cancel, and you get $\frac{1}{\sqrt{x-4}+3}$.

3. **Calculate the limits.**

   When you plug 13 into the function, you get $\frac{1}{6}$, which is your limit.

Finding the lowest common denominator

For the fourth and final limit-finding technique using algebra, you'll be given a complex rational function. The technique of plugging fails, because you end up with a 0 in the denominator somewhere. The function isn't factorable, and you have no square roots to rationalize. This lets you know to move on to the last technique. With this method, you combine the functions by finding the least common denominator (LCD). The terms will cancel, at which point you can find the limit. For example, to find $\lim_{x \to 0} \frac{\frac{1}{x+6}-\frac{1}{6}}{x}$, follow these steps:

1. **Find the LCD of the fractions on the top.**

   This gives you $\lim_{x \to 0} \frac{\frac{6 \cdot 1}{6(x+6)}-\frac{1(x+6)}{6(x+6)}}{x}$.

2. **Distribute the numerators on the top.**

   You now have $\lim_{x \to 0} \frac{\frac{6}{6(x+6)}-\frac{x+6}{6(x+6)}}{x}$.

3. **Add or subtract the numerators and then cancel terms.**

   Subtracting the numerators gives you $\lim_{x \to 0} \frac{\frac{6-x-6}{6(x+6)}}{x}$, which then cancels to $\lim_{x \to 0} \frac{\frac{-x}{6(x+6)}}{x}$.

4. **Use the rules for complex fractions to simplify further.**

   This boils down to $\lim_{x \to 0} \frac{-x}{6(x+6)} \div x = \lim_{x \to 0} \frac{-x}{6(x+6)} \cdot \frac{1}{x} = \frac{-1}{6(x+6)}$.

5. **Substitute the limit value into this function and simplify.**

   You want to find the limit as x approaches 0, so the limit here is $-1/36$.

Operating on Limits: The Limit Laws

If you know the limit laws in calculus, you'll be able to find limits of all the crazy functions that calc can throw your way. Thanks to limit laws, for instance, you can find the limit of combined functions (addition, subtraction, multiplication, and division of functions, as well as raising them to powers). All you have to be able to do is find the limit of each individual function separately.

If you know the limits of two functions (see the previous sections of this chapter), you know the limits of them added, subtracted, multiplied, divided, or raised to a power. If the $\lim_{x \to b} f(x) = L$ and $\lim_{x \to b} g(x) = M$, you know the following:

- **Addition law:** $\lim_{x \to b} (f(x) + g(x)) = L + M$
- **Subtraction law:** $\lim_{x \to b} (f(x) - g(x)) = L - M$
- **Multiplication law:** $\lim_{x \to b} (f(x) \cdot g(x)) = L \cdot M$
- **Division law:** $\lim_{x \to b} \left(\frac{f(x)}{g(x)} \right) = \frac{L}{M}$
- **Power law:** $\lim_{x \to b} (f(x))^p = L^p$

For example, if the $\lim_{x \to 3} f(x) = 10$ and $\lim_{x \to 3} g(x) = 5$, you find $\lim_{x \to 3} \left[\frac{2f(x) - 3g(x)}{g(x)^2} \right]$ with the following calculations:

$$\frac{2 \cdot 10 - 3 \cdot 5}{5^2} = \frac{20 - 15}{25}$$

$$= 5/25$$

$$= 1/5$$

It really is that easy!

Exploring Continuity in Functions

The more complicated a function becomes, the more complicated its graph becomes as well. A function can have holes in it, it can jump, or it can have asymptotes, to name a few variations (as you've seen in previous examples in this chapter). However, a graph that's smooth without any holes, jumps, or asymptotes is called *continuous*. We often say, informally, that you can draw a continuous graph without lifting your pencil from the paper.

Any polynomial function, exponential function, or logarithmic function will always be continuous at every point (no holes or jumps). If your textbook or teacher asks you to describe the continuity of one of these particular groups of functions, your answer is that it's always continuous!

Also, if you ever need to find a limit for any of these functions, you can use the first technique we mention in the previous section because the functions are all defined at *every* point. You can plug in any number, and the y value will always exist.

You can look at the continuity of a function at a specific x value. You don't usually look at the continuity of a function as a whole, just at whether it's continuous at certain points. Even discontinuous functions are only discontinuous at certain spots. In the following sections, we show you how to determine whether a function is continuous. You can use this information to tell whether you're able to find a derivative (something you'll get very familiar with in calculus).

Determining whether a function is continuous

Three things have to be true for a function to be continuous at some value x in its domain:

- **$f(c)$ must be defined.** The function must exist at an x value (c), which means you can't have a hole in the function (such as a 0 in the denominator).
- **The limit of the function as x approaches the value c must exist.** The left and right limits must be the same, in other words, which means the function can't jump or have an asymptote. The mathematical way to say this is that $\lim_{x \to c} f(x)$ must exist.
- **The function's value and the limit must be the same.** $f(c) = \lim_{x \to c} f(x)$.

For example, you can show that $f(x) = \frac{x^2 - 2x}{x - 3}$ is continuous at $x = 4$ because of the following:

1. **$f(4)$ exists.** You can substitute 4 into this function to get an answer: 8.
2. **$\lim_{x \to 4} f(x)$ exists.** If you look at the function algebraically (see the earlier section on this topic), it factors to $\frac{x(x-2)}{x-3}$. Nothing cancels, but you can still plug in 4 to get $\frac{4(4-2)}{4-3} = \frac{f(2)}{1}$, which is 8.
3. **$f(4) = \lim_{x \to 4} f(x)$.** Both sides of the equation are 8, so it's continuous at 4.

If any of the above situations aren't true, the function is discontinuous at that point.

Dealing with discontinuity

Functions that aren't continuous at an x value either have a *removable discontinuity* (a hole) or a *nonremovable discontinuity* (an asymptote):

- If the function factors and the bottom term cancels, the discontinuity is removable, so the graph will just have a hole in it.

  For example, $f(x) = \frac{x^2 - 4x - 21}{x + 3}$ factors to $\frac{(x+3)(x-7)}{(x+3)}$, which, after canceling, leaves you with $x - 7$. This means $x + 3 = 0$ (or $x = -3$) is a removable discontinuity — there's a hole in the graph, like you see in Figure 15-3a.

- If a term doesn't factor, the discontinuity is nonremovable, and the graph will have a vertical asymptote.

  If you look at the function $g(x) = \frac{x^2 - x - 2}{x^2 - 5x - 6}$, it factors to $\frac{(x-2)(x+1)}{(x+1)(x-6)}$. Because the $x + 1$ cancels, you have a removable discontinuity at $x = -1$ (you'd see a hole in the graph there, not an asymptote). But the $x - 6$ didn't cancel in the denominator, so you have a nonremovable discontinuity at $x = 6$. This creates a vertical asymptote in the graph at $x = 6$. Figure 15-3b shows the graph of $g(x)$.

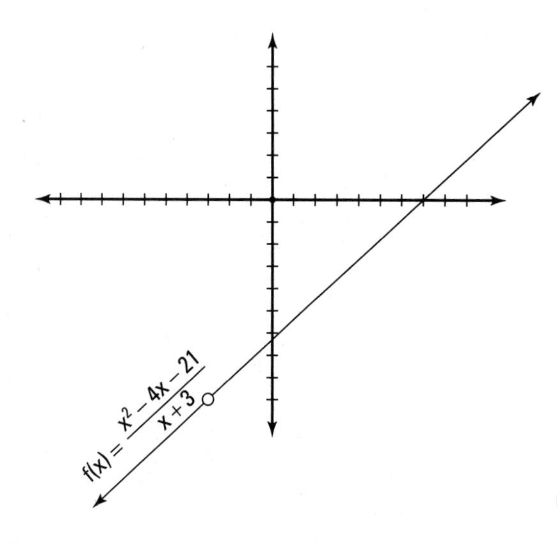

a.

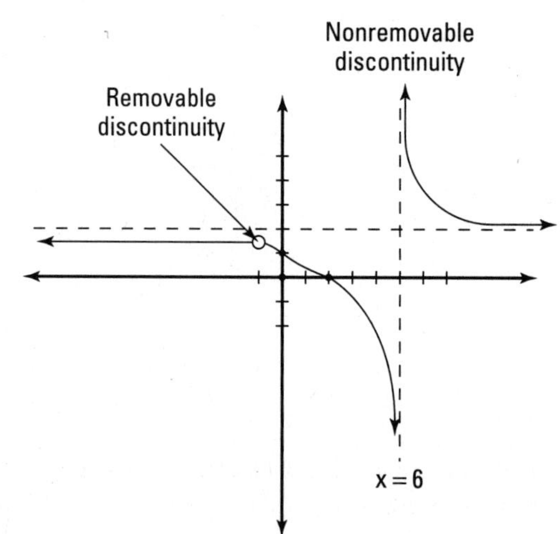

b.

Figure 15-3: The graph of a removable discontinuity leaves you feeling empty, while the graph of a nonremovable discontinuity makes you feel jumpy.

Part IV
The Part of Tens

"Okay, ma'am, I'm going to ask you to walk a straight line, then I'm going to ask you to bisect that line with a perpendicular line that slopes to the equation $y = 3x + 5$."

In this part . . .

This part presents two opposite sides of the spectrum in terms of preparation for calculus: good math habits to take into calculus, and bad habits to break before calculus. Both ends of this spectrum are critical for success because the problems get longer and the sympathy from teachers for algebra errors gets shorter.

Chapter 16

Ten Habits That Help You Attack Calculus

In This Chapter

- Preparing to solve a problem
- Working through a problem
- Verifying your accuracy after solving a problem
- Going the extra mile to ensure pre-calc success

Adopting certain tasks as habits will surely help your brain as you gear up to tackle calculus. In this chapter, we outline ten habits that should be a part of your daily math arsenal. Perhaps your teachers have been singing the praises of certain tasks since elementary school — such as "show all your work" — but other tricks may be new to you. Either way, we're confident that if you remember these ten pieces of advice, you'll be ready for whatever calculus throws your way.

Figure Out What the Problem Is Asking

Often, math teachers test students' reading comprehension (we know, unfair!) and ability to work with multiple parts that comprise a whole, which is the essence of the concepts behind mathematics. When faced with a math problem, start by reading the whole problem or all the directions to the problem. Look for the question inside the question. Keep your eyes peeled for words like "solve," "simplify," "find," and "prove." These are common buzz words in any math book. Don't begin working on a problem until you're certain of what it wants you to do.

For example, take a look at this problem:

> The width of a rectangular garden is 24 inches longer than the garden's length. If you add 3 inches to the length, the width is 8 inches more than twice the length. How long is the new, bigger garden?

If you miss any of the important information, you may start to solve the problem to figure out how wide the garden is. Or you may find the length but miss the fact that you're supposed to find out how long it is with 3 inches *added* to it. Look before you leap!

It's often helpful if you underline key words and information in the question. We can't stress this enough. Highlighting important words and pieces of information will solidify them in your brain so that as you work, you can redirect your focus if it veers off-track. When presented with a word problem, for example, first turn the words into an algebraic equation. If you're lucky and are given the algebraic equation from the get-go, you can move on to the next step, which is to create a visual image of the situation at hand.

Draw Pictures (And Plenty of 'Em)

Your brain is like a movie screen in your skull, so it's your job to project what you see onto the paper. When you visualize math problems, you're more apt to comprehend them. Draw pictures that correspond to a problem and label all the parts so you have a visual image to follow that allows you to attach mathematical symbols to physical structures. This process works the conceptual part of your brain and helps you remember important concepts. As such, you'll be less likely to miss steps or get disorganized.

If the question is talking about a triangle, for instance, draw a triangle; if it mentions a rectangular garden filled with daffodils for 30 percent of its space, draw that. In fact, every time a problem changes and new information is presented, your picture should change, too. (Among the many in this book, Chapters 6 and 10 illustrate how drawing a picture in a problem can greatly improve your odds of solving it!)

If you were asked to solve the rectangular garden problem from the previous section, you'd start by drawing two rectangles: one for the old, smaller garden and another for the bigger one. These pictures get labels in the next section, where we begin to plan how to get the solution (see Figure 16-1).

Plan Your Attack

When you know and can picture what you must find, you can plan your attack from there. The equations that you'll be working with come from this. If you follow the path we've shown below, you'll be solving word problems in a jiffy!

Try making a "let x =" statement to start. In the garden problem from the last two sections, you're looking for the length and width of a garden after it has been made bigger. Start by defining some variables:

- Let x = the garden's length now
- Let y = the garden's width now

Now add those variables to your picture of the old garden (see Figure 16-1a).

You know that the new garden has had 3 inches added to its length, so

- Let $x + 3$ = the garden's new length
- Let y = the garden's width (which doesn't change at all)

Now add these labels to the picture of the new garden (see Figure 16-1b).

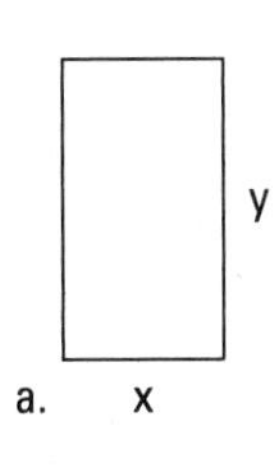

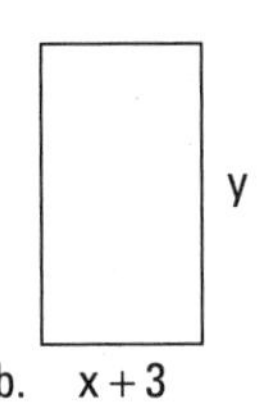

Figure 16-1: Picturing the old garden and the new garden helps you plan your attack.

Be sure to write down the given information somewhere near the question. You could write the following information for our example problem:

- **The width is 24 inches more than the length.**

  This becomes the algebraic equation $y = x + 24$.

- **When the length has 3 more inches added on, the width will then be 8 inches more than twice the length.**

  This becomes the algebraic equation $y = 2(x + 3) + 8$.

This technique helps you identify the equation that you need to solve.

Write Down Any Formulas

If you start your attack by writing the formula needed to solve the problem, all you have to do from there is plug in what you know and then solve for the unknown. A problem will always make more sense if the formula is the first thing you write when solving. Before you can do that, though, you need to figure out which one to use; that should be evident in the words of the problem.

In the case of the garden problem from the previous section, the two equations you write — $y = x + 24$ and $y = 2(x + 3) + 8$ — become the formulas that you need to work with. However, there are times when you won't have to do so much thinking to come up with the formula you need, though you should still write down the formula.

For example, if you need to solve a right triangle, you may start by writing down the Pythagorean Theorem (see Chapter 6) if you know two sides and are looking for the third. For another right triangle, perhaps you're given an angle and the hypotenuse and need to find the opposite side; in this situation, you'd start off by writing down the sine ratio (also in Chapter 6).

Show Each Step of Your Work

Yes, you've been hearing it forever, but your third-grade teacher was right: Showing each step of your work is vital in math. Writing each step on paper will minimize silly mistakes that you can make when you calculate in your head. It's also a great way to keep a problem organized and clear. It takes precious time to write every single step down, but it's well worth your investment.

Neglecting to show your steps can keep you from getting partial credit. If you show all your work, your teacher may reward you for the knowledge you show, even if your final answer is wrong. But if you get a wrong answer and show no work, you'll receive no credit.

Know When to "Quit"

What's worse than a pop quiz or snoring after you've fallen asleep in class? Glaring at a multiple-choice question that doesn't list your answer as a possibility. It will leave you screaming. No, the test isn't wrong, though you really want to believe it. Sometimes, a problem has no solution. If you've tried all the tricks in your bag and you haven't found a way, consider that there may be no solution at all.

Some common problems that may not have a solution include the following:

- Absolute value equations
- Equations with the variable under a square root sign
- Quadratic equations (which may have solutions that are complex numbers; see Chapter 11)
- Rational equations
- Trig equations

On the other hand, you may get a solution for some problems, but it just won't make sense. Watch out for the following:

- If you're solving an equation for a measurement (like length or an angle in degrees) and you get a negative answer, either you made a mistake or there is no solution. Measurement problems include distance, and distance can't be negative.
- If you're solving an equation to find the number of things (like how many books are on a bookshelf) and you get a decimal answer, that just doesn't make any sense. How could you have 13.4 books on a shelf?

Check Your Answers

Even the best mathematicians make mistakes. When you hurry through calculations or work in a stressful situation, you tend to make mistakes more frequently. So, check your work. Usually, this is a very easy process: You take your answer and plug it back into your equation to see if it really works. It takes very little time to make the check, and if it guarantees you got the question right, why not do it?

Now we'll go back and solve the garden problem from earlier in this chapter by looking at its system of equations:

$$y = x + 24$$

$$y = 2(x + 3) + 8$$

Solving this system (using the techniques we describe in Chapter 13), you get the ordered pair (10, 34). Plug $x = 10$ and $y = 34$ into *both* of the original equations, just to be sure; when you do, you'll see that they both work:

$$34 = 10 + 24 \quad ✓$$

$$34 = 2(10 + 3) + 8 \quad ✓$$

Practice Plenty of Problems

You're not born with the knowledge of how to ride a bike, play baseball, or even speak. The way you get better at difficult tasks is to practice, practice, practice. And the best way to practice math is to work the problems. You can seek harder or more complicated examples of questions that will stretch your brain and make you better at a concept the next time you see it.

Along with working along with us on the example problems in this book, you can take advantage of the *For Dummies* workbooks, which include loads of practice exercises. Check out *Trigonometry Workbook For Dummies,* by Mary Jane Sterling, *Algebra Workbook* and *Algebra II Workbook For Dummies,* both also by Mary Jane Sterling, and *Geometry Workbook For Dummies,* by Mark Ryan (all published by Wiley), to name a few.

Even your textbook from class is great for practice. Why not try some (gulp!) problems that your teacher didn't assign, or maybe go back to an old section to review and make sure you've still got it. Typically, textbooks show the answers to the odd problems, so if you stick with those you can always double-check your answers. And if you get a craving for some extra practice, just search the Internet for "practice math problems" to see what you can find! For example, to see more problems like the garden problem from previous sections, we searched the Internet for "practice systems of equations problems" and found over 3 million hits. That's a lot of practice!

Make Sure You Understand the Concepts

Most math classes build on previous material so students can understand new concepts. So, typically, if you miss an idea in Chapter 1 of a book, this problem will affect you for the rest of the course. For this reason, you should always remember not to move on in math until you've mastered each concept.

Your teacher is your best resource, because he or she knows what's important and what's coming up in the rest of the course. Find your teacher after class and get the help that you need until the concept makes sense to you. Because these tutorial types of meetings generally are one-to-one, your teacher can take personal time to explain the concept to *you.*

In addition, given two students — one who isn't doing well but is trying to seek out assistance, and one who isn't doing well but doesn't ask questions after class — who do you think the teacher will have more time, patience, and sympathy for? That's right. Showing your face equals showing you care. The next section talks more in-depth about questions and how to ask them, so we urge you to keep reading.

Pepper Your Teacher with Questions

There's no such thing as a silly math question. If you don't understand a concept a teacher has presented, or if you simply can't figure out a problem on your own, seek guidance. Ask 20 questions about one problem until you completely understand it. *Trust us,* your teacher won't care that you're asking a lot of questions if you're earnestly trying to understand! (If you're asking questions just to goof off in class, though, your teacher *will* know that and will indeed care.)

Here are just a few examples of really good math questions you may want to ask your teacher:

- *Why* does this work the way that it does?
- *Why* am I taking this step here?
- *What* are the steps, in general, to solving this type of equation?
- *Can* you give me another example to try?
- *When* will I use this information in the future of this class? In the real world? In my life?

Chapter 17

Ten Habits to Break before Calculus

In This Chapter

- Recognizing all the bad habits
- Remembering all the correct habits

You've come this far in math, for which you deserve high praise. But we can say with confidence that only 1 percent of you have few to no bad math habits. So, it makes us a little nervous to include them, out of fear that you may actually pick them up! For the other 99 percent of you, we've created a chapter focused on breaking bad math habits in hopes that you'll recognize a mistake that you commonly make and learn to not make it anymore. The 1 percent of you who are perfect can now leave the building!

Operating Out of Order

Don't fall for the trap that many students do by performing operations in order from left to right. For instance, $2 - 6 \cdot 3$ doesn't become $-4 \cdot 3$, or -12. Why? Because you didn't do the multiplication first like you're supposed to. Focus on PEMDAS every time, all the time:

Parentheses (and other grouping devices) first

Exponents

Multiplication and **D**ivision from left to right

Addition and **S**ubtraction from left to right

Don't ever go out of order, and that's an order!

Squaring without FOILing

Always remember to FOIL. Here's a common mistake we see: $(x + 3)^2$ doesn't equal $x^2 + 9$. Why? Because you forgot that to square something means to multiply it times itself. Foiled again! Here's the correct process: $(x + 3)^2 = (x + 3)(x + 3)$. You should reach the answer of $x^2 + 6x + 9$ for this one. Don't be a square and forget to FOIL. (See Chapter 4 for a review of the FOIL method.)

Splitting Up Denominators

We're here to tell you that $\frac{3}{2x+1}$ doesn't equal $\frac{3}{2x} + \frac{3}{1}$. If you look at this situation with numbers only — like $\frac{3}{2+1}$ — you'd get $3/2 + 3/1$, which should be $9/2$. But notice that you forgot to do PEMDAS (see the first section of this chapter). That division bar is a grouping symbol. You have to simplify what's on the top and what's on bottom before doing the division. The answer should be $3/3$, or 1. The same warning applies to variables, too, like in $\frac{3}{2x+1}$. Because variables represent numbers (albeit unknown ones), they have to follow the same rules as numbers do.

Combining the Wrong Terms

Recognizing like terms can be fairly simple or annoyingly difficult. On the simple side, you must realize that you can't add a variable and a constant — $4x + 3$ isn't $7x$, no matter how hard you try. These are not like terms. (That's like saying that four chickens plus three dogs is seven chogs. Not quite!)

The more complicated the polynomial, the more likely you are to combine unlike terms, so be especially careful when dealing with doozies. For example, $4a^2b^3 - a^3b^2$ is in simplified form. Even though the two terms look remarkably similar, they're not like terms so you can't combine them. The expression $4a^2b^3 - a^2b^3$, however, is a different story. These are like terms, and you can find their difference: $3a^2b^3$.

Forgetting the Reciprocal

The chances of making mistakes increase when you're working with complex fractions — particularly when it comes to dividing them. You're moving right along, and then $\frac{\frac{4}{x}}{\frac{2}{x+2}}$ suddenly becomes $\frac{4}{x} \cdot \frac{2}{x+2}$. You forgot to do the division! The big bar between the two fractions means "divide!", so you really should do this: $\frac{4}{x} \div \frac{2}{x+2}$.

Dividing a fraction means to multiply by its reciprocal, or $\frac{4}{x} \cdot \frac{x+2}{2}$. When this simplifies, you get $\frac{2(x+2)}{x}$.

Losing Track of Minus Signs

When subtracting polynomials, most people tend to lose minus signs. For example, when you work on this problem:

$$(4x^2 - 6x + 3) - (2x^2 - 4x + 3)$$

You may have a tendency to write

$$4x^2 - 6x + 3 - 2x^2 - 4x + 3\text{, or } 2x^2 - 10x + 6$$

But you forgot to subtract the whole second polynomial; you only subtracted the first term. Make sure that the minus sign in front of the second polynomial comes out to play in each subsequent term. The right way to work this problem is $4x^2 - 6x + 3 - 2x^2 + 4x - 3$, which becomes $2x^2 - 2x$.

One common mistake like this occurs when subtracting rational functions — $\frac{4x-1}{x+3} - \frac{2x-6}{x+3}$, for example. Most people will turn this into $\frac{4x-1-2x-6}{x+3}$, but if they do, they forget to subtract the whole second polynomial in the numerator. The problem should become $\frac{4x-1-(2x-6)}{x+3}$, which eventually simplifies to $\frac{2x+5}{x+3}$.

Oversimplifying Radicals

People tend to oversimplify radicals when progressing through the stages of math. Some common mistakes include losing the root sign altogether, so that $\sqrt{3}$ becomes just 3. Some people even toss the index so that $\sqrt[3]{4}$ becomes $\sqrt{4}$, which is 2 . . . except that it's not. The cube root of 4 is between 1 and 2.

Other mistakes include adding roots that shouldn't be added — like $\sqrt{3} + \sqrt{5} = \sqrt{8}$, or $2\sqrt{2}$. These two roots aren't like terms, so you can't add them (see the earlier section "Combining the Wrong Terms").

When working with radicals, be sure to always simplify, but make sure you don't fall for the common traps described here.

Erring in Exponential Dealings

Multiplying monomials with exponents doesn't mean you multiply their exponents — you add them instead. For instance, $x^3 \cdot x^4$ isn't x^{12}; it's x^7. Similarly, when finding the power of a product, don't forget to apply the power to every term by multiplying the exponents. For example, $(3x^4y^2)^2$ isn't $3x^8y^4$, because you forgot the power on the 3. It should be $9x^8y^4$.

Watch for negatives — especially on calculators. The monomials -3^2 and $(-3)^2$ represent -9 and 9, respectively, because the order of operations says that you must take an exponent first.

Canceling Out too Quickly

People cancel incorrectly in so many different ways, we could write a whole *For Dummies* book to cover them all. Instead, we just look at the most common mistakes in the following list:

- **When dealing with rational expressions, you can't simplify constants by throwing the distributive property out the window.** When canceling terms on the numerator and denominator, the bottom term must go into *every* term on the top. Division is a lot like multiplication; if you have an expression with more than one term in the numerator, you must make sure that the term on the bottom divides evenly into all terms on the top. For example, $\frac{4x-1}{2}$ doesn't equal $2x - 1$, because the 2 on the bottom doesn't divide evenly with both $4x$ and -1. Therefore, this rational expression doesn't simplify; you leave it alone as $\frac{4x-1}{2}$. If you're going to divide the 2, the result had better be $2x - \frac{1}{2}$.
- **You can't get cancel happy and cross out variables from terms.** $\frac{3x^2-6x+2}{2x}$ doesn't simplify to $3x - 6 + 1$, or $3x - 5$, by canceling the x terms and the 2s. You can rewrite the fraction $\frac{3x^2-6x+2}{2x}$ as $\frac{3x^2}{2x} - \frac{6x}{2x} + \frac{2}{2x}$, which simplifies to $\frac{3x}{2} - 3 + \frac{1}{x}$.
- **Don't cancel multiple terms that can't be cancelled on the top and bottom of a rational expression.** After you've factored any factorable polynomials and cancelled like terms, you're done. For example, we've seen students cancel expressions like $\frac{x^2-2x-3}{x^2-x-6}$ by doing this: $\frac{\cancel{x^2}-2\cancel{x}-\cancel{3}^1}{\cancel{x^2}-\cancel{x}-\cancel{6}_2} = \frac{-2}{2} = -1$. You have to factor first to get $\frac{(x-3)(x+1)}{(x+2)(x-3)}$, which reduces by canceling $\frac{x+1}{x+2}$. Now you're done. Please don't fall for the trap and simplify it any further.

Even if there's no factoring to do in a rational expression, don't cancel if you can't. For instance, $\frac{x^2-8}{x+2}$ has no factoring to do. You can't start canceling like this: $\frac{\cancel{x}^{2\,x} - \cancel{8}^{4}}{\cancel{x}+\cancel{2}} = x - 4$. Remembering that variables represent numbers, you have to follow the order of operations to fully simplify the numerator and the denominator before you can divide (or cancel). $\frac{x^2-8}{x+2}$ can't simplify on the top or the bottom in any way, so you can't do any canceling.

Distributing Improperly

When distributing, don't forget to multiply the term you're distributing to every single term within the parentheses. (Think of how the negative sign in a subtraction applies to every term.) For instance, $2(4x^2 - 3x + 1)$ doesn't equal $8x^2 - 3x + 1$ or even $8x^2 - 6x + 1$; it equals $8x^2 - 6x + 2$.

Index

Symbols and Numerics

• A •

• B •

• C •

• E •

• F •

• G •

• *H* •

• I •

• K •

• L •

• N •

• O •

• P •

• Q •

• R •

• S •

• U •

• V •

• W •

• X •

• Y •

Z

Notes

Notes

Notes

Notes

BUSINESS, CAREERS & PERSONAL FINANCE

0-7645-9847-3

0-7645-2431-3

Also available:

- Business Plans Kit For Dummies 0-7645-9794-9
- Economics For Dummies 0-7645-5726-2
- Grant Writing For Dummies 0-7645-8416-2
- Home Buying For Dummies 0-7645-5331-3
- Managing For Dummies 0-7645-1771-6
- Marketing For Dummies 0-7645-5600-2
- Personal Finance For Dummies 0-7645-2590-5*
- Resumes For Dummies 0-7645-5471-9
- Selling For Dummies 0-7645-5363-1
- Six Sigma For Dummies 0-7645-6798-5
- Small Business Kit For Dummies 0-7645-5984-2
- Starting an eBay Business For Dummies 0-7645-6924-4
- Your Dream Career For Dummies 0-7645-9795-7

HOME & BUSINESS COMPUTER BASICS

0-470-05432-8

0-471-75421-8

Also available:

- Cleaning Windows Vista For Dummies 0-471-78293-9
- Excel 2007 For Dummies 0-470-03737-7
- Mac OS X Tiger For Dummies 0-7645-7675-5
- MacBook For Dummies 0-470-04859-X
- Macs For Dummies 0-470-04849-2
- Office 2007 For Dummies 0-470-00923-3
- Outlook 2007 For Dummies 0-470-03830-6
- PCs For Dummies 0-7645-8958-X
- Salesforce.com For Dummies 0-470-04893-X
- Upgrading & Fixing Laptops For Dummies 0-7645-8959-8
- Word 2007 For Dummies 0-470-03658-3
- Quicken 2007 For Dummies 0-470-04600-7

FOOD, HOME, GARDEN, HOBBIES, MUSIC & PETS

0-7645-8404-9

0-7645-9904-6

Also available:

- Candy Making For Dummies 0-7645-9734-5
- Card Games For Dummies 0-7645-9910-0
- Crocheting For Dummies 0-7645-4151-X
- Dog Training For Dummies 0-7645-8418-9
- Healthy Carb Cookbook For Dummies 0-7645-8476-6
- Home Maintenance For Dummies 0-7645-5215-5
- Horses For Dummies 0-7645-9797-3
- Jewelry Making & Beading For Dummies 0-7645-2571-9
- Orchids For Dummies 0-7645-6759-4
- Puppies For Dummies 0-7645-5255-4
- Rock Guitar For Dummies 0-7645-5356-9
- Sewing For Dummies 0-7645-6847-7
- Singing For Dummies 0-7645-2475-5

INTERNET & DIGITAL MEDIA

0-470-04529-9

0-470-04894-8

Also available:

- Blogging For Dummies 0-471-77084-1
- Digital Photography For Dummies 0-7645-9802-3
- Digital Photography All-in-One Desk Reference For Dummies 0-470-03743-1
- Digital SLR Cameras and Photography For Dummies 0-7645-9803-1
- eBay Business All-in-One Desk Reference For Dummies 0-7645-8438-3
- HDTV For Dummies 0-470-09673-X
- Home Entertainment PCs For Dummies 0-470-05523-5
- MySpace For Dummies 0-470-09529-6
- Search Engine Optimization For Dummies 0-471-97998-8
- Skype For Dummies 0-470-04891-3
- The Internet For Dummies 0-7645-8996-2
- Wiring Your Digital Home For Dummies 0-471-91830-X

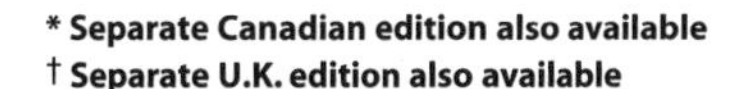
*** Separate Canadian edition also available**
† Separate U.K. edition also available

Available wherever books are sold. For more information or to order direct: U.S. customers visit www.dummies.com or call 1-877-762-2974. U.K. customers visit www.wileyeurope.com or call 0800 243407. Canadian customers visit www.wiley.ca or call 1-800-567-4797.

SPORTS, FITNESS, PARENTING, RELIGION & SPIRITUALITY

0-471-76871-5

Baby Massage For Dummies

0-7645-7841-3

Also available:

- Catholicism For Dummies 0-7645-5391-7
- Exercise Balls For Dummies 0-7645-5623-1
- Fitness For Dummies 0-7645-7851-0
- Football For Dummies 0-7645-3936-1
- Judaism For Dummies 0-7645-5299-6
- Potty Training For Dummies 0-7645-5417-4
- Buddhism For Dummies 0-7645-5359-3
- Pregnancy For Dummies 0-7645-4483-7 †
- Ten Minute Tone-Ups For Dummies 0-7645-7207-5
- NASCAR For Dummies 0-7645-7681-X
- Religion For Dummies 0-7645-5264-3
- Soccer For Dummies 0-7645-5229-5
- Women in the Bible For Dummies 0-7645-8475-8

TRAVEL

0-7645-7749-2

0-7645-6945-7

Also available:

- Alaska For Dummies 0-7645-7746-8
- Cruise Vacations For Dummies 0-7645-6941-4
- England For Dummies 0-7645-4276-1
- Europe For Dummies 0-7645-7529-5
- Germany For Dummies 0-7645-7823-5
- Hawaii For Dummies 0-7645-7402-7
- Italy For Dummies 0-7645-7386-1
- Las Vegas For Dummies 0-7645-7382-9
- London For Dummies 0-7645-4277-X
- Paris For Dummies 0-7645-7630-5
- RV Vacations For Dummies 0-7645-4442-X
- Walt Disney World & Orlando For Dummies 0-7645-9660-8

GRAPHICS, DESIGN & WEB DEVELOPMENT

0-7645-8815-X

0-7645-9571-7

Also available:

- 3D Game Animation For Dummies 0-7645-8789-7
- AutoCAD 2006 For Dummies 0-7645-8925-3
- Building a Web Site For Dummies 0-7645-7144-3
- Creating Web Pages For Dummies 0-470-08030-2
- Creating Web Pages All-in-One Desk Reference For Dummies 0-7645-4345-8
- Dreamweaver 8 For Dummies 0-7645-9649-7
- InDesign CS2 For Dummies 0-7645-9572-5
- Macromedia Flash 8 For Dummies 0-7645-9691-8
- Photoshop CS2 and Digital Photography For Dummies 0-7645-9580-6
- Photoshop Elements 4 For Dummies 0-471-77483-9
- Syndicating Web Sites with RSS Feeds For Dummies 0-7645-8848-6
- Yahoo! SiteBuilder For Dummies 0-7645-9800-7

NETWORKING, SECURITY, PROGRAMMING & DATABASES

0-7645-7728-X

0-471-74940-0

Also available:

- Access 2007 For Dummies 0-470-04612-0
- ASP.NET 2 For Dummies 0-7645-7907-X
- C# 2005 For Dummies 0-7645-9704-3
- Hacking For Dummies 0-470-05235-X
- Hacking Wireless Networks For Dummies 0-7645-9730-2
- Java For Dummies 0-470-08716-1
- Microsoft SQL Server 2005 For Dummies 0-7645-7755-7
- Networking All-in-One Desk Reference For Dummies 0-7645-9939-9
- Preventing Identity Theft For Dummies 0-7645-7336-5
- Telecom For Dummies 0-471-77085-X
- Visual Studio 2005 All-in-One Desk Reference For Dummies 0-7645-9775-2
- XML For Dummies 0-7645-8845-1